NASA SP-4023

ASTRONAUTICS AND AERONAUTICS, 1978

A Chronology

The NASA History Office

The NASA History Series

 Scientific and Technical Information Branch 1986
National Aeronautics and Space Administration
Washington, DC

NASA maintains an internal history program for two principal reasons: (1) Sponsorship of research in NASA-related history is one way in which NASA responds to the provision of the National Aeronautics and Space Act of 1958 that requires NASA to "provide for the widest practicable and appropriate dissemination of information concerning its activities and the results thereof." (2) Thoughtful study of NASA history can help agency managers accomplish the missions assigned to the agency. Understanding NASA's past aids in understanding its present situation and illuminates possible future directions. The opinions and conclusions set forth in this book are those of the author; no official of the agency necessarily endorses those opinions or conclusions.

Preface

Astronautics and Aeronautics for 1978 is the 18th volume in a series of annual chronological digests of principal events in the fields of astronautics and aeronautics. Initiated in 1961 with a compendium of events from 1915-1960, this series is designed to serve as a reference for policymakers, reference librarians, and researchers. Each entry is followed by one or more citations of sources to which users may turn for further information on the particular event or development being chronicled.

The *Astronautics and Aeronautics* series is one way in which the NASA History Office has attempted to implement its charter to disseminate information about the activities of the National Aeronautics and Space Administration, and to provide support for historians and other researchers in the area of United States aeronautical and space programs. As a result of staff and budgetary limitations, the NASA History Office has been unable to maintain the annual currency of this series. We expect, however, to bring the series up to date with a forthcoming compendium volume covering principal events over a six-year period. This volume was written by Ms. Bette R. Janson.

Reliability is essential to the usefulness of any reference work of this scope. As with previous volumes, readers can contribute by notifying our office of any errors so that we can publish them in subsequent volumes.

Sylvia D. Fries
Director, NASA History Office

July 1986

Four varieties of spacecraft launched by NASA in 1978: top left, scientific international ultraviolet explorer IUE 1, *launched Jan. 26; top right, earth-resources satellite* Landsat 3, *launched March 5; bottom left, environmental meteosat* Tiros N, *launched Oct. 13; bottom right, comsat NATO 3C, launched Nov. 18.*

Contents

Preface	iii
January	1
February	25
March	49
April	77
May	103
June	129
July	155
August	181
September	209
October	235
November	259
December	289

Appendixes

A. Satellites, Space Probes, and Manned Spaceflights, 1978	317
B. Major NASA Launches, 1978	347
C. Manned Spaceflights, 1978	351
D. NASA Sounding Rocket Launches, 1978	353
E. Abbreviations of References	361
Index	367

Illustrations

First women named by NASA as astronaut candidates	9
Space Shuttle orbiter Enterprise	72
Six astronauts awarded Congressional Medal of Honor	237
Photo mosaic from the Viking orbiter	267
NASA's second high-energy astronomy observatory	269
Artist's concept, Pioneer Venus orbiter and multiprobe	281

January

January 2: Aviation Week and Space Technology magazine reported that President Carter would soon receive from Secretary of Defense Harold Brown a plan for a unified U.S. space policy covering all government and private-sector programs. The plan was the product of a policy-review committee headed by Brown and consisting of representatives from NASA, the Dept. of Agriculture, Joint Chiefs of Staff, CIA, DOD, the office of the President's national security adviser, and the Office of Science and Technology Policy; the Interior Dept. and NOAA had participated in a steering committee under the main group. After evaluating the effect of the U.S. space program on foreign policy, national security, and overall economic benefit, the group had agreed that national security considerations were the main hindrances to formulating a national space policy. The group had considered space-arms control (maximum pacification of space, with limitations on killer satellites); availability of reconnaissance-satellite data (greater federal agency access); regulation of private-sector remote sensing (resolution limits); interagency management (a coordinating committee, to be chaired either by DOD and NASA, or by the President's science adviser); broadcast satellite policy (no restrictions on signals crossing national borders); and certain intelligence, DOD, NASA, NOAA, and civilian programs. (*AvWk*, Jan 2/78, 14)

• *AvWk* reported NASA's announcement that it would send *Pioneer 11* to fly just outside the rings of Saturn as a pathfinder for two Voyager spacecraft scheduled to reach Saturn in 1980 and 1981. Passing 15 000mi rather than 3700mi from Saturn's surface, the new course would place *Pioneer 11* at about the same distance from Saturn as the trajectory of *Voyager 2*, if NASA should decide to use Saturn's gravity to hurl *Voyager 2* toward Uranus. The outside trajectory would permit NASA to achieve the maximum science return for all three spacecraft. *Pioneer 11* should reach Saturn in Sept. 1978. (*AvWk*, Jan 2/78, 15; *Lewis News*, Jan 6/78, 2)

• President Carter's $100 million reduction in MX-missile FY 1979 funding would not prevent DOD's continuing advance MX development for another yr, *AvWk* reported. The remaining $160 million would cover costs of assessing basing modes and alternatives for the missile. The President's decision to continue MX without starting engineering development would delay initial operational capability until 1987. Key administration officials doubted that MX would ever be deployed, said *AvWk*, because of its links to future SALT agreements preventing deployment of mobile-based systems. (*AvWk*, Jan 2/78, 13)

January 5: NASA reported a board of inquiry's findings on the Sept. 9, 1977, failure of an Atlas-Centaur launch vehicle destroyed 55sec into the flight. Launched from Kennedy Space Center, the vehicle had carried an Intelsat IV-A. The board attributed failure to a rupture in the high-pressure "omega joint" in the Atlas-stage gas-generator discharge line, caused by contamination during production. Technicians had tested joints on all completed vehicles, and all those at KSC had passed; a few in the factory that were corroded or questionable had been replaced. Changing the fabrication process should prevent further contamination, the report stated. (NASA Release 78-3)

January 6: Lewis Research Center reported it had signed a $24 million 1 yr contract with Honeywell, Inc., Avionics Division, St. Petersburg, Fla., for management and engineering support of Centaur inertial guidance. This automatic self-contained navigation and guidance system with a digital computer unit would permit Centaur to compute and adjust its flight for delivery of a spacecraft to a preselected orbit without ground command. The nation's first high-energy liquid-hydrogen liquid-oxygen rocket, the Centaur, as an upper stage on both Atlas and Titan III boosters, had helped launch *Voyager 1* and *2*, scheduled to fly by Jupiter and Saturn in 1979 and 1980. (*Lewis News*, Jan 6/78, 2)

• Preliminary analyses of *Viking Orbiter 1* (VO-1) data on the Mars satellite Phobos revealed low density and a spectral reflectance and low albedo that suggested a composition similar to type I carbonaceous chondrites, reported *Science* magazine. VO-1 had obtained a spectacular picture sequence at 100km, an encounter by far the most demanding navigation performance of any interplanetary-spacecraft exploration to date. Since Phobos had a mean radius of only 11km and was only partially illuminated when viewed during the flyby, the VO-1 position relative to Phobos had required accuracy of within a few kilometers. Imaging data had revealed small positive features (hummocks), small craters with bright ejecta blankets, dark material on the floors of some craters, large variations in topography, and an abundance of linear features resembling crater chains. Further analysis would refine density estimates to determine Phobos's origin. (*Sci*, Jan 6/78, 61)

January 6-7: NASA launched *Intelsat IVA F-3*, fifth of a series of improved commercial communications satellites, at 7:15pm EST Jan. 6 from Cape Kennedy on an Atlas-Centaur into a transfer orbit. ComSat-

Corp had fired the apogee kick motor Jan. 7, putting the satellite into near-geosynchronous orbit. After initial positioning at 74°E the satellite would be moved to 60°E as backup to *Intelsat IVA F-6*, and was scheduled to be on station at 63°E by mid-June.

Intelsat IVA F-3 had an overall height of 6.99m and a diameter of 2.38m. Height of the solar panel was 2.82m. Liftoff weight was approximately 1515kg (3340lb); in-orbit weight after apogee motor firing, 825.5kg (1820lb). Design life was 7yr.

Ten Intelsat IV and IVA satellites were currently in synchronous equatorial orbit over the Atlantic, Pacific, and Indian Oceans, providing full-time telephone, teletypewriter, data, and television services to 107 countries, territories, and possessions. The Intelsat IVA had almost two-thirds greater communications capability than the Intelsat IV series, with about 6250 two-way voice circuits and 2 television channels in its system configuration. The number of operating transponders had increased from 12 to 20. Improvement of the antenna system had produced greater effective bandwidth through frequency reuse by beam separation: for Indian Ocean service, the east-pointed beam of *Intelsat IVA F-3* would illuminate Japan and Southeast Asia using a 320MHz bandwidth, and would reuse the same 320MHz bandwidth in the west beam illuminating Europe and Africa. In addition to eight east-beam transponders and eight west-beam transponders, four transponders connected to an earth-coverage antenna would illuminate all stations in the Indian Ocean region.

The Communications Satellite Corp. (ComSatCorp), U.S. member of INTELSAT, would be management services contractor for the satellite system; ComSatCorp on behalf of INTELSAT would reimburse NASA under a launch-services agreement for costs of the Atlas-Centaur and the launch. The Intelsat IVA program represented an investment by 101 nations of approximately $295 million (US); this launch had cost approximately $47 million ($18 million for the satellite and $29 million for the launch vehicle and related services). (NASA Release 77-250; MOR E-491-633-78-5 [prelaunch] Jan 5/78, [postlaunch] June 23/78; *W Star*, Jan 7/78, A-5)

January 9: NASA announced selection of 16 experiments in space technology for the Long Duration Exposure Facility (LDEF) mission, scheduled as payload for the first Space Shuttle operational flight in 1980. LDEF, managed by Langley Research Center, was a reusable unmanned free-flying structure carrying various technical and scientific experiments in special trays, providing an easy and economical means to conduct primarily passive experiments in earth orbit.

NASA had selected the 16 experiments from 190 candidates that responded to an opportunity notice distributed in June 1976. Investigators selected to date represented 6 universities, 6 private organizations or research institutes, and 5 NASA centers. Seven experiments were from France, 2 from England, and 1 from Canada. R&D value, compatibility with LDEF and other experiments, and effort cost had been selection criteria. Technical areas included materials, thermal-control coatings, detectors, power, micrometeoroids, electronics, lubrication, optics, and space-debris detection. Eighty percent of LDEF's experiment trays had been filled; the remaining space would contain micrometeoroid-detection panels. LDEF would remain in orbit 6 to 12mo for experiment exposure to space environment. The Space Shuttle orbiter would retrieve and return it to earth; experiments would go to their investigators for data analysis. (NASA Release 78-1)

• NASA announced appointment of Kenneth Chapman as associate administrator for external relations, effective Jan. 8. Chapman had been acting in the position since its creation Nov. 8, 1977, and had been assistant administrator for DOD and interagency affairs since joining NASA April 1, 1977. Chapman would be senior policy official responsible for integrating NASA activities in public, legislative, university, community, interagency, and international affairs. Chapman had come to NASA from the Nuclear Regulatory Commission. He retired from the Air Force in 1975 after a 29-yr career. (NASA Release 78-4)

• KSC had begun preparation for launching Shuttle Orbiter 102 in Oct., *Spaceport News* reported. Construction of new facilities and modifications to existing ones were proceeding on schedule. The orbiter-landing facility and orbiter-processing facility had been the only new facilities built at KSC to support Shuttle operations. Other preparations included modifications of existing facilities originally designed and built to support Apollo/Saturn operations. KSC would transform hanger AF at Cape Canaveral Air Force Station into a solid-fuel rocket-booster recovery and disassembly facility. (*Spaceport News,* Jan 9/78, 1)

• The Commerce Department had predicted the total value of U.S. aerospace-industry shipments of complete aircraft, engines, and space vehicles at $22.2 billion this year, reported *Aviation Week*. This was a 21% increase over the estimated value of $18.3 billion for 1977, resulting from a surge of orders for large commercial transports in the first half of 1977 for 1978 delivery. A hundred data collectors working for DOC concluded that the U.S. would maintain worldwide preeminence in aerospace; that aerospace exports would continue to provide the largest U.S. manufacturing trade surplus; that sales, shipments, and profits would move up with the demand for large transports causing the greatest optimism; that general aviation would continue to grow in 1978; and that economic growth in 1978 would increase passenger traffic about 6% over 1977, to 200 billion passenger miles. Aerospace imports, the magazine said, would probably remain low—about 2.2% of U.S. consumption in 1977—and would probably cost the U.S. about $648 million in 1978. (*AvWk*, Jan 9/78, 15)

- NASA announced plans for 25 launches in 1978 and support of 3 Atlas-F space launches from Vandenberg Air Force Base. The 1978 missions would be almost equally divided between those directly benefiting people (for communications and geodetic, environmental, navigation, meteorological, and earth-resources studies) and those primarily for scientific and exploratory research. Fifteen launches would be for paying customers, including the European Space Agency (ESA), ComSatCorp, the U.S. Navy, Japan, NOAA, the United Kingdom, and Canada. Cape Canaveral had been scheduled for 10 Delta launches and all the Atlas-Centaur launches; one Delta would be launched from the Western Test Range at Vandenberg AFB. (*Spaceport News*, Jan 9/78, 1; *Goddard News*, Jan/Feb 78, 1; *AvWk*, Jan 9/78, 48)

- Lockheed-Georgia Co. announced it would test its USAF C-5A transport as a cruise missile launch platform, *Aviation Week* reported. Lockheed noted that the platform could accommodate the USAF/Boeing air-launched cruise missile (ALCM), the Navy/General Dynamics Tomahawk (TALCM), and air-to-air self-defense missiles with equal facility. Key features would include no need for structural modifications to the aircraft or its systems or subsystems; operation essentially by gravity; and flexibility that would allow installation on any C-5, Lockheed C-141, or C-130 aircraft. The launch system would maximize flight safety, as missiles would be loaded in launch position, secured, and released and moved only on a launch command, Lockheed officials emphasized. (*AvWk*, Jan 9/78, 40)

January 10: The Federal Aviation Administration would demonstrate in Europe and Africa in early 1978 a family of advanced all-weather landing systems, Administrator Langhorne Bond announced. The FAA would airlift three versions of the time-reference scanning-beam microwave landing system (TRSB/MLS) developed by the U.S. and Australia to four overseas sites in Jan. and Feb. to demonstrate the new approach-and-landing aid at various airport environments. TRSB/MLS versions designed for small community airports would be demonstrated Jan. 23 and 24 at Kristiansand's Kjevik Airport in Norway, and in Feb. at Dakar's Yoff Airport in Senegal; at the Nairobi, Kenya, airport; and at Gosselies Airport outside Brussels, Belgium. FAA would demonstrate an expanded TRSB/MLS system Feb. 1-3 at the Brussels airport. The

U.S. and Australia had proposed TRSB/MLS to the International Civil Aviation Organization (ICAO) as the new worldwide standard all-weather approach-and-landing guidance system. (FAA Release 3-78)

January 11: KSC announced award of a $3.5 million contract to Federal Systems Division, Honeywell Information Systems, for delivery of a central data subsystem for Shuttle launch processing, consisting of two large scale computers sharing a common memory that would store test procedures, a master-program library, pre- and post-test data analyses, and other types of data to support Shuttle checkout, countdown, and launch. KSC had procured the subsystem for Vandenberg AFB for a second Shuttle launch and processing facility. (KSC Release 3-78)

• Marshall Space Flight Center (MSFC) announced that Corning Glass Works, Canton, N.Y., had begun work on a foot-thick primary-mirror blank for the Space Telescope, for which the center would have overall management responsibility. The telescope had been scheduled for launch in 1983. Corning would deliver two 8ft discs to Perkin-Elmer Corp. for grinding and polishing and telescope assembly; Lockheed Missiles and Space Co. would build the primary spacecraft and support system, and integrate the complete telescope. Glass had been used for the mirror blank as it would change shape only slightly with temperature changes during viewing; the titanium-silicate mirror blank would have zero expansion on a scale used by physicists. Aperture of the mirror blank was about 94in; diameter of the center hole was about 2ft. (*Marshall Star*, Jan 11/78, 4)

• MSFC announced successful completion of the first test of the Space Shuttle external tank, a major milestone leading to firing tests of the Shuttle's main propulsion system. The Dec. 21 test assured that the main propulsion test article (the test facility and associated hardware) could handle super-cold propellants, liquid hydrogen, and liquid oxygen used by the Shuttle main engine. Propellants used to fill the external tank had flowed through connecting piping to the 3 engines until stopped by the main-engine valve; test engineers then made sure the engines had cooled down to operating temperature, several hundred degrees below zero. A series of static firings of the external tank, a simulated orbiter with a flight aft fuselage, three main engines, and associated piping would validate system operation. (*Marshall Star,* Jan 11/78, 4)

• A simulated Space Shuttle orbiter had represented the Enterprise during practice hoists into the tall tower where the latter would soon be tested, the *Marshall Star* reported. The real orbiter had recently completed a series of landing tests in Calif. and would next undergo a series of tests at MSFC to see how the orbiter and other Shuttle components would maintain guidance and control under vibration during launch and powered flight. MSFC engineers would use the simulated orbiter to rehearse tests; for example, they had towed the simulator along center roadways to check clearances. The team had then practiced hoisting the simulator into the 131m (430ft) test stand. After completing tests at MSFC, the simulator would go by barge to KSC to check out procedures for launch. Vandenberg AFB had asked to borrow the simulator for its Shuttle launch facilities. (*Marshall Star,* Jan 11/78, 1)

• Tass reported that the USSR launched *Soyuz 27* from Tyuratam on Jan. 10, 1978, at 7:26pm Moscow time to dock with the already occupied orbital station *Salyut 6*. Crewmen were Lt. Col. Vladimir Dzhanibekov, commander, and Oleg Makarov, civilian flight engineer. *Soyuz 26* had been docked with *Salyut 6* since Dec. 11, 1977; the 2 crews would carry out research and experiments on board. (FBIS, Tass in English, Jan 11/78)

January 12: ESA announced that the third (upper) stage of the heavy launcher Ariane, designed to use highly advanced liquid-hydrogen and liquid-oxygen propulsion techniques, had been successfully tested Jan. 10 at the Société Européenne de Propulsion's Vernon test center. The firing had lasted 250sec, as scheduled. With previous successful tests, this test had demonstrated European mastery of the technology, ESA said. ESA had developed Ariane to launch communications and other applications satellites weighing up to 1000kg into geostationary orbit 36 000km above the equator; ESA had plans to launch about 200 geostationary satellites between 1980 and 1990. (ESA Release Jan 12/78)

January 13: NASA announced that the National Remote Sensing Agency of India (NRSA) had agreed to build a ground station at Hyderabad in south central India to receive data directly from NASA earth resources satellites—*Landsat 2,* as well as Landsat C (scheduled for launch in March 1978) and Landsat D (scheduled for launch in 1981)—and to offer India and neighboring countries the practical benefits of remote-sensing technology. (NASA Release 78-6)

January 16: NASA announced it had selected 15 Space Shuttle pilot-astronaut candidates and 20 mission-specialist candidates as its eighth group of astronauts, including 6 women, 3 blacks, and an American of Japanese descent. The first women and minority astronauts could be in orbit within 3yrs. Selection criteria for the more than 8000 applicants had been motivation, educational background, and work experience. All but one of the 15 pilot candidates and 7 of the 20 mission-specialist candidates had come from the U.S. military services. Of the 35 selectees, 10 were from the Air Force, 9 from the Navy, and 1 each from the Marine Corps and the Army. The mission specialists selected had come from medical, biological, earth sciences, physical and chemical sciences, and engineering disciplines. The number of selections had been keyed to the expected need for Shuttle crews to supplement the 17 pilot and 10 scientist astronauts now on flight status in Houston. The new astronauts would begin 2 years of training July 1 at Johnson Space Center in Houston.

Selected to train as mission specialists were Anna Fisher, Shannon Lucid, Judith Resnick, Sally Ride, Margaret Seddon, Kathryn Sullivan, USAF Maj. Guion Bluford, USMC Capt. James Buchli, USAF Maj. John Fabian, Navy Lt. Dale Gardner, Terry Hart, Steven Hawley, Jeffrey Hoffman, Ronald McNair, USAF Capt. Richard Mullane, George Nelson, USAF Capt. Ellison Onizuka, Army Maj. Robert Stewart, Norman Thagard, James Van Hoften; as pilot astronauts, Navy Lt. Cdr. Daniel Brandenstein, Navy Lt. Cdr. Michael Coats, USAF Maj. Richard Covey, Navy lt. Cdr. John Creighton, Navy Lt. Robert Gibson, USAF Maj. Frederick Gregory, Stanley Griggs, Navy Cdr. Frederick Hauck, Navy Lt. Cdr. Jon McBride, USAF Capt. Steven Nagel, USAF Maj. Francis Scobee, USAF Capt. Brewster Shaw, Jr., USAF Capt. Loren Shriver, Navy Lt. Cdr. David Walker, and Navy Lt. Cdr. Donald Williams. (NASA press conference Jan 16/78; *Marshall Star*, Jan 17/78, 1; *W Star*, Jan 17/78, A-5; *AvWk*, Jan 23/78, 19)

• A NASA internal assessment chartered by the Carter Administration had found areas "where clearly there are too few people to do the job that needs to be done," reported *AvWk*. The study would aid in defining the proper size and structure of NASA through the early to mid-1980s, based on tasks assigned. NASA would continue to analyze its alternatives, including agency employment levels and the outlook on agency responsibilities. The resulting recommendations would be designed to accommodate a variety of funding levels. (*AvWk*, Jan 16/78, 21)

• FAA questions about isolation of hydraulic systems in the Anglo-French Concorde had delayed the plane's certification, *Aviation Week* reported. The British Aerospace Corp. had expressed confidence that the FAA would approve certification; an FAA team concluded earlier that 129 out of 149 certification items had complied with U.S. standards. Later, the engine had been certified, and cabin and crew compartments inspected. An FAA memorandum to the British noted questions on "the possibility of single-hydraulic system failure causing loss of all hydraulic systems, including flight control." The British and French would have to show that their fail-safe type of system provided a degree of safety

First women to be named by NASA as astronaut candidates (left to right): Margaret R. (Rhea) Seddon, Anna L. Fisher, Judith A. Resnik, Shannon W. Lucid, Sally K. Ride, and Kathryn D. Sullivan. (NASA photo 78-H-81)

equivalent to that of the usual dual system, which observers expected them to do. (*AvWk*, Jan 16/78, 31)

• The Congressional battle over President Carter's efforts to delete funds for B-1 bomber aircraft 5 and 6 would resume after Congress reconvened in Jan., said *Aviation Week*. The outcome would depend on a Senate rescission move led by Sen. John Stennis (D-Miss.), chairman of the Senate Armed Services Committee, to uphold the administration's stand against B-1 funding. Last year the House had voted to oppose the Senate rescission stand after Senate-House conferees failed to resolve the issue. Senate B-1 proponents led by Sen. Alan Cranston (D-Calif.), Senate majority whip, had joined forces to sustain the House action and return the question to the conference. The administration might move quickly to win endorsement of the Stennis move to rescind while many senators were still away from Washington. (*AvWk*, Jan 16/78, 15)

January 17: The *Washington Post* reported that the U.S. and USSR had opened a more reliable "hotline" using satellites, instead of telephone wires and undersea cables, to link the White House and the Kremlin. The direct line had been available since 1963 for use in an international emergency. The direct link would use the synchronous-communications satellite Intelsat and the Soviet Molniya system of satellites. The previous hotline had been subject to unexpected interruptions, a potentially grave problem in event of tension between the two governments; it had been cut at least three times, once when a Finnish farmer cut a buried cable while cultivating land. (*W Post,* Jan 17/78, A-10)

January 18: The 16th Aerospace Sciences Meeting at MSFC of the American Institute of Aeronautics and Astronautics with nearly 700 registrants concluded with posthumous presentation of the space systems award to Dr. Werner von Braun, first MSFC director. Former astronaut Maj. Gen. Thomas Stafford had accepted the Haley spaceflight award on behalf of the Apollo-Soyuz Test Project crew. (*Marshall Star,* Jan 18/78, 1)

• MSFC announced that NASA had instituted a NASA End-to-End Data Systems (NEEDS) project to ease the enormous burden of storing data collected by its satellites and aircraft. NASA's public archives had been growing by 100 000 000 000 bits of data per day; during the 1980s and beyond, spacecraft like Spacelab, earth-orbiting satellites, and planetary probes would produce a staggering quantity of data, creating a major problem of storage and availability to users. NEEDS would decrease processing time and storage space, reduce maintenance costs, and provide better user access. Four field centers would participate in the project, MSFC being responsible for developing a medium to replace magnetic tape and microfilm. Working with Harris Corp. of Melbourne, Fla., MSFC had developed the optical mass memory system, using a dry-process film that could store as much data as on magnetic tapes or microfilm in 100 times less space, and could make access 100 times faster. The goal of NEEDS would be a tenfold reduction in the present $300-million-per-year storage/maintenance cost, and a thousandfold increase in information return to users. (*Marshall Star,* Jan 18/78, 4)

January 19: INTELSAT announced it had awarded Nippon Electric Co., Tokyo, Japan, a $175 000 contract for development of a multimode phase-shift-keying (PSK) modem for a nonlinear satellite channel. INTELSAT had established full-time global communications using 197 antennas at 161 earth stations in 87 countries. (INTELSAT Release 78-3-M)

January 20: NASA announced successful completion of the second full-duration ground-static firing of a huge solid-fuel rocket motor to be used in Space Shuttle flights beginning in 1979. Thiokol Corporation's Wasatch Division, prime contractor for MSFC development of the motor, had conducted the firing at Brigham City, Ut. The 125ft-long (38.1m) motor on the test fixture had fired for more than 2min, reaching a thrust level of 2832lb (12 596 736 newtons). The test firing had used 1 102 400lb (500 048kg) of propellants. The contractor had gimballed the motor nozzle early in the test and again later in the firing for about half the total burn time; the solid-fuel rocket booster's motor nozzles during Shuttle launches would provide directional control for the entire Shuttle assembly. MSFC had scheduled the next test for early summer. (MSFC Release 78-6; *Marshall Star,* Jan 25/78, 1)

• *Spaceport News* reported a spacecraft problem had delayed launch of FltSatCom-A, first of a new highly versatile 3-satellite military communications system in which each satellite in geosynchronous orbit would provide 23 operating channels. A FltSatCom (fleet satellite communications) system would provide reliable secure communications between naval units anywhere except in extreme polar regions, and would become part of the Air Force Satellite Communications System (AFSatCom) linking the national command authority with Strategic Air Command (SAC) and other USAF units. DOD had reserved one wideband 500KHz channel on FltSatCom-A. (*Spaceport News,* Jan 20/78,1)

January 21: NASA Administrator Robert Frosch in a briefing announced that NASA's FY79 budget would propose an 8% increase, providing "buying power" at about the 1978 level. The space science request had gone up about 27%, largely because of the Space Telescope and the Galileo (Jupiter-orbiter) project. Applications and aeronautics were each up about 16%. The proposed FY79 budget was $4.371 billion; the FY78 budget was $4.064 billion. The FY79 budget requested money for only four Space Shuttle orbiters; Frosch explained that the four orbiters would use two launch sites (KSC and Vandenberg AFB) and that NASA and the Air Force would phase out the existing stable of expendable boosters. Omitted was OV-101, the Enterprise, which would be used for approach and landing tests or for spare parts, or could be reconfigured in an emergency to spaceflight status.

The major new start NASA had included in the budget was the "Out of Ecliptic" (OOE) or Solar-Polar, a cooperative venture with ESA to fly two satellites to look at the structure of the sun's surface and at sunspots and poles. Other new starts NASA had requested were the HALO (halogen-occultation) experiments to study upper-atmosphere pollution, and the earth radiation-budget satellite system to study incoming and outgoing radiation. Key application continuations would be Landsat C and D, the Nimbus-G environmental satellite, Tiros-N next-generation weather satellite, and Seasat-A. Space Transportation Systems would receive $1 827 700 000; Space Science $513 200 000; Space and Terrestrial Applications $283 400 000; Aeronautics and Space Technology $375 400 000; Tracking and Data Acquisition $305 400 000. Manpower would remain at the FY78 level, with 45 positions shifted from MSFC to

JSC, KSC, and the Natl. Space Technology Laboratories. (NASA budget briefing, Jan 21/78; *Marshall Star,* Jan 25/78, 1; *W Star,* Jan 23/78, A-8; *AvWk,* Jan 23/78, 13; Jan 30/78, 28)

January 22: FBIS reported that, according to *Tass,* the automatic cargo-transport ship *Progress 1* launched Jan. 20 had docked with the manned scientific complex *Salyut 6/Soyuz 27* at 1:12pm Moscow time, first use of an automatic spaceship to supply fuel, equipment, and materials to a manned orbital station. The shipment would prolong considerably the functioning and effectiveness of manned orbital complexes. Linkup was to the docking gear in the instrumented compartment of the *Salyut 6* station. The mutual search, rendezvous, tethering, and linkup was automatic, using onboard radio-technical and computing devices. To increase cargo capacity and cut transportation costs, the USSR had designed *Progress 1* not to return to earth but to burn up on reentry. (FBIS, Tass in English, Jan 22/78)

January 23: NASA had formally designated as Project Galileo a 1982 mission to orbit a spacecraft around Jupiter for at least 20mo and send a probe into its atmosphere, NASA Administrator Robert Frosch announced. The designation had honored the 16th century Italian astronomer, first to observe Jupiter by telescope and discover its moons. Galileo would be the first planetary spacecraft carried by the Space Shuttle and the first planetary spacecraft named for a person, although NASA had christened the *Orbiting Astronomical Observatory 3 (OAO 3)* "Copernicus" after it was launched. (NASA Release 78-11)

• NASA announced it had retired *Landsat 1,* first spacecraft designed to discover and monitor earth's natural resources, on Jan. 16 after 5.5yr of operation in outer space. The multispectral scanner carried on *Landsat 1* had revolutionized earth observation from space. Launched in 1972 with a life expectancy of 1yr, *Landsat 1* had taken more than 300 000 pictures and had demonstrated remote-sensing applications in geology, oceanography, agriculture, forestry, hydrology, urban planning, crop prediction, and other disciplines. *Landsat 2,* launched in 1975, was still in orbit; a third Landsat would be launched in March. Three U.S. ground stations had received Landsat data, as had stations in Canada, Brazil, and Italy. Iran and Japan were building stations; Argentina planned one. (NASA Release 78-9; *Goddard News,* Jan-Feb 78, 5)

• *AvWk* reported that the third instead of the fifth Space Shuttle mission would attempt to raise the orbit of the abandoned Skylab space station, bettering chances of Skylab's remaining in orbit until the Shuttle could come to the rescue. NASA had scheduled the third Shuttle flight for Oct. 1979; the fifth, for Feb. 1980. A Martin Marietta teleoperator propulsion unit carried on the Shuttle to boost Skylab to a higher orbit would be left in space for retrieval by the fifth Shuttle flight. MSFC would build a solar panel for the teleoperator to keep it in orbit for as long as necessary. (*AvWk*, Jan 23/78, 22)

• The USSR had decided to keep a permanent manned space station in earth orbit as a result of the current *Salyut 6* space station mission, *AvWk* reported. Staffing this and follow-on Salyuts without major intervals between missions would be a milestone in manned spaceflight, and would demonstrate the USSR's decision to invest major resources in space.

The USSR had claimed the *Salyut 6/Soyuz 27* mission was more scientific than military; its activities had included photographing the Soviet Union to explore for natural resources, detailed medical studies of cosmonaut blood circulation, and upper-atmosphere and micrometeorite monitoring. The crew had also devoted more time to system checkout and monitoring, because the design of *Salyut 6* was a significant change from that of previous Salyut space stations.

The USSR had not previously flown a spacecraft configuration like *Salyut 6/Soyuz 26/Soyuz 27* to test the interaction of three large vehicles. It had tried various spacecraft attitudes to evaluate stresses during maneuvers on the docked system, and had transferred food and equipment from the *Soyuz 27* resupply vehicle and film, biological samples, and records to the *Soyuz 27*.

The Soyuz/Salyut mission had achieved three manned-spaceflight firsts: four-man space station, simultaneous docking of two transport spacecraft with an orbiting space station, and manned resupply of an occupied station. (*AvWk*, Jan 2/78, 20; Jan 9/78, 25; Jan 16/78, 20; Jan 23/78, 19)

• The National Oceanic and Atmospheric Administration (NOAA) announced it had awarded Eppley Laboratory, Inc., Newport, R.I., a $79 000 contract for specialized solar-tracking devices to measure solar radiation reaching earth, to extend its solar-radiation network. The Department of Energy had provided NOAA with funds for 50 pyrheliometers, one for each of NOAA's solar-radiation collecting facilities, to measure increased temperatures proportional to the intensity of incoming radiation. Architects, engineers, and others interested in solar-energy applications would use this information with other temperature, precipitation, and cloud cover data. (NOAA Release 78-8)

January 24: Tass announced that the USSR had launched a *Molniya 3* comsat carrying relay apparatus to operate in the centimeter waveband, into an orbit with 40 631km apogee in the northern hemisphere and 661km perigee in the southern hemisphere, period of 12hr 16min, and 62.8° inclination. The satellite would ensure operation of a long-distance telephone, telegraph, and radio-communications system, and program relay from USSR central television to points in the Orbita network. (FBIS, Tass in English, Jan 24/78)

January 25: NASA announced results of two November 1977 meetings with the USSR Academy of Sciences on further cooperation in space. An agreement reached in May 1977 focused on ways to benefit from the different attributes of Salyut and a reusable-Shuttle type of spacecraft. Dr. Noel Hinners, associate administrator for space science at NASA, had headed the U.S. delegation; Dr. Boris Petrov, chairman of the Intercosmos Council of the Soviet Academy of Sciences, had led the Soviet delegation. The groups had agreed to meet in the U.S. in late March or early April, in Moscow in July, and again in the U.S. in October.

The eighth annual meeting at Wallops Flight Center Nov. 19-25 of the NASA-Soviet Space Biology and Medicine Working Group had discussed biomedical findings: preliminary results of the Soviet *Cosmos 936* unmanned flight carrying U.S. experiments, and of the *Salyut 5/Soyuz 21* and *24* missions. The USSR had invited the U.S. to participate in Soviet biosatellite flights in 1980 and 1981. A workshop on simulated weightlessness held in Bethesda, Md., had preceded the meeting. Dr. David Winter, NASA director for life sciences, had headed the U.S. life sciences delegation and Dr. Rufus Hessberg, NASA director of space medicine, the workshop delegation. Soviet leader at both meetings was Dr. Nikolai Gurovsky of the USSR Ministry of Health. The ninth meeting would take place in the USSR in the second half of 1978. The workshop and meeting had been part of a program set up under a 1971 Science and Applications Agreement between NASA and the Soviet Academy of Sciences. (NASA Release 78-10)

• LeRC announced it had worked with Kresge Eye Institute in Detroit to reduce and regulate pressure inside the eye during glaucoma surgery using a device based on fluid-systems and components technology developed by NASA. The pressure-reduction system could mechanically lower intraocular pressure (inside the eye) to any level desired by the physician over a set time and in a controlled manner. A fluid supply with pressure matched to existing intraocular pressure would enter the anterior (front) chamber of the eye through a tiny tube inserted near the edge of the cornea; controlled reduction of fluid-supply pressure would reduce intraocular pressure by the same amount and at the same rate. The system would permit investigation and assistance in reduction of some post-eye surgery complications such as cataract formation, vascular-membrane detachment, flat anterior chamber, and malignant glaucoma. NASA planned to continue development of the device under controlled clinical conditions. (NASA Release 78-12)

• MSFC announced plans for two runs by a high-speed sled in Feb. at the Sandia sled track, Albuquerque, N.M., to test the "broadside deployment phase" of the Space Shuttle's solid-fuel rocket booster parachute-recovery system. The sled tests would deploy only the pilot chute to determine if the nosecap when ejected would clear the vehicle without becoming entangled, as boosters would come in flat, not sideways like an arrow, and air would flow across the nose cone at right angles. MSFC had developed the SRB and its recovery system. *(Marshall Star*, Jan 25/78, 4)

• INTELSAT announced appointment of 3 deputy directors and 4 division directors. New deputy director generals were Andrea Caruso, administration; Reginald Westlake, finance; and H. William Wood (former NASA deputy associate administrator for networks, OSTDS), operations and development. New division directors were Francis Latapie, procurement; Emeric Podraczky, engineering; William Geddes, operations; and NandKishore Chitre, systems planning. After INTELSAT's contract with ComSatCorp for management services expired in Feb., INTELSAT would assume responsibility for all technical, operational, and procurement functions previously performed by ComSatCorp, which would continue to provide certain technical tasks under two service contracts. (INTELSAT Release 78-4-I)

January 26: NASA announced it had launched the International Ultraviolet Explorer (IUE) from KSC at 12:36pm EST on Jan. 26, 1978, aboard a Delta rocket. KSC had delayed the launch for 4hr to secure an access door on the Delta heatshield. The Delta had injected IUE into a transfer orbit with 46 342km (28 784mi) apogee, 167km (104mi) perigee, and 28.7° inclination. IUE was part of a joint program by NASA, ESA, and the United Kingdom's Science Research Council (SRC); more astronomers (200 from 17 countries) would use IUE than had used any previous satellite. The satellite's telescope, aimed by ground controllers, would examine the ultraviolet spectrum between 1 150 and 3 200A, a region inaccessible to study from the ground.

Studies would range from planets in earth's solar system to some of the most distant objects in the universe, including quasars, pulsars, and black holes in space. Astronomers would study a wide range of stars to learn how they were born, live, and die; the material between these stars from which they are formed and are still forming; strange objects emitting radio waves or x-rays, or both; the sedate nearby galaxies and the violent distant quasars; and earth's planetary neighbors and their satellites. IUE would complement and extend current NASA observations from the two orbiting astronomical observatories, *OAO 2* and *Copernicus*, and ESA's *TD 1* satellite. IUE's follow-on would be the 10-ton Space Telescope (ST) to be launched by the Shuttle in 1983.

The IUE spacecraft was an octagonal cylinder approximately 1.5m tall and 1.3m between opposing faces. Overall length of 4.2m included a telescope tube at the forward end and an apogee motor extending from the rear. Solar arrays on opposite sides of the spacecraft when unfolded in space would measure 4.3m (14ft) wide. Spacecraft weight was 1479lb, of which 272lb was scientific instrumentation. After stabilization, the spacecraft would continuously point one of its two solar arrays at the sun.

Goddard Space Flight Center would have a constant view of IUE; ESA's Villafranca station near Madrid, Spain, a view at least 10hr per day. GSFC had designed, integrated, and tested IUE and provided U.S. ground-support facilities; ESA had built the solar array and the Madrid ground facilities. SRC, in collaboration with Univ. College, London, had provided four television camera detectors to transform spectral displays into video signals for transmission to the ground. The joint agreement had allocated 16hr of viewing time a day to NASA, 8hr to be shared jointly by ESA and the United Kingdom. The U.S. had spent $57 million on the spacecraft, Delta rocket, and other items; ESA had spent $21 million; SRC and Univ. College had spent $9 million. (NASA Release 78-8; MOR S-868-78-01 [prelaunch] Jan 19/78, [postlaunch] Feb 2/78; *NYT*, Jan 27/78, A-8; ESA Release Jan 20/78; *Goddard News*, Jan-Feb 78, 1)

• NASA announced it would celebrate Jan. 31 the 20th anniversary of the first U.S. artificial earth satellite, *Explorer 1,* which had ushered the western world into the space age. Principal scientific achievement had been discovery of the Van Allen radiation belts surrounding the earth. Weighing 13.9kg (30.8lb), measuring 15.2cm (6in) in diameter and 203.2cm (80in) long, and shaped like a stovepipe, *Explorer 1* had been the product of two other organizations now elements of NASA: the development group of the Army Ballistic Missile Agency (ABMA) at Huntsville, Ala., and the Jet Propulsion Laboratory (JPL) of CalTech at Pasadena, Calif. In 1960 ABMA had become the nucleus of MSFC; JPL had gone under contract to NASA in Dec. 1958. ABMA had provided the modified Redstone booster and the basic satellite design, and JPL had furnished the solid-propellant upper stages of the carrier vehicle and had packaged and tested the payload.

Explorer 1's launch from Cape Canaveral, Fla., on a Jupiter C vehicle at 10:48pm EST, Jan. 31, 1958, had come 45 days after DOD had given the Army the go-ahead to prepare a backup to the existing U.S. Navy Vanguard project. *Explorer 1* had continued to transmit information on the space environment until May 23, 1958, when its batteries were exhausted; it had reentered March 31, 1970. Dr. James Van Allen had designed the scientific package that discovered the radiation belts named for him. The *Explorer 1* launch had been part of U.S. participation in the International Geophysical Year. (NASA Release 78-13; MSFC Release 78-9; JPL Release Jan 27/78; NRC *News Report*, Feb. 78, 1; *Spaceport News,* Jan 9/78, 2; Jan 20/78, 3; *JPL Universe,* Jan 20/78, 1; *Marshall Star*, Jan 25/78, 2)

• Although the U.S. and USSR had signed separate treaties in 1963 and 1967 prohibiting test and deployment of nuclear weapons in space, western observers had interpreted the USSR's Cosmos series as an antisatellite development program, Farooq Hussain reported in a *Nature* article. Killer satellites had exploited the poor resistance of satellite components (particularly solar cells) to intense heating and radiation damage; they could aim at satellites' high energy-lasers or ion beams to cause arcing and discharge through instrumentation. Precision-guided missiles could destroy satellites either by colliding with them or by detonating conventional explosives close to a target. One explanation of recent Soviet tests had been China's development of reconnaissance satellites. The only U.S. test of an antisatellite had occurred in 1963 when a modified McDonnell Douglas Thor missile had been fired against a missile booster in low earth orbit.

Recent advances had indicated that killer satellites might carry high-energy continuous-wave or pulsed lasers to destroy target satellites; both sides had used ground-based lasers for several years to interrogate reconnaissance satellites. In Sept. 1977 the Air Force had awarded to Vought Corp., Dallas, Tex., a $58.7 million contract for development of a killer satellite. To avoid the cost of this arms race, the article said, both the U.S. and USSR had been willing to agree on limiting development of killer satellites. (*Nature*, Jan 26/78, 293)

January 27: NASA announced that scientists from NASA's Ames Research Center and NOAA's Environmental Research Laboratory in Boulder, Colo., had begun flight tests of an instrument designed by NOAA scientist Dr. Peter Kuhn to warn pilots of impending turbulence (see *A&A 1976,* Nov. 11, 277). Dr. Kuhn, an experimenter on NASA's Gerard P. Kuiper Observatory (a C-141 aircraft equipped with a large telescope for astronomy research), had noticed that variations in atmospheric water vapor as measured by an infrared (IR) water-vapor radiometer had often accompanied extreme turbulence encountered by the aircraft; he had succeeded in using IR radiometer to define atmospheric conditions in the telescope's field of vision. Dr. Kuhn then suggested mounting an IR radiometer aimed forward in an aircraft; this system had successfully predicted occurrences of turbulence in clear air from 2.5 to 5.5min before encounter. However, researchers needed more encounters from various directions and altitudes to test the instrument, and wanted to experiment with filters and to fine-tune the instrument for optimum results.

The joint NASA-NOAA program would consist of exhaustive tests using a NASA Learjet research aircraft to develop a low-cost system that would enhance flight safety and comfort, be suitable for any aircraft, be able to operate unattended and require minimum maintenance, and produce a cockpit visual alert 4 to 5min before a turbulence encounter. (NASA Release 78-15)

• LaRC reported on its Jan. 17-19 seminar on Large Space Systems Technology (LSST) attended by more than 200 persons who had exchanged ideas on the technology needed to build large space systems in the future. The group had tried to guess what research and technology the multimission needs of the Space Shuttle would be in 1985 to 2000; NASA had scheduled no definite programs for 1985 to 2000, but wanted to ensure performance of economical and well-organized future missions.

Seminar subjects included predictions of structural behavior in space, particularly under loads and stresses; advanced materials and techniques for joining various kinds of structures; distribution of controls for large structures; and electronics equipment needed for operation and control of structures. Those attending had represented six NASA centers, 24 aerospace firms, 8 universities, and 4 DOD centers. To continue cooperation between government agencies and industry, LaRC planned to hold a similar seminar yearly. (LaRC Release 78-3; *Langley News*, Jan. 27/78, 1)

January 28: A Canadian Air Force transport searching for radioactive contamination from a Soviet satellite that disintegrated over northwest Canada had reported a "highly probable" radiation contact, the *W. Star* reported, first to be discovered since Jan. 24 when the crippled *Cosmos 954* satellite powered by a nuclear reactor containing 100lb of highly radioactive uranium 235 had reentered earth's atmosphere. The occurrence had sparked controversy over hazards of radioactive space debris; the U.S. space program had used radioisotope power systems, but had launched only one fission reactor, that on the *Snap 10A* test vehicle. Fission reactors such as that on *Cosmos 954* would create strontium, cesium, and cerium byproducts potentially lethal if radiation dosages were high or prolonged.

The USSR had launched at least 10 reactors on spacecraft like *Cosmos 954*, 16 with the same operating parameters having flown since Dec. 27, 1967; it had put all the reactor craft except *Cosmos 954* in safe high orbit. The Soviet and U.S. governments had kept close contact during the incident. National Security Affairs Assistant Zbigniew Brzezinski, quoted by TASS, said, "I would like to point out two things: first that this incident which, to my mind, is nothing out of the ordinary in the space age, is fully controlled so we believe there is no danger. . . . Second, that the governments of the U.S., the Soviet Union, and other governments concerned cooperated intensively concerning this matter." (FBIS, Tass in English, Jan 25/78, Jan 28/78; *W Star*, Jan 26/78, A-1; *AvWk*, Jan 30/78, 33; *Economist*, Jan 28/78, 75)

• NASA announced that it had appointed Dr. John McCarthy director of LeRC, effective Oct. 1, 1978. Currently director of Massachusetts Institute of Technology's Center for Space Research and a professor of aeronautics and astronautics, Dr. McCarthy had been widely recognized as an expert in systems engineering and vehicle design. He had served as a member of the Air Force Scientific Advisory Board and chairman of the Aeronautical Systems Division Advisory Group, Air Force Systems Command, and had been a member of the Joint Strategic Target Planning Staff Scientific Advisory Group for the Joint Chiefs of Staff. (NASA Release 78-16)

January 30: FBIS reported that the People's Republic of China had launched on Jan. 26 an earth satellite that had functioned normally in orbit and returned to earth "after fulfilling tasks of scientific experimentation." The report added that the achievement had resulted from the guidance of the wise leader Chairman Hua Kuo-feng. (FBIS, Tass in English, Jan 30/78)

• Thiokol Corp.'s Wasatch Div. had used an airfilm material-handling system at its Brigham City, Utah, plant to move and position Space Shuttle solid-fuel rocket-motor components weighing up to 300 000lb, *AvWk* reported. The size of components for the Shuttle's solid-fuel rocket motors and the large number of Shuttle missions expected had led Thiokol to try new types of material-handling systems.

Heaviest load moved had been a solid-fuel rocket-motor casting segment loaded with propellant weighing about 300 000lb on a transporter dolly weighing an additional 50 000lb, for a combined total of about 175 tons moved. (The motor would consist of four segments individually cast with solid rocket motor fuel in one of several pits at the Thiokol facility.) The airfilm system used compressed air to inflate urethane pads (bearings) attached to the bottom of specially-designed dollies; air forced into the bearings and seeping between them and the floor had created a space of two or three hundredths of an inch—enough to "float" the transporter and its load on a film of air. (*AvWk*, Jan 30/78, 56)

January 31: ESA announced that it had successfully tested the second stage of its Ariane launcher in flight configuration, at the Deutsche Forschungs- und Versuchsanstalt fur Luft- und Raumfahrt e.V. test center in Hardthausen, Germany. The 138sec test had checked compatibility between the structure (fuel tanks) and propulsion system (engines) and that of the pressurization system, propulsion-system performance, and structure resistance. ESA had scheduled two further tests in the next few weeks; all three stages of the launcher had reached complete-stage testing. The Centre National d'Études Spatiales, France, had developed Ariane for ESA; more than 50 firms in 10 ESA-member countries had participated in the program. (ESA Release Jan 31/78)

• Federal Aviation Administrator Langhorne Bond said the British Civil Aviation Authority and the British Plessey Corporation had been "using the press and other news media to deliberately mislead and confuse the world aviation community" about a U.S.-backed all-weather approach and landing system. The FAA administrator said the time reference scanning-beam/microwave landing system (TRSB/MLS) had been developed in a "goldfish bowl" from the outset with wide public participation; he labelled as "simply not true" British allegations that the U.S. had conducted a misleading test program or issued false data, and that the TRSB system was unsafe. Bond said the International Civil Aviation Organization (ICAO), at its worldwide meeting in April, would probably select TRSB as the international system and that the campaign to discredit TRSB had been a last-minute effort by the British and Plessey to influence ICAO's decision. (FAA Release 9-78; *W Star*, Jan 7/78, A-1)

During January: NASA announced it had tentatively selected 18 scientists to participate in the design and early operational phases of the Space Telescope project. The 10-ton observatory to be launched by the Space Shuttle in 1983 would make astronomical observations deeper into space and in more detail than ever before. The scientists would head teams representing investigations definition (design and develop the focal plane instruments and carry out major scientific investigations); astrometry science (ensure that a fine guidance system could perform the required astrometric functions); faint-object camera science (U.S. members to assist ESA's team); telescope science (interpret performance requirements in terms of telescope design specifications); data and operations (establish instrument specifications for control system, flight operation, and ground data handling); and interdisciplinary science (provide a broad scientific overview of ST development). A Space Telescope science working group to give scientific guidance to the project would consist of team leaders, 2 telescope scientists, interdisciplinary scientists, and appropriate project personnel. MSFC had management responsibility for the telescope; GSFC would manage operations and the development of scientific instruments. *(NASA Activities,* Jan 78, 9)

• NASA announced it would hold a symposium Jan. 16 and 17 in Washington, D.C., on the proper role NASA should be prepared to fulfill during the early to mid-1980s. The symposium would host 20 nationally-known authorities in fields including federal research policy, economics, education, science, aircraft development, and space applications. The symposium was a part of its current assessment of its institutional base. (NASA Release 78-5)

• NASA reported that DOD had lent it 5 Navy divers to support simulated weightlessness tests in the Space Shuttle program. The divers, all fleet-salvage men selected for their expertise in erecting or dismantling underwater structures or equipment, had been assigned to the neutral buoyancy simulator at MSFC. Working underwater in full-scale segments of the Space Shuttle, they would assist test engineers with contingency procedures, equipment operation, evaluation of space-construction techniques, and other activities scheduled for Shuttle and Spacelab missions. *(NASA Activities,* Jan 78, 5)

• GSFC reported that, as a result of its concern about possible volcano damage to a Spaceflight Tracking and Data Network (STDN) station, Mt. Cotopaxi in Ecuador, tallest volcano in the world, almost 6km (3.75mi) high, had been transmitting continuous information on its seismic activity. Dr. Richard Allenby of the Geophysics Branch at GSFC had installed a geophysical monitoring system consisting of a single-axis vertical seismometer and two tiltmeters. The system had transmitted data to the STDN station for quick review and relay for analysis to Dr. Minard Hall, geology professor at Ecuador's Escuela Politecnia National. Any unusual volcanic activity identified by Hall would be immediately relayed back to the station and to the Ecuadorian authorities.

The seismometer and its preamplifier, the only permanent volcano-monitoring system in South America, had been encased in a sealed plastic tube and buried about a meter (3ft) underground for fear of an eruption within the next few years. Other remaining field instruments and batteries had been buried in 50-gal oil drums. Although the equipment had not detected major volcanic activity, it had detected ground motions along a nearby active fault line. (*Goddard News*, Jan/Feb 78, 4)

• MSFC had issued to industry a request for proposals to study an automated longwall guidance and control system for coal mining and to build a prototype, *NASA Activities* reported. Studies by MSFC engineers over the past 2 yrs had indicated the potential practicality of a system using gamma rays, radar beams, impact devices, or reflected light to guide a longwall shear's cutting edge. The engineers had applied automated guidance and control-system technology developed for the lunar roving vehicle to a "longwall machine" that would grind coal from a wall or face of a coal seam in a more or less straight line, send it out on a conveyor, and advance with mechanized roof supports. Automating such a machine could eliminate the need for humans to work next to the cutting operation. Automated guidance and control would permit much faster removal of virtually all the coal present, increasing annual production by millions of tons of purer coal. The automated system should reduce wear on cutters and minimize machine downtime. (*NASA Activities*, Jan 78, 9)

• The Air Force Systems Command reported it had successfully concluded a 2-yr space-communications test program 4mo ahead of schedule, saving the Air Force more than $500 000. Directed by the Satellite Communications Terminal Program Office, SAMTEC's 6555th Aerospace Test Group had launched the Lincoln experimental satellites *LES 8* and *9* from Cape Canaveral 22 000mi into space to demonstrate advanced technology that would provide dependable jam-proof communications between satellites spaced thousands of miles apart and terminals scattered over more than 75% of the earth's surface. The satellites and terminals had worked essentially flawlessly.

In the test, a receiver and transmitter on each satellite had relayed messages for the first time directly between two spacecraft without ground or airborne relay. Acting as earth-orbiting relay stations during the test program, the satellites exchanged voice and data information between airborne, ground, and shipboard terminals. Instead of batteries charged by solar panels (susceptible to radiation damage), each satellite had used two nuclear-powered thermal generators for long-term dependable power services. Some tests had run continuously for 2 or 3 days, requiring 24-hr manning. Nerve center for the program had been the test-management facility at the Mitre Corp., Bedford, Mass. (*ASFC Newsreview*, Jan 78, 6)

• NASA announced personnel changes at Hq.
—Herbert Rowe, acting deputy associate administrator for external relations, would leave to become vice president of the Electronic Industries Association. Rowe had joined NASA in 1975.
—NASA had appointed Frank Penaranda director of the institutional Operations Division, Office of Management Operations, effective im-

mediately. He would be responsible for managing the agency's R&PM budget and institutional management system, including program support negotiations between centers and program offices, coordinating agency-wide civil service and contractor manpower planning, and overseeing the center's aircraft operations.

—NASA had appointed C. Ronald Hovell director of the Resource Management/Administration Office, Office of Space Transportation Systems. He would be responsible for budget and control support of each STS program directorate having responsibility for STS funds.

—Effective Jan. 1, Dr. S. Ichtiaque Rasool had become chief scientist in the Office of Space and Terrestrial Applications. Rasool had maintained an active relationship with academic, industrial, and government organizations working on development and use of scientific and technological instrumentation, theory, and techniques for study earth and its nearby space environment. (NASA anno Jan 11, 16, 20, 30/78)

February

February 1: Fifteen scientists who would manage earth-orbital investigations for the third Spacelab in mid-1981 had met at Marshall Space Flight Center to discuss preparations for the mission, the *Marshall Star* reported. Known as the investigators' working group for Spacelab mission 3 (IWG-3), the group would work with NASA mission management to develop a payload offering the greatest possible scientific return and to select payload specialists who would perform experiments aboard Spacelab 3. Those at the meeting agreed on professional qualifications for payload specialists and on criteria for their selection, beginning in the second half of 1978.

Sponsored by the NASA Hq Office of Space and Terrestrial Applications, Spacelab mission 3 would include materials processing in space, the atmospheric cloud-physics laboratory, the drop-dynamics module, and vestibular-function research. Experiments would obtain basic knowledge about crystal-growing processes in earth orbit, mechanisms for generating small latex spheres, dynamic behavior of free-floating drops of liquid, cloud formation in the atmosphere, red blood cell aggregation, and adaptation of life forms to space. MSFC would manage integration and operation of the scientific payloads flown on the first 3 Spacelab missions. (*Marshall Star*, Feb 1/78, 1)

• MSFC reported that the USAF had redesignated an interim upper stage (IUS) for Space Shuttle as the "inertial upper stage," to reflect its inertial-guidance capabilities as compared to the NASA-sponsored spinning solid upper stage. The new name would also indicate that DOD payloads would use the stage well into the 1980s. (*Marshall Star*, Feb 1/78, 1; *Spaceport News*, Feb 17/78, 3)

February 2: NASA announced plans to launch on Feb. 9, 1978, the first spacecraft in a new worldwide military communications system. DOD's Fleet Satellite Communications (FltSatCom) satellites would provide vital communications for the Navy, Air Force, DOD, and the Presidential Command Network [see Jan. 20]. The spacecraft would go into synchronous-transfer orbit on an Atlas Centaur from ETR's launch complex 36.

FltSatCom, a highly versatile military-communications system built by TRW Defense and Space Systems, would offer 23 operating channels:

the Navy would use 1 fleet-broadcast and 9 fleet-relay channels for communications between naval aircraft, ships, submarines, and ground stations, ship-to-ship and ship-to-shore. The system could offer reliable secure communications between naval units, even relatively small vessels equipped with inexpensive antennas and simple equipment, anywhere in the world except in extreme polar regions. The Air Force would use 12 narrowband channels as a part of the Air Force Satellite Communications System (AfSatCom) linking the national command authority with Strategic Air Command (SAC) units and other arms of the Air Force; DOD had reserved 1 channel. The Naval Electronic Systems Command would manage the FltSatCom program.

The spacecraft would consist of two major hexagonal elements (a payload module and a spacecraft module), with the majority of electronic equipment mounted on 12 panels enclosing the payload and spacecraft modules. The payload module would contain UHF and X-band communications equipment, and antennas—the UHF-transmit antenna, built of ribs and mesh to open like an umbrella; and the receive antenna, a separate deployable helix. The spacecraft module would contain earth sensors, an apogee kick motor, attitude and velocity control, telemetry tracking and command, electrical-power distribution, and a solar array folded around the spacecraft module until reaching final position in orbit, to be deployed by spring-loaded hinges. The 1875kg (4132lb) spacecraft weighed about 390kg (860lb), more than any satellite previously launched into synchronous orbit on an Atlas Centaur, NASA's standard launch vehicle for intermediate-weight payloads. (NASA Release 78-17; MOR M-491-202-78-01 [prelaunch], Feb 3/78)

• Former NASA Administrator Dr. James C. Fletcher received NASA's Distinguished Service Medal at a ceremony at NASA Hq on Jan. 30, 1978. Dr. Frank Press, President Carter's science adviser, made the presentation in recognition of Dr. Fletcher's "distinguished leadership . . . outstanding scientific and administrative abilities . . . and contribution to the nation's significant achievements in the exploration of space and utilization of space to manage the earth's resources." Dr. Press noted, "His decision to proceed with the development of the Shuttle orbiter exemplified the perceptive leadership which foresaw space as a giant laboratory where men will work to improve the future of all mankind." Dr. Fletcher, who was NASA administrator from April 27, 1971 until May 1, 1977, was a consulting engineer and vice president of the National Space Institute in Washington, D.C. (NASA Release 78-18, *Langley Researcher,* Feb 24/78, 1)

• A detailed article on manned spaceflight in a recent Chinese technical journal had raised speculation that the People's Republic of China might attempt to put a man in orbit, reported FBIS. *Navigation Knowledge,* published by the China Navigation Society, had printed a description of manned-flight problems such as weightlessness and satellite design, the effects of weightlessness on the body, and the growing of food in outer space. The article suggested that the PRC, having retrieved three unmanned satellites from orbit, was now ready to experiment with living in space.

Much of the article concerned physical changes resulting from a spacecapsule environment, and the food, clothes, and waste facilities appropriate for an astronaut, information apparently based on completed experiments. By using the direct rays of the sun, the article claimed, a space voyager could grow rice and wheat the size of a Chinese date tree and eggplant and peppers the size of watermelon. Observers believed Chinese space technology had developed considerably since the initial space launches and that the next (No. 8) launch might have an animal aboard. (FBIS, Kyodo in English, Feb. 2/78)

February 3: Kennedy Space Center reported it had begun a 3-mo exercise called Spacelab Critical Design Review, whose results would go to a final board meeting on design changes needed for Spacelab to fly. ESA's prime contractor, ERNO, had assembled Spacelab data consisting of 34 volumes and approximately 3000 drawings and had packaged 5 sets, weighing a total of 500lb, for shipment to the U.S. KSC review groups would look for discrepancies to ensure that the data reflected the hardware design and that the design was compatible with KSC operations. The reviewers would submit any discrepancies to a screening group, which would select those to be forwarded (along with those identified by other NASA centers) to MSFC for consolidation and decision on which would be sent to Europe. Those finally selected would be consolidated with ESA-identified discrepancies and forwarded to ERNO. *(Spaceport News,* Feb 3/78, 1)

• John Pierce of the CalTech department of electrical engineering wrote in a *Science* magazine editorial that space exploration and technology had been remarkably rewarding by any standards, because the American public and government had been willing to try new and promising things without any guarantee of success. "Some areas of space," he said, "appear to be reserved perpetually for our government or competing governments. We must continually spend government money to ensure that our launch capabilities keep ahead of those of the rest of the world. Without this, even the most lucrative uses of space will pass into other hands." He advocated continued funding for planetary exploration and other radically new space science, even without immediate financial return: "If space science, including planetary exploration, is not adequately funded, we will lose an art which, having led us to glory, may lead us much farther. A future without adequate support in this area of great national success would be dismal indeed." (*Sci*, Feb 3/78, 4328)

February 4: FBIS reported that the Japanese Defense Agency had deciphered the "identification, friend or foe" (IFF) code of the Soviet MiG-25 flown by a defecting pilot to Japan's Hakodate airport in Sept. 1976. IFF, a top-secret device, would enable a pilot flying at Mach 2 to identify a plane appearing on a radar screen. At that speed, since identification could not be made by sight, the pilot would send an electric wave in cipher; a plane that failed to return a prearranged signal would be attacked. The decoding was the result of intensive study of data observed from the MiG-25. Western military experts had discovered the Soviet approach to constructing secret codes, the basic units of pulse signals emitted by the IFF, and the sequence of Soviet codes. The USSR would now have to revise all IFF codes for the MiG series, including 300 MiG-25s and 115 MiG-25Bs. One aviation expert noted that the IFF had provided vital information; ". . . it is not an overstatement to say that the IFF was the only secret about the MiG-25 that landed at Hakodate." Japanese and U.S. authorities had returned the MiG to the USSR after study. (FBIS, Kyodo in English, Feb 4/78)

February 6: NASA announced it had appointed Dr. John Klineberg deputy associate administrator in the Office of Aeronautics and Space Technology (OAST), where he would be responsible under the OAST associate administrator for general management and direction of OAST programs. Klineberg had come to Ames Research Center in 1970 from CalTech; while at ARC, he had done fundamental research on transonic flow, particularly on viscous effects and boundary-layer separation. After joining the OAST Aerodynamics and Vehicle Systems Division at Hq in 1974, Klineberg had managed aeronautics research on short- and reduced-takeoff-and-landing (STOL), vertical-takeoff-and-landing (VTOL), and rotorcraft vehicles. Appointed in 1976 to the OAST Aircraft Energy Efficiency Office, he had been responsible for aerodynamics and active-controls elements of that program. (NASA anno Feb 6/78)

• *Av Wk* reported that the Carter administration had planned an intensive $400- to $500-million program of killer satellite technology to overcome the USSR's 10-yr lead. The USAF Flight Test Center at Edwards AFB, Calif., announced it had sought proposals for an air-launched space booster usable in an antisatellite program. Vought Corp. had undertaken work on killer-satellite technology under a $58.7-million contract. Such a non-nuclear warhead system could consist of small rocket tubes clustered around an optical-sensor package containing a small digital computer;

this device would use the rocket-propulsion tubes as the kill mechanism. In the antiballistic-missile mode, a warhead's terminal homing device would detect the target, and the computer would govern maneuvers by firing the tube-mounted rockets. Once the device approached within range of the target, the rocket tubes would separate and fire into the path of the object, destroying it.

President Carter revealed that the U.S., in addition to the "hot-metal" kill mechanism, had been working on lasers as possible spaceborne defense against Soviet killer-satellite attacks. At the same time the USSR had pushed forward with laser-equipped killer-satellite systems, the U.S. had jeopardized its negotiating position in upcoming killer-satellite limitation talks by failing to convene a cabinet-level interagency group to decide between State Dept. and DOD negotiating strategies, marked by differences in opinion on verification and on impact on U.S. systems proposals. (*AvWk*, Feb 6/78, 18)

• *Av Wk* reported that the Research Ministry of the Federal Republic of Germany had sent NASA a financial deposit to reserve space on Spacelab flights in 1981 and 1982. ESA, also interested in space on two near-term flights, had not yet made the small deposit required for inclusion in mission planning. FRG also had placed an order with NASA for space on the Shuttle for 25 "getaway special" small payloads, most of them space-processing experiments that would lead to larger processing activities on later Shuttle/Spacelab missions. (*AvWk*, Feb 6/78, 15; Feb 27/78, 13)

February 8: MSFC announced that its Boeing 747 modified for Shuttle-orbiter transport had arrived at Redstone Army Airfield in Fla. for a dry run in preparation for its March landing with an orbiter atop. Crews from MSFC, KSC, and Dryden Flight Research Center had checked out handling gear and procedures to be used at MSFC for receiving the orbiter and offloading it from its carrier. MSFC had installed a large lifting derrick at the airfield with lateral-restraint and other devices needed to lift the Enterprise clear of the 747 and position it on a trailer-like transporter. When the 747 had departed, handling crews at the airfield would continue verification of the mate-demate device, using the orbiter ground transporter and a full-size orbiter simulator. (*Marshall Star*, Feb 8/78, 1)

• Tass announced that the automatic supply-transport ship *Progress 1* had reentered and burned in the atmosphere Feb. 8. The vehicle had a mass of 7020kg and could carry into orbit 2300kg of equipment, materials, fuel, and gas. Principal new features of *Progress 1* had been the refueling system and a means of carrying an increased air supply. Its automatic operation on command from either the ground or from the orbiting Salyut had been without precedent in spaceflight. *Progress 1* had also recorded dynamic characteristics of the *Salyut 6/Soyuz 27/Progress 1* complex and had corrected the station's orbit. (FBIS, Tass in English, Feb 8/78; Feb 6/78)

February 9: Dr. Frank Press, director of the White House Office of Science and Technology Policy, had launched a campaign to reduce paperwork required by the government from university scientists receiving federal support, *Nature* reported. Press said money saved in this way might mean millions of dollars for every university, and wrote to NASA and NSF asking what steps they could take to reduce paperwork and to streamline grants regulations. Press added that, although there was little support in the administration for a Dept. of Science, interest was growing in establishing a Dept. of Technology and Industrial Development. (*Nature*, Feb 9/78, 497)

February 10: NASA announced it had selected Fairchild Space and Electronics Co., Germantown, Md., to negotiate a fixed-price incentive contract for integrating, testing, and delivering 2 multimission modular spacecraft (MMS) for Landsat-D at a target cost of about $10.3 million. The MMSs would serve as basic spacecraft buses for the Landsat-D mission and a backup.

Under the contract managed by GSFC, Fairchild would integrate and test the spacecraft buses; design and test the propulsion module; fabricate and test a signal-conditioning and control unit; fabricate a spacecraft structure; design and fabricate a spacecraft harness for the thermal and electrical systems; acquire ground-support systems and a simulator for spacecraft integration and testing, and for delivery to spacecraft users; integrate these elements with 3 government-furnished subsystem modules (power, communications and data-handling, and attitude-control subsystems); test the integrated systems; and provide 24 man-months of instrument support. The contract would include options for integrating and testing up to 4 additional MSSs for future programs through approximately 1982. (NASA Release 78-21)

• NASA announced plans to develop hardware systems to support materials-processing experiments aboard the Space Shuttle in mid-1981. The materials-processing in space (MPS) program would take advantage of near-weightlessness in earth orbit to eliminate sedimentation in liquid materials and to reduce movement of liquid caused by heat, allowing precise control of processes such as casting and crystal growth.

MSFC had asked industry to submit proposals by Feb. 23 for support of 8 materials-processing experiments on 1981 Shuttle/Spacelab missions, specifically a fluids-experiments system and a solidification-experiments system using a high-temperature furnace; other hardware systems to be developed would support individual experiments and related earth-based systems. The fluid-experiments system would use an area in the Spacelab for optical observation of fluid phenomena; it would accommodate a variety of cells for use in a broad range of experiments. The solidification experiments system would require a multiuse system for processing metals, alloys, eutectics, semiconductors, crystals, ceramics, glasses, etc., in orbit by melting, refining, and resolidification in near weightlessness. Scientists at MSFC's ground control laboratory would process "control samples" under conditions identical (except for gravity) to those on the spacecraft. (MSFC Release 78-20; *Marshall Star*, Feb 22/78, 2)

• To alleviate natural gas shortages and other disruptions caused by unusually harsh weather, the U.S. government would increase efforts to understand and predict seasonal variations in climate by starting a national program on climate research described in President Carter's budget proposals for 1979, *Science* magazine reported. Climate research was not new: the Depts. of Defense, Commerce, Agriculture, Energy, and Interior, as well as NASA, EPA, and NSF, would spend $76 million on it in 1978. Next year, however, would be the first time a single agency—NOAA—would coordinate all research under a national plan. Cost of climate research would rise 37%, to $104 million in FY79.

NOAA's Climate Program Plan would use a fourfold approach: assessment of the relationship between climate variations and national activities through studies of carbon dioxide, crop yields, land and water resources, and coal use; early warnings of short-term climate fluctuations through a variety of observation systems; better understanding of climate processes through research carried out on the oceans and in the atmosphere; and study from space of long-term climate variations by monitoring subsurface ocean temperatures and snow levels. NOAA would not have authority to approve or reject any agency's climate-program proposals, but could consolidate them into a single budget proposal for review by the White House each yr. (*Sci*, Feb 10/78, 671)

February 12: Space station *Salyut 6* would complete a thousand orbits on this date, Tass announced. During the day, cosmonauts Yuri Romanenko and Georgy Grechko had installed and switched into operation a new life-support-system air filter delivered by *Progress 1*. Of great interest to Soviet scientists were the crew's regular reports on natural calamities: dust storms, forest fires, and volcanic activity. Orbital complex *Salyut 6/Soyuz 27* would continue its flight. (FBIS, Tass in English, Feb 12/78)

February 13: NASA announced it had tentatively selected 4 more experiments for its Long Duration Exposure Facility (LDEF) scheduled as a major Space Shuttle payload in 1980. The four scientific experiments—to study hazards to man of ion particles in space, the chemistry of micrometeoroids, the interstellar wind, and cosmic-ray nuclei—would join 23 technology experiments selected for LDEF earlier in 1978 [see Jan. 9]. LDEF, an unmanned reusable free-flying facility carrying technical and scientific experiments mounted in special trays, would offer an easy and economical way to expose primarily passive experiments to the space environment during 6 to 9 mo of LDEF orbit.

Selected experiments and principal investigators were: free-flyer biostack experiment, Dr. Horst Bucker, Universitat Frankfurt-am-Main, W. Germany; interstellar gas experiment, Dr. Don Lind, Johnson Space Center; high-resolution study of ultra-heavy cosmic ray nuclei, D. V. Domingo and Dr. K. P. Wenzel of ESA (the Netherlands), and Prof. C. O. Ceallaigh, Dr. D. O'Sullivan, Dr. A. Thompson of the Dublin Inst. for Advanced Studies, Ireland; and chemistry of micrometeoroids, Dr. Fred Horz, JSC. (NASA Release 78-20; LaRC Release 78-5; *JPL Universe,* Feb 3/78, 1; ESA newsletter, Feb 78, 3)

• Airport and aircraft noise-reduction legislation faced a rocky future, said *Av Wk*, because of congressional concern that safety needs were taking a back seat to retrofit and replacement for aircraft compliance with federal noise standards, and that the proposed scheme to fund safety measures was unacceptable. One bill under consideration would set up a voluntary program to assist airports and surrounding communities to reduce noncompatible land uses near airports. Main target of congressional concern was Title 3 of the bill, lowering by 2 percentage points the current domestic passenger-ticket and waybill taxes and replacing them with (among other charges) a 2% surcharge on tickets and waybills. Each airline would use surcharge funds for retrofitting, reengineering, or replacing aircraft to comply with recent federal noise standards. Witnesses testified, according to *Av Wk*, that about 75% of the 2100 jet engine aircraft in the U.S. airline fleet today had not met those standards.

"We believe," testified Transportation Secretary Brock Adams, "there are significant benefits to the public at large to be achieved by providing incentives for replacement of noisy aircraft with new technology aircraft, rather than merely retrofitting or reengineering." Among those with substantial concern over the surcharge was Rep. Al Ullman (D-Ore.), who said: "Everyone is trying to impose taxes that are really outside the tax system. How do you keep it in the tax system?" Representatives of the Aerospace Industries Association had testified in favor of the bill. (*Av-Wk,* Feb 13/78, 32)

• Robert Kirk, president and chief executive officer of Vought Corp., in a speech to the Aviation/Space Writers Association, had called for a comprehensive multinational treaty on international operations in space, reported *Av Wk*. "Why not a United Nations conference to work out international protection—not unlike the 'freedom of the seas' philosophy which has worked so well throughout the world for centuries?" Kirk asked. He said a treaty had to be approached as a totality, because space defense was only one element of a freedom-of-space policy. And he noted that treaty negotiators should consider all space systems on a broad international basis. "As you know," he pointed out, "Communist China and European Space Agency countries are already entering the picture, so issues are no longer confined to U.S. and Soviet interests. Therefore, in seeking an enduring treaty, a broad international forum is required. Most important," Kirk concluded, "any agreement must guarantee that we are free to operate—that any nation is free to operate—in space and to share its benefits. . . . A good treaty is an important step in this direction." (*AvWk*, Feb 13/78, 7)

• Aeritalia and the Italian ministry of research and technology had discussed a new Italian telecommunications-satellite program, *Av Wk* reported. Most of Italy's annual space budget had been devoted to ESA programs, but Italian government and industry had continued a limited number of strictly domestic projects, including the *Sirio* experimental comsat launched from KSC and the Alfa 2-stage booster that had qualified in three successful test firings from Italy's Saito di Quirra experimental test range. Current discussions of future satellite programs could lead to detailed telecommunications project studies this year, and development work beginning in 1979. A major space undertaking for Aeritalia was the ESA Spacelab; Aeritalia had been responsible for the structure of the pressurized Spacelab module. (*AvWk*, Feb 13/78, 41)

• INTELSAT announced it had signed agreements with Saudi Arabia and Zaire to lease capacity on global-coverage Intelsats for domestic telecommunications in those countries. Saudi Arabia had leased a total of 2.25 global-beam transponders, extending domestic service to 8 cities in addition to Jeddah, Riyadh, and 2 mobile antennas served since 1977. The 5-yr agreement with Zaire had earmarked a full global-beam transponder in a spare-capacity satellite to serve 13 locations in that nation. Fifteen countries had begun (or would soon begin) to use the Intelsat system for domestic telecommunications. (INTELSAT Release 78-5-I)

February 14: NASA announced two promotions: James Morrison to the position of European representative, and Louis Lushina to director of the information systems division, Office of Management Operations. For the past 3.5yr Morrison had managed the earth resources survey program in NASA's Office of Applications; before that, he had served for

12yr in NASA's Office of International Affairs. He had been an F-86 and F-100 pilot and served as a flight commander, operations officer, and squadron commander. As European representative, Morrison would be liaison between NASA and European space agencies.

In his new position, Lushina would be responsible for managing agencywide information-systems efforts and ADP resources. He had come to NASA in 1968 from the Army Materiel Command and had served since then as deputy of financial management in the Office of the Comptroller. Previously he was deputy comptroller of the U.S. Army Communications Zone in Europe and of the U.S. Army Supply and Maintenance Command, and assistant comptroller of the U.S. Army Ordnance Corps. He held the NASA exceptional service medal and the Army's meritorious civilian service medal. (NASA Release 78-23; NASA anno Feb 14/78, Feb 15/78)

• ESA reported first results from the international ultraviolet explorer launched Jan. 26. A European team working Feb. 9 at ESA's Villafranca station had located and observed Capella (a star in the Milky Way galaxy) for the first time in the complete ultraviolet spectrum. Like the sun, the strong-magnitude (bright) star appeared to be surrounded by an area called the chromosphere, in which chemico-gaseous exchanges occurred. IUE's program of scientific and technological observation would continue until the end of March. (ESA Release Feb 14/78)

February 15: MSFC announced NASA had extended the mission lifetime of the first earth-orbiting high-energy astronomy observatory *Heao 1*, originally scheduled to end Feb. 12, to at least mid-November when the second of the series was due for launch. The high quality of scientific data received from *Heao 1* had prompted the decision to extend the mission. Launched Aug. 12, 1977, *Heao 1* had been the first of three planned large unmanned orbital observatories managed by MSFC; the third would be launched in 1979. *Heao 1* had been a scanning mission for fully mapping the celestial sphere; the second Heao mission would point very precisely at x-ray sources identified by *Heao 1*, which had already significantly advanced knowledge of celestial x-ray and gamma-ray sources. By extrapolating these results, the scientists expected to increase the number of x-ray objects identified by the first 6 mo of data from 200 to 1000 or more, with further increases during the extended operation.

Heao 1, although basically a scanning instrument, had been versatile enough to stop its rotation and point briefly at items of particular interest; resolution and sensitivity would increase tremendously in the pointing mode. *Heao 1* was performing about 5 daytime pointings lasting about 3hr during a week; extended mission lifetime would enable the mission-control crew to add night pointings. Dr. Fred Speer, mission director, explained: "Astronomers would like to have their optical viewings and optical photography of stellar objects—which of course can only be done at night with optical telescopes—coincide with our pointings with the X-ray sensors aboard *Heao 1*, so that data can be correlated." (*Marshall Star*, Feb 15/78, 1)

• MSFC announced that the American Inst. of Astronautics and Aeronautics (AIAA) had installed Dr. William Lucas, MSFC director, as a fellow at the 14th AIAA annual meeting Feb. 8. AIAA had selected Dr. Lucas "in recognition of your many outstanding contributions and acknowledged leadership in the professional aerospace community," according to a letter from AIAA president R. H. Miller. Others named AIAA fellows were former astronauts Sen. Harrison Schmitt (R-N.M.) and David Scott, Dr. Donald Hearth of Langley Research Center, and Harvard Lomax of ARC. Elected an honorary fellow was former NASA Administrator Dr. James C. Fletcher. (*Marshall Star*, Feb 15/78, 1)

February 16: Scientists in NASA's Viking project this spring would award grants totaling $5000 to support undergraduate research projects in planetology or astronomy, NASA announced. The money had been part of the annual Newcomb Cleveland Award presented Feb. 15 by the American Assn. for the Advancement of Science (AAAS) to 150 Viking scientists, authors of scientific papers that had appeared in 3 special Viking issues of *Science* magazine. This had been the first Cleveland Award to a group. The grants, ranging from $500 to $2000, would fund a student's (or group of students') ideas or would supplement student work in progress, selected on the basis of originality and feasibility. (LaRC Release 78-7; *Langley Researcher*, Feb 24/78, 3)

• Modification and flight of U.S. Air Force jet cargo/tanker aircraft delivered to DFRC would allow NASA to research the possibility of significant fuel savings, NASA announced. Winglets (2.7m [9ft]-long airfoil sections attached to the wingtips of a KC-135 aircraft) could improve performance in cruise flight by approximately 8%, for an annual fuel saving of 11.9 million liters (45 million gal) based on 1975 use rates for KC-135 aircraft. Other NASA studies had indicated that use of the winglet could benefit existing medium-weight civil-transport aircraft.

LaRC had developed the winglet concept in its wind tunnels. Winglets, made of aluminum 0.6m (2ft) wide at the tip and 1.8m (6ft) wide at the base, could weigh almost 68kg (150lb). The Boeing Co. would install them on outer wing panels under an Air Force contract estimated at $3 million. Varying the incidence and cant angle of the winglets between flights would demonstrate their effectiveness at various positions. After installation of recorders, the KC-135 would make the first flights in a joint NASA-USAF program scheduled after winglet installation in early

1979. Comparison of data would demonstrate winglet effectiveness. (NASA Release 78-25)

• Using a multichamber furnace delivered to *Salyut 6* by the *Progress 1* unmanned tanker/freighter, Soviet cosmonauts had begun major space-processing experiments, Tass reported. Operated in a Salyut airlock to vent into the vacuum of space, the furnace had three areas: one maintaining heat up to 1100C, a "cold" area maintaining 600 to 700C, and a "gradient" area allowing linear temperature change between maximum and minimum furnace-heating capabilities. The Salyut space-processing system, designated "Splav" (alloy), would use material samples in capsules, each containing 3 crystal ampules to fuse when heated in the furnace and form monocrystals in the "gradient" area, with 3-dimensional crystallization in the hot and cold sections of the furnace. Combinations of materials heated to form new substances not possible in earth gravity would be aluminum/antimony, aluminum/tungsten, molybdenum/gallium, copper/indium, and indium/antimony. Some of the alloys might have semiconductor properties, suggesting that Soviet interest would be in specific applications in earth orbit rather than in theory.

The cosmonauts had also reported a major technology experiment using a closed-cycle cryogenic system designed to provide temperatures comparable to liquid helium (−269C) but did not say whether or not the test actually used liquid helium. Cosmonauts had also used cryogenics to cool the receivers of a "submillimeter telescope," Tass reported. (FBIS, Tass in English, Feb 15/78, Feb 16/78; Moscow Dom Svc in Russian, Feb 21/78; *AvWk*, Feb 27/78, 21)

February 17: NASA's *Skylab* workshop launched in May 1973 might descend to 150 nautical miles (173 statute miles, 278km) altitude and begin reentry into earth's atmosphere as early as late summer of 1979, *JSC Roundup* reported, citing orbit data from the North American Air Defense Command's satellite trackers, the Smithsonian Astrophysical Observatory, and the Swiss federal observatory. NASA would try two ways of postponing *Skylab* reentry: first, to reactivate the workshop's thruster attitude-control system in the spring of 1978 and put it into a very slow tumble, decreasing atmospheric drag and perhaps adding several months to its orbital lifetime; second, to use a teleoperator-retrieval system (TRS) launched on an early Space Shuttle mission (about Oct. 1979) and flown by remote control to dock with *Skylab*. Once docked, the TRS propulsion system could either raise *Skylab*'s orbit or control its reentry.

Although *Skylab* would break up and burn during descent, some debris would probably survive reentry, and probably would land in an ocean, because 80% of the *Skylab* orbit had been over water (between 50°N and 50°S). *Skylab*, largest payload in earth orbit at 85 tons, and about 96ft in length, had been manned during 3 missions by 3 different astronaut crews. When the last crew had left *Skylab*, NASA estimated the workshop would remain in orbit until 1983; however, its altitude had dropped more rapidly than expected because of increased atmospheric drag caused by sunspot activity. (*JSC Roundup*, Feb 17/78, 1; *DFRC X-Press*, Feb 10/78, 2; *Marshall Star*, Feb 8/78, 4; *W Post*, Feb 2/78, A-2; *C Trib*, Feb 6/78, Sec. 1; *Nature*, Feb 9/78, 499)

• A laser-heterodyne radiometer (LHR) developed by Dr. Robert Menzies of the Jet Propulsion Laboratory had obtained evidence of ozone destruction in the upper atmosphere, JPL reported. The unique instrument, carried on a gondola suspended from a helium-filled polyethylene balloon, had gathered extremely accurate data by "looking" at the sun through the stratosphere during sunset at an altitude of 25km (120 000ft) where ultraviolet light breaks apart carbon molecules, releasing chlorine. Atoms of chlorine normally existed in quantities so small that their presence was difficult to detect; however, scientists had determined the effect of chlorine in destroying ozone, through a chain reaction that would form chlorine monoxide molecules. Later flights would measure the effect of seasonal variations and would look for other molecules responsible for ozone destruction. The research would assist federal regulators investigating danger from fluorocarbons used in refrigeration and air-conditioning; regulators might decide to ban aerosol propellants by June of this yr. (*JPL Universe*, Feb 17/78, 1)

• JSC reported it had begun a new procedure for receiving and processing rare and pristine meteorite samples. Such a program had not been necessary previously because available meteorite fragments had been contaminated by reaction with the soil where they fell, by weather, and by often unsterile handling conditions. The new meteorite facility would contain a near-pristine meteorite collection found in the winter of 1977 near McMurdo Station in Antarctica; the first two meteorite samples were 200- to 300-gram (baseball-size) fragments believed by NSF to be rare carbonaceous chondrites that might include primordial matter from the solar system.

Dr. William Cassidy, Univ. of Pittsburgh geologist working under an NSF grant, who had found the samples on an Antarctic ice shelf, had hypothesized that large accumulations of meteorites might exist in areas of the Antarctic, where ancient "blue ice" rose to the surface. Meteorites that fell on Antarctica centuries ago might appear on the surface along with the ice. Although the movement of the ice to the surface was not well understood, within 2 mo (last Dec. and Jan.) Cassidy had found 310 fragments believed to represent between 20 and 50 different meteorites, the most concentrated find known so far. Because of the exceptional preservation possible in the extreme cold and dry environment, Cassidy's find of 310 samples was probably the least contaminated near-pristine meteorite group ever collected. To ensure proper handling, the NSF and NASA's Lunar Curatorial Facility in Houston had equipped Cassidy with special sterile equipment to safeguard his collection.

Researchers at the meteorite-processing facility, using glove boxes first used to handle material from the moon, would document the samples and make initial characterizations before freeze-drying them to remove any trapped water-ice and making a mold, finally storing the samples in dry nitrogen gas. A special committee representing the NSF, the Smithsonian Museum of Natural History, NASA, and Dr. Cassidy's team would plan systematic examination of the samples. (*JSC Roundup*, Feb 17/78, 1)

February 20: AvWk reported that NASA had been struggling with institutional and funding arrangements for launching a Hughes-built comsat, Syncom-4, without giving Hughes the unfair advantage of a free Space Shuttle launch. The problem required prompt attention, because NASA had to make decisions within a few weeks if Syncom-4 was to fly on an early Shuttle mission.

Hughes had designed and offered Syncom-4 specifically to fit the Space Shuttle payload-bay design, since NASA had wanted to deploy a test spacecraft on a Shuttle orbital mission. Under the Hughes proposal, the spacecraft would have no significant communications capability, and NASA would launch the unit free of charge; however, the satellite-user community (the state of California, Agency for International Development, and the Interior Dept.) had asked that the spacecraft include some communications capability instead of merely verifying a design. NASA had to decide who would pay for the added capability and whether Hughes would have an unfair advantage.

Without a decision, NASA Administrator Dr. Robert Frosch said the whole exercise could revert to the original concept, with no significant communications channeled through the satellite. NASA might decide to run an open competition for systems to demonstrate Shuttle-launched communications spacecraft in the same manner as Syncom-4. However, Frosch was concerned that in an open competition Hughes could be "competed out of it in a way that is not quite proper," since the Syncom-4 concept had been a Hughes innovation. "As it has developed," Frosch explained to a House appropriations subcommittee hearing, "it is not completely clear that there is a proper solution to this dilemma." (*AvWk*, Feb 20/78, 60)

• The Canadian government had asked the USSR to pay possibly more than $2 million as the cost of an extensive search to recover debris from *Cosmos 954*, AvWk reported. Canada's request was in accord with a

1968 United Nations agreement that "expenses incurred in fulfilling obligations to recover and return a space object or its component parts . . . shall be borne by the launching authority." The Canadians had delivered an official statement to the Soviet embassy in Ottawa that pieces of *Cosmos 954* had landed on Canadian soil Jan. 24, and that Canada would return the debris to the USSR on receipt of the desired payment. The Canadians had also sent the UN a note reserving the right to claim compensation under the 1968 UN agreement, and announcing their intent to request compensation from the USSR. The report did not specify the amount of compensation requested. (*AvWk*, Feb 20/78, 24)

• A Soviet-French complex of thermostatic equipment on *Soyuz 27* had passed the test of terrestrial and space conditions, Tass reported. Soviet and French biologists had prepared biological cultures and chilled them in an ordinary refrigerator before putting them into the semiconductor instrument "Termokont-2." On the night before the *Soyuz 27* launch, Termokont-2 had handed over its biological "load" to another Soviet instrument, the thermostat "Bioterm-8," to maintain a temperature of 8°C during the launch and orbit of *Soyuz 27* until it docked with *Salyut 6*. After docking, the biological insert had gone into the French semiconductor thermostat "Cytos," which thawed the cultures and began to study spaceflight influence on cellular division of microorganisms under the strictly controlled temperature of 25°C. (FBIS, Tass in English, Feb 20/78)

February 21: NASA announced results of a demonstration of curriculum-sharing planned by LaRC using television through the communications technology satellite *Cts* at North Carolina A&T State Univ., Greensboro, N.C. This was the first time a predominantly black college had participated in such an experiment. NASA's portable earth terminal (PET), a ground-based communications system contained in a trailer, had transmitted 2 days of telecasts originating at North Carolina A&T to *Cts* 22 500 mi up, which relayed the programs to Jackson State Univ. in Miss.; Bell-Northern Research Ltd., Ontario, Canada; and Stanford Univ. and Rockwell Intl. in Calif. Participants in each location could see and communicate with each other. The program, funded under NASA contract to demonstrate the use of satellites in sharing curriculum information, research techniques, and results, included a question-and-answer session with Frederick Gregory, black Shuttle astronaut recently selected by NASA. ARC had made available almost 7 hr of satellite time for the project. (LaRC Release 78-9; NASA photo L-78-1209; *Langley Researcher*, Feb 24/78, 1)

• The number of airports, heliports, and other landing facilities available to civil aircraft in the U.S. and its possessions had increased in

1977 to 14 117, a net gain of 347 from the previous year, the Department of Transportation reported. Figures included civilian landing facilities and military facilities that permitted civil-aircraft operations. Privately-owned landing facilities accounted for more than 300 of the increased number, from 9103 in 1976 to 9409 in 1977. The 1977 year-end total included 11 713 airports, 1850 heliports, 513 seaplane bases, and 41 short-takeoff-and-landing (STOL) airports or runways, as well as 48 facilities in U.S. possessions. Landing facilities abandoned in 1977 numbered 294. Of the total, 747 were served by air carriers holding certificates of public convenience and necessity issued by the Civil Aeronautics Board. (FAA Release 12-78)

February 22: NASA announced plans to launch Landsat-C, an improved satellite to monitor earth resources, from WTR March 5 aboard a 2-stage Delta vehicle. Third of a series designed to orbit earth at 917km (570mi) altitude and scan earth's surface in a systematic resources and environment study, Landsat-C would be placed in a near-polar orbit circling earth every 103 min. Its remote sensors would view a 185km (115mi)-wide strip of earth running nearly north-to-south at a 99° angle to the equator. Earth-surface coverage would proceed west, with a slight overlap, covering the globe every 18 days. Synoptic repetitive coverage under consistent observation conditions was required for maximum use of the multispectral imagery.

Most important uses of Landsat data would correspond to three of the world's major problems: energy supplies, food production, and global large-scale environment monitoring. Innovations in the multispectral scanner system (MSS) would permit detection of day and night temperature differences in vegetation, bodies of water, and urban areas. The return beam vidicon (RBV) sensor improvements would increase recorded image resolution by 50%. Areas as small as half an acre would be identified and studied. Landsat-C also carried a data-collection system (DCS) that would collect data radioed directly from as many as 1000 remote ground platforms and relay them to a Landsat data-acquisition station. The DCS would collect data on volcano activity, stream flow, water and snow depth, water temperature, and sediment density.

Goddard Space Flight Center was responsible for project management of the Landsat spacecraft, Delta launch vehicle, NASA image-processing facility, and worldwide tracking network. KSC would supervise launch. General Electric Co.'s Space Division was Landsat spacecraft prime contractor. (NASA Release 78-22)

• NASA announced it had selected Teledyne Brown Engineering, Huntsville, Ala., for negotiations leading to award of a contract for payload/missions integration for the sixth Space Shuttle orbital flight test and for the first 3 Spacelab missions to be managed by MSFC. An experiment-integration contractor, Teledyne Brown would ensure that the group of experiments for each mission was compatible with the Space Transportation System scheduled to carry it into orbit. The contract would provide for manpower, materials, and associated services required for integration and operation of Spacelab payloads. Flights scheduled during 1980 to 1981 would offer opportunities for scientific investigations on Spacelab missions in the planning stage with scientists from throughout the world participating. These missions would give scientists the first chance to go into space as payload specialists and conduct their investigations in earth orbit. (*Marshall Star*, Feb 22/78, 1)

• MSFC reported a two-wk visit by 13 Japanese scientists doing initial design review of the SEPAC (space experiments with particle accelerators) package to be flown on Spacelab 1. Heaviest and most complex of the experiments scheduled for that mission, the SEPAC had 16 scientific objectives in addition to proving its performance and compatibility with the Spacelab. It had been designed for multiple reuse, and would accommodate a number of scientific disciplines. The particle accelerator, for example, would shoot an electron beam to create a small-scale artificial aurora in the upper atmosphere, timed to occur at a specific location for simultaneous observations from space and from earth for later comparison.

Professor Tatsugo Obayashi of the Institute of Space and Aeronautical Science at the Univ. of Tokyo said that nature had provided ideal laboratory conditions in space for many experiments. "There we can actively interact with nature and, by artificially exciting the electrons and ions of plasma and studying the results, we can eventually define the dynamics of space." The Japanese scientists had been working on the SEPAC experiment since early 1976. (*Marshall Star*, Feb 22/78, 4)

• NASA announced that LaRC would sponsor a government-industry conference Feb. 28-March 3 on technology advances in design and operation of conventional takeoff-and-landing (CTOL) transport aircraft. Technical sessions would present new technology generated by NASA in-house and contract efforts, including the ongoing Aircraft Energy Efficiency Program (ACEE). Speakers would represent organizations in the airframe and engine industry, universities, and 4 NASA centers. The 45 presentations would be organized into 6 disciplinary sessions: propulsion, structures and materials, laminar-flow control, advanced aerodynamics and active controls, operation and safety, and advanced systems. (LaRC Release 78-10)

• FBIS, in an unattributed commentary in English, reported that the USSR had been testing a laser system called "Glissada" that might permit safe airline takeoffs and landings in any weather. Questioning Professor Igor Berezhnoy, the commentator said: "The new system enables the pilot, in any weather without any additional aids, to find his airfield and the landing strip." Berezhnoy explained: A few dozen km from the airfield the pilot would see 3 red laser beams motionless, like pillars, over the airfield and would discern other beams, 2 red beams clearly indicating runway boundaries and a third directly down its middle; it is this axial beam that the plane would follow to land. The system would enable the pilot to detect even a half-meter deviation from course. The laser system would require little additional training of pilots and no additional equipment on the planes. The Soviets claimed the cost of the system and its installation and maintenance would be a fraction of the cost of present radio systems and airport-lighting equipment. (FBIS, unattributed commentary in English, Feb 23/78)

February 23: LaRC reported NASA had appointed astronaut Donald Slayton to manage the orbital-flight test (OFT) program for Space Shuttle at JSC. One of the original 7 Mercury astronauts and a member of the U.S. crew of the Apollo-Soyuz test project, Slayton would be responsible in his new position for all mission-unique activities associated with the 6 Shuttle orbital-flight tests scheduled to begin in the spring of 1979. He would represent program manager Robert Thompson across the total program to ensure timely and appropriate resolution of all OFT mission-unique issues. (LaRC Release 78-12)

February 24: NASA announced that *Nimbus 6*, a meteorological research satellite, would track the 6000km (3728mi) journey by dogsled of a lone Japanese explorer from northern Canada to the North Pole and back. The Smithsonian Institution, which had a scientific stake in the venture, had asked GSFC to manage the tracking. The 6-mo journey would begin March 4, 1978. The explorer, 37-yr-old Naomi Uemura of Tokyo, had driven a dogsled solo some 12 000km (7457mi) from Greenland to Alaska over an 18mo period in 1975 and 1976. During his upcoming Arctic journey, Uemura would carry a 4.5kg (10lb) battery-powered satellite beacon on his dogsled to transmit once per min an automatic radio signal, including local temperature and atmospheric pressure. *Nimbus 6* overflying the pole every 108min would monitor the signals and collect the data for relay to GSFC by a NASA tracking station in Fairbanks, Alaska. Uemura would systematically collect snow, ice, and air samples for Japan's Natl. Inst. of Polar Research and the Water Research Inst. of Nagoya Univ. Researchers needed as accurate a record as possible of Uemura's daily positions to correlate with the collected data. Satellite-tracking data would also verify Uemura's dead-reckoning and celestial-navigation readings, particularly difficult in polar regions. NASA's expenditures on the venture would be minimal because both the *Nimbus 6* and its ground-control computer would be routinely operating on a 24hr basis. (NASA Release 78-29)

• DFRC reported that it had awarded the Ames Industrial Corp., Bohemia, N.Y., a $218 000 fixed-price contract to develop and fabricate a small lightweight manned oblique-wing aircraft. NASA had begun a low-cost study of piloting an oblique-wing aircraft. At lower flight speeds, the wing would orient itself perpendicular to the fuselage, producing efficient quiet operation for takeoff and landing and for low-speed cruise flight. This would substantially reduce engine thrust required for takeoff, resulting in quieter operations during takeoff and landing. In high-speed flight, the wing would pivot fore and aft, forming oblique (up to 60°) angles with the aircraft fuselage.

Studies had suggested the "scissor-wing" concept would offer better flight performance at high speeds; as an aircraft went faster, pivoting the wing to an oblique angle would decrease air drag, giving increased speed and longer range for the same fuel expenditure. Computer and wind-tunnel studies at ARC had shown that an oblique-wing transport aircraft flying at 1000mph might realize twice the fuel economy of either the current British-French Concorde or the Soviet SST, and also might alleviate the sonic-boom problem. DFRC expected delivery of the aircraft in late 1978 and planned its first flights early in 1979. (*DFRC X-Press,* Feb 24/78)

February 26: NASA announced that scientists working for NASA and NSF had discovered living organisms inside rocks on the frozen deserts of the Antarctic. The discovery, made in a region called the Dry Valleys whose harsh climate resembled conditions on Mars, would significantly extend the known limits of life on earth, and would have important implications in a search for extraterrestrial life.

Dr. E. Imre Friedmann and Dr. Roseli Ocampo-Friedmann, a husband-wife team who had searched for more than 15yr for microbial life in rocks, had made the discovery. They had isolated the newly discovered microorganisms (bacteria, algae, and fungi) and were growing them in laboratory cultures for clues to their endurance. Although the Dry Valleys of Antarctica had appeared lifeless, and the region's freedom from snow and ice, combined with constantly blowing winds, had produced an environment among the world's harshest, the Friedmanns had found in the Dry Valleys a widespread rich microbial vegetation under the surface of rocks, in the airspaces of porous rocks, or in fissures. Rocks broken open had revealed the presence of organisms as a dark-green layer a few millimeters deep.

Dr. Richard Young, NASA's chief of planetary biology, said the Dry

Valleys in many ways resembled the environmental extremes found on Mars by the 1976 Viking landers, which searched the Martian soil unsuccessfully for signs of microbial life or organic molecules. "If Martian life forms exist only in the interior of Martian rocks, as is principally the case in the Antarctic, that could easily serve as an explanation for the lack of evidence on Mars," Young said. "This interesting (if speculative) analogy is of considerable interest to NASA in designing future attempts to study planetary surfaces for evidence of life." NASA and NSF research grants had supported the Friedmanns' work. (NASA Release 78-14)

February 27: Dr. Robert Frosch, NASA administrator, announced a reorganization at NASA Hq to significantly restructure organization elements such as Aeronautics and Space Technology, Space and Terrestrial Applications, External Relations, and Management Operations, with lesser changes in the Equal Opportunity, Procurement, and Audit organizations. Frosch's Feb. 22 approval of the new organization structures had authorized office heads to proceed with final placement of people.

In explaining the reorganization, Alan Lovelace, deputy administrator, said: "In keeping with the announced policy of President Carter, each Headquarters employee affected by the reorganization will be made a good-faith offer of a position and to the extent possible the offer will be made at the employee's current grade level. As I announced previously, there will be no involuntary reduction in force at NASA Headquarters. . . . The classification survey of Headquarters positions has been completed. Based on tentative aggregate findings in the survey thus far, plus the potential impact of the reorganization, Dr. Frosch has asked the Civil Service Commission for authority to delay any demotions for people in the General Schedule resulting from survey results and the impact of the reorganization. NASA was granted such authority on Feb. 22. . . . As to future classification activity, it is important that we maintain a rigorous system for the accurate classification and grading of all Headquarters positions. . . . I would like to assure you that we are taking all of the steps possible to provide appropriate job opportunities for all Headquarters employees and, to the extent we can, we shall do our best to avoid adverse impacts." (NASA special anno Feb 27/78)

• *Av Wk* reported that Congress might override President Carter's decision not to fund a fifth Space Shuttle orbiter in NASA's FY79 budget. Congressmen of both parties had said that a 4-orbiter fleet would limit the development of a civilian space program, making the U.S. extremely vulnerable in strategic use of space. Officials who considered addition of a fifth orbiter inevitable had pointed out that Carter's decision to defer acquisition would increase Shuttle program cost by $235 to $250 million if the administration should later change its mind. Some congressmen had also expressed displeasure with the administration's apparent inability to provide stated goals and directions for NASA's space capabilities and with the attempt to cut NASA advance-planning funds. (*AvWk*, Feb 27/78, 20)

• *Av Wk* reported award to Lockheed Missiles and Space Co. of a $34.5 million USAF contract to design and build a space platform using long wavelength infrared radiation for space surveillance of relatively cool objects. Designated space test program mission P80-2, to be managed by USAF's Space and Missile Systems Organizations (SAMSO), the platform would go into orbit from WTR in Jan. 1981. Principal payload for the P80-2 mission would be a satellite infrared experiment (SIRE) developed for USAF by Hughes Aircraft Co., to operate in the 8- to 12-micrometer region of the spectrum, detecting and tracking objects such as a spacecraft or booster after burnout. The platform would also carry an autonomous navigation system built for USAF by Martin Marietta and a solar-flare isotopic-composition experiment designed by the Univ. of Chicago. The contract provided that Lockheed would integrate the payload (for which it had no responsibility) with the space platform. (*AvWk*, Feb 27/78, 39)

• Congress had strongly supported legislation establishing a power satellite research and development and demonstration program scheduled to become a major issue among changes to President Carter's FY79 solar energy proposals, *Av Wk* reported. Rep. Ronnie Flippo (D-Ala.) had introduced a bill directing the Dept. of Energy and NASA to begin research and development on technical problems in implementing a solar power satellite. However, Dale Myers, undersecretary at DOE, who had described the joint solar energy projects contemplated by NASA and DOE at a House Science and Technology space science and application subcommittee hearing, said he questioned the need for such a bill. (*Av Wk*, Feb 27/78, 47)

February 28: The *Congressional Record* reported that President Carter had nominated former astronaut Maj. Gen. Thomas Stafford to be Air Force deputy chief of staff for research and development. Stafford, who had spent more than 500hr in space, was serving as commander of Edwards AFB Flight Test Center. The nomination, which carried a promotion to the rank of Lt. Gen., was subject to Senate confirmation. (*CR*, Feb 28/78, S 2580; *NYT*, Feb 26/78, (27), 14)

During February: Gossamer Condor, the manpowered aircraft that won the $85 000 Kramer prize in 1977, was on display at the National Air and Space Museum in Washington, D.C., the National Aeronautic Association reported. The 96-ft wingspan had been cut into 4 pieces for shipping across country with the fuselage and stabilizer. Gossamer Condor had the largest wingspan and was the lightest of any plane in the museum, the wing-covering accounting for only 4.5lb of the craft's total 70lb. An intricate web of piano wire held the structure together. Flight speed, reaching 10mph, had required the manpower equivalent of one-third of a horsepower. Designer Paul MacCready had substantiated his belief in the potential of ultralight craft. (NAA newsletter, Feb 78, 4)

• The Carter administration, calling for national commemoration of the 75th anniversary of the Wright brothers' first powered flight, had designated DOT lead agency to coordinate programs and materials, the National Aeronautic Association reported. The task force for the 75th anniversary, which had held its first meeting at DOT in Jan., would be a clearinghouse to avoid duplication, facilitate innovation, and develop ideas suitable for agencies and private organizations. Peter Clapper, FAA assistant administrator for public affairs, noted that the American people had taken for granted the tremendous advances in aviation over the last 75yr and that the anniversary offered an opportunity to bring these advances to public attention. (NAA newsletter, Feb 78, 1)

• Rep. Melvin Price (D-Ill.), chairman of the House Armed Services Committee, had written President Carter expressing concern that recent administration decisions could seriously impair the strategic defense posture of the U.S., the National Aeronautic Association reported. Price wrote: "I have attempted to comprehend the rationale behind the combination of decisions leading to cancellation of the B-1, cancellation of production of the Short Range Attack Missile (SRAM), terminating production of the Minuteman III Missile, and the reduction in funds for the development of the advanced MX missile system." He emphasized that, despite the consensus of defense experts that U.S. silos would be vulnerable within 3 to 5yr, the President had reduced MX funds and stretched out its schedule.

Price also voiced concern over the apparent lack of correlation or adequate interchange between the military departments principally responsible for strategic capability and others in the administration principally responsible for preparing proposals for and participating in SALT negotiations. Price concluded: "It is my hope that working with the administration we can be prepared (at the time the fiscal 1979 defense program is considered) to begin to provide those systems necessary to ensure a halt to the trend that sees the United States falling behind the strategic might of the Soviets." (NAA newsletter, Feb 78, 1)

• The USAF announced it had selected McDonnell Douglas Corp.,

Long Beach, Calif., as primary contractor for the advanced tanker cargo aircraft (ATCA). Award of the $28 million contract would initiate production-engineering tooling and other nonrecurring activities. The aircraft would be funded over the next 5yr, with the intent of eventually ordering 20 of them. Primary objective of ATCA would be to increase mobility of U.S. forces in contingency operations by refueling fighters and simultaneously carrying fighter-support equipment and personnel on overseas deployments; refueling strategic airlifters (such as C-5 and C-141) during overseas deployments and resupply missions; refueling strategic offensive and reconnaissance aircraft during long-range conventional operations; and selectively augmenting cargo-carrying capability. (AFSC *Newsreview*, Feb 78, 1)

• The USAF announced it had tested at Ramstein, W. Germany, a new "brain" for the imaging infrared (IIR) Maverick missile, a digital centroid tracker developed for a joint Air Force-Navy IIR program by Hughes Aircraft Co., to prevent loss of target lock-on, a problem inherent in the older analog system. The centroid tracker would locate a target such as an industrial complex or tank using a sensor that picked up infrared or heat images, calculated boundaries of the target, and guided the missile to the approximate center. The IIR Maverick (designated AG65D), most advanced version of this new missile, had demonstrated in tests earlier this year improved performance in daytime or nighttime and adverse weather conditions over earlier TV-guided Mavericks. (AFSC *Newsreview*, Feb 78, 4)

• The Jan/Feb issue of *Satellite Pathways* reviewed COMSAT accomplishments, noting that satellite communications technology had advanced through research and development programs at COMSAT Laboratories, engineering support it had provided to current satellite systems, and its work on satellite systems of the future. COMSAT's research and engineering programs had aimed primarily at more powerful and long-lived multiple-beam satellites operating with smaller earth stations in higher communications frequencies; these programs would lead to more efficient use of the frequency spectrum allocated to satellite communications by simultaneous multiple use of the same frequency bands and transmitting communications in high-speed digital rather than conventional analog modes.

Among achievements in 1977, the article mentioned new nickel-hydrogen batteries developed by COMSAT under INTELSAT sponsorship; data collected from COMSAT-built 13- and 18-gigahertz transponders aboard the *Ats 6* satellite and the 19- and 28-gigahertz beacons aboard the COMSTAR satellites; a simulator that could handle communications for the Intelsat V satellites that would form a global system in the 1980s; and new communications processing equipment to convert conventional analog signals into digital signals, and vice versa. (*Satellite Pathways*, Jan/Feb 78, 12)

• *Science Indicators 1976*, ninth annual report of the National Science Board recently transmitted to Congress by the President, noted an increase in U.S. research and development expenditures during 1976 to an estimated $38.1 billion in current dollars; however, constant-dollar spending of $28.5 was only 2.5% above the 1974 level. Measured in current dollars, federally supported R&D expenditures in 1976 had climbed to an estimated new high of $20.1 billion; however, constant-dollar levels stood at $15 billion, 18% below the peak reached in 1967. National spending for basic research in current dollars had climbed substantially since 1960, to a new high in 1976. However, estimated constant-dollar expenditures in 1976 had advanced only about 2% above the 1975 level, remaining nearly 11% below the peak year of 1968. The federal government was the source of most support for basic research in 1976, as it had been in past years. The federal share amounted to 68% of the total since 1971, compared with approximately 60% in 1960, whereas industry's share had remained stable at approximately 15% throughout the 1970s, in sharp contrast to its 28% share in 1960. (NSF report, Feb 78)

March

March 1: Marshall Space Flight Center engineers reported completion of tests simulating checkout of a solid-fuel rocket booster (SRB) using flight-type hardware interfaced with the launch-processing system (LPS) and tests run by computer. The tests would verify compatibility of SRB hardware and demonstrate use of the LPS to checkout the SRB system. Flight equipment removed from simulated SRB hardware had been relocated in equipment racks to simulate a left-hand SRB configuration; next use of the LPS would be to checkout the left-hand SRB. United Space Boosters, Inc., would perform the checkouts, then move the equipment to Kennedy Space Center for use in future SRB checkouts. (*Marshall Star*, Mar 1/78, 2)

- NASA announced appointment of Norman Terrell as director of international affairs and Arnold Frutkin as deputy associate administrator for external relations. Terrell, who joined NASA in 1977 as chief of the International Program Policy Office where his responsibilities included U.S.-Soviet space relations and United Nations space affairs, had worked for the Nuclear Regulatory Commission and as foreign service officer in the State Department, where he received an award in 1975 for his work on the threshold test ban treaty. Frutkin, who had headed NASA's Office of International Affairs for the past 18yr, had served as deputy director of the U.S. National Committee for the International Geophysical Year. (NASA anno Mar 1/78)

March 2: The USSR had launched *Soyuz 28* at 4:28pm Moscow time, Tass announced, with cosmonaut Vladimir Remek of Czechoslovakia accompanying Soviet cosmonaut Aleksey Gubarev on the first of a series of international cooperative flights that would include representatives from Bulgaria, Hungary, the GDR, Cuba, Mongolia, Poland, and Romania. Soyuz launches later in 1978 would carry Polish and GDR cosmonauts. Since 1967 when the Intercosmos program began, scientists and specialists of participating countries had been using Soviet satellites and research rockets to study the physical properties of outer space, space meteorology, communications, biology, medicine, and natural resources. (FBIS, Tass in English, Mar 3/78)

March 3: NASA announced award of a contract worth approximately $10.5 million to RCA for a high-quality television camera system to transmit live color and black and white TV during manned orbital Space Shuttle flights. The closed-circuit TV camera system installed on the Space Shuttle would photograph earth-orbital missions starting in 1979 through the 1980s. Under contract to Johnson Space Flight Center, RCA would provide up to 50 cameras for approximately 500 Shuttle flights over the next decade. Each Shuttle orbiter might carry up to 6 cameras; the system would consist of several TV cameras, a video control unit, pan and tilt mechanisms, and various monitors. Cameras installed in the crew compartment, cargo bay, and on a remote manipulator arm would use a 525-line standard compatible with broadcast television. A portable TV camera used during astronaut extravehicular activity (EVA), for example, could look at areas of the orbiter not visible to the permanently mounted cameras. The portable camera equipped with its own viewfinder would allow astronauts to focus accurately on an object such as the moon or a free-flying satellite in space. The TV system should help the crew in retrieving satellites from space, removing spacecraft from the cargo bay of the Shuttle, and repairing or replacing parts of a satellite on the servicing platform mounted in the orbiter's payload bay. (*JSC Roundup,* Mar 3/78, 1)

• Jet Propulsion Laboratory announced it had fabricated and displayed a prototype roving ball planetary explorer, an instrument-laden reinforced-fabric ball capable of inflating and deflating automatically or on remote command from earth during possible Mars exploration. Driven by high winds known to prevail on the planet's surface, or powered and steered by an inner drive system much like that of a bicycle, such balls or balloons could roam freely and conduct scientific experiments and measurements over a vast area. A ball could partially deflate if necessary for additional study of a particular area. This "bouncing ball" would offer the advantages of low cost, long life, and extended traverse of the planet's landscape. As a Mars explorer, a ball would constitute a payload of 20 to 30kg (about 44 to 66lb) capable of a traverse of 100 to 200km (60 to 120mi). The "skin" of the ball, "sandwiched" or laminated Mylar and Kevlar (both registered DuPont products), would be strong enough to cover some of Mars's most difficult terrain. Schjeldahl Corp. designed and built the ball for the Aerospace Corp. for the ATMOSAT project. Jacques Blamont of France's Centre National d'Études Spatiales (CNES) had proposed the concept. (JPL *Universe,* Mar 3/78, 1)

• Remote control might soon operate the two drawbridges linking KSC with the Florida mainland over the main channel of the Intracoastal Waterway, NASA announced. An experimental remote-control system tested on the 170-ft-long twin-leaf bridge spanning Haulover Canal would be a prototype for a permanent system both at the canal and at the dual-bridge span of the NASA causeway across the Indian River. The experimental system should demonstrate the feasibility and cost-effectiveness of remote bridge operation; in actual use, a duty officer in the launch-control center would use TV cameras, loudspeakers, and microphones mounted on the bridge to monitor and talk with canal and road traffic. A small microwave dish atop the bridge-control tower would transmit TV signals to a receiver on top of the 525ft-tall vehicle assembly building at KSC, and audio signals would travel by telephone wire. An operational system could save approximately $200 000 per yr and pay for itself over 4yr. (KSC Release 22-78; *Spaceport News,* Mar 3/78, 1)

• Langley Research Center announced it would host Mar. 7-9 an advanced technology airfoil research conference offering the aviation community a comprehensive review of the latest technical developments in airfoil computational analysis and design methods, airfoil test facilities and techniques, and various applications of new airfoil technology. Presented by NASA, supporting contractors, and university grantees, the technical sessions would cover progress in mathematical codes for use in electronic digital computers for airfoil analysis and design, coupled with progress in analog computers through wind-tunnel test facilities and special techniques of investigation; general aviation aircraft; transports; rotorcraft; propellers; windmills; and agricultural aircraft. (LaRC Release 78-12)

March 6: NASA announced plans to launch from Cape Canaveral no earlier than Mar. 23 an experimental Japanese direct-broadcast communications satellite capable of sending television programs directly into an individual's home. The satellite, technically named Medium-Scale Broadcasting Satellite for Experiment Purposes (BSE), had been a project of Japan's National Space Development Agency (NASDA). As agreed by the U.S. and Japan, NASA would launch BSE on a Delta rocket; NASDA would reimburse NASA for the cost of the launch vehicle, launch services, and other administrative costs. NASA's worldwide Spaceflight Tracking and Data Network (STDN) would track BSE until it achieved final orbit, when NASDA would take over tracking. Present Japanese TV, though covering 97% of the main islands, had not been economical for remote islands and mountain regions; also, tall buildings had degraded TV reception in large cities. NASDA had designed BSE to transmit high-quality color television to the Japanese islands and Okinawa at lower cost; user antennas could be as small as 1 to 1.6 meters (3.3 to 5.2ft), with production costs estimated at $200 per unit. The high power of BSE could transmit directly to individual TV sets even in bad weather, making TV available to Japanese living on the offshore islands and in inaccessible mountain areas.

Goddard Space Flight Center would manage the Delta vehicle for NASA's Office of Space Transportation Systems, and KSC would manage launch operations. General Electric Co., Valley Forge, Pa., had built the BSE under contract to Tokyo Shibaura Electric Co. The Japanese Broadcasting Corp., the Japanese Ministry of Post and Telecommunications, and the Radio Research Laboratories of Japan would sponsor or participate in the program. (NASA Release 78-33; MOR M-492-212-78-01 [prelaunch], Feb. 28/78; *Spaceport News*, Mar 17/78, 1)

• The House subcommittee on space science and applications had added funding of $4 million to the NASA FY79 budget for a fifth Space Shuttle orbiter and a new stereosat earth resources spacecraft, *AvWk* reported. The subcommittee had also added $7 million to NASA's Office of Space Transportation funds for advanced planning, plus $3 million each for large space-structures work and solar power satellite studies. (NASA's initial budget request had not included solar power satellite funding, because of Dept. of Energy priority in this area.) The subcommittee had added $40.5 million to the NASA request and cut $25 million, resulting in a net $15.5 million increase.

Funds for a stereosat came with a subcommittee call for development of the $40- to $45-million spacecraft based on significant funding from the user community; primary users would be geologists exploring for petroleum and minerals. Other subcommittee additions were $10 million for distribution of Landsat data and training programs to states for crop-prediction research, encouragement of microwave development, and work on an operational Landsat system; $4 million to stimulate bioengineering work; $1.5 million to support industrial application of NASA technology; and $4 million for modifying NASA's tracking system to facilitate full and timely handling of Spacelab data early in Spacelab's flight program. (*AvWk*, Mar 6/78, 13)

• *AvWk* reported that the Navy's Electronic Systems Command planned to issue a request for proposals (RFP) to establish and operate a follow-on system to the Navy/TRW fleet satellite communications (FltSatCom) system. The Navy would lease from the contractor capacity on the new system, designated Leasat, which could be either a hybrid (providing service to merchant shipping as well as to the Navy) or dedicated (serving only the Navy). The RFP would require 3 satellites positioned over the Atlantic, Pacific, and Indian oceans. The first satellite should be available Oct. 1, 1981; the second, Apr. 1, 1982; the third, Oct. 1, 1987. A satellite that failed would have to be replaced within 30 days. The contractor would also build and operate a telemetry, tracking, and control system with central headquarters for coordination at 4 Navy communications stations in Va., Hawaii, Calif., and Guam. (*AvWk*, Mar 6/78, 14)

• Upon successful completion of the Spacelab Critical Design Review [see Feb. 3], ESA announced it would authorize manufacture of the Spacelab flight unit and confirm technical orientation already given in certain critical areas. Plans called for delivery of the engineering model to NASA in mid-1979; ESA would deliver the flight unit to NASA in two shipments (in autumn 1979 and early 1980) to satisfy requirements of the first 2 Spacelab missions, scheduled for Dec. 1980 and Apr. 1981. (ESA Release Mar 6/78)

March 8: NASA announced the successful launch of *Landsat 3* (formerly Landsat-C), third of a series, from WTR at 12:54pm EST on March 5. Orbital parameters were: apogee, 913.96km; perigee, 897.3km; period, 103.108min; and inclination, 99.134°. Landsat was part of a U.S. program to develop remote sensing for improved earth resources management. The total program included development of remote-sensing instruments; data-analysis research using data from spacecraft, aircraft, and ground-truth sites; and a spaceflight program. *Landsat 3* carried two piggyback payloads, the larger (approximately 34kg) being a plasma-interaction experiment (PIX) by Lewis Research Center to measure plasma-coupling current and negative-voltage breakdown of a solar-array segment and a gold-plated steel disk. Objective of PIX was to establish design guidelines, materials, devices, and test methods for controlling detrimental interactions between high-voltage systems and the space-plasma environment. The PIX would remain attached to the second stage of the Delta vehicle. The second piggyback payload, the 27.3kg *Amsat Oscar* communications-relay satellite built by radio amateurs, ejected from the second stage approximately 12.3min after Delta separation from *Landsat 3*, would be available to amateur operators for science education, communications experiments, and search-and-rescue systems tests. The *Oscar D* carried two communications transponders having 1 to 2 watts of power output, and was magnetically stabilized.

Changes in the *Landsat 3* payload included alteration of the return-beam vidicon system to produce higher resolution panchromatic images for detailed ground mapping, and inclusion in the multispectral scanner (MSS) of a fifth band in the emitted terrestrial infrared radiation region. Objective of the MSS was to acquire multispectral high-spatial-resolution (80m) images of solar radiation reflected from earth's surface, emitted infrared-radiation images of 240m resolution, and panachromatic RBV images of 40m resolution. The MSS would obtain contiguous images about once every 18 days, weather permitting. The multispectral images would aid research and operations demonstrations in agriculture and forestry resources, mineral and land resources, land use, water resources, marine resources, cartography, and the environment. Applications would include surveying land use; measuring factors causing stress on crops and forest; inventorying crops and forests; assessing crop vigor and health; classifying areas by geological or geomorphic characteristics; delineating promising areas for mineral exploration; determining water-runoff patterns and extent of snowcover; monitoring population movements and environmental hazards to man; mapping sea ice; and analyzing shorelines. Demonstration projects called Applications Systems Verification and Transfer (ASVT), focusing on landcover inventory for federal, regional, state, and private users, would also use *Landsat 3* data.

The Landsat-C spacecraft (basically a Nimbus design modified to accommodate Landsat instruments and mission requirements) had the following parameters: weight, 90kg; mission margin, 51.4kg; stabilization type, 3-axis earth-oriented; stabilization accuracy, 0.7° pointing error, 0.04° per sec rate; average power available for experiments, 375 watts; thermal range, 10 to 30C; data storage capacity, 60min of video data (25.2 million bits). Launch vehicle was a 2-stage Thor Delta approximately 116ft long and 8ft in maximum body diameter. The first stage was a McDonnell Douglas extended long-tank Thor booster incorporating 9 strap-on Thiokol Castor II solid-fuel rocket motors; second stage was powered by the TRW TR-201 liquid-fuel pressure-fed engine gimbal-mounted to control pitch and yaw through second-stage burn. A nitrogen gas-system using 8 fixed nozzles provided roll control during powered and coast flight, as well as pitch and yaw control after second-stage cutoff. Two fixed nozzles fed from the propellant-tank helium-pressurization system provided retrothrust after spacecraft separation. Total cost of the Landsat program (excluding costs of the launch vehicles, and including tracking and data network-systems support) would be $43.9 million through 1979. (MOR E-641-78-03 [postlaunch] Mar 8/78, [prelaunch] Feb 22/78)

• MSFC had been responsible for a variety of ground tests around the U.S. to prove flightworthiness of the Space Shuttle's solid-fuel rocket booster, the center newsletter reported. Each Shuttle mission would use 2 of the 578 097kg (600ton) SRBs, 45.46 meters (149ft) long and 3.7 meters (12ft) in diameter, each providing 12 232 550 newtons (2.75 million lb) of thrust for about 2min from the launch pad to burnout and separation at an altitude of 45.3km (27mi). Tests at MSFC had put loads on a "short" version of the booster to duplicate load conditions encountered on the launch pad, in flight, during parachute deployment, and in water impact and recovery. Other MSFC engineers had tested the flight system designed to gimbal (or swivel) the motor nozzles at the lower end of each booster to guide the Shuttle on a proper course.

Engineers at United Technologies Corp.'s Chemical System Div. near San Jose, Calif., had tested 8 small solid-fuel booster-separation motors, fired simultaneously to separate each SRB and move it away from the Shuttle during flight. Each end of the two boosters would carry 4 motors, a total of 16 on each flight. These would have about 88 960 newtons (20 000lb) of thrust and would fire only about 1 sec each to accomplish separation. The separation motor had completed development testing; the Calif. test series would qualify it for flight.

Thiokol Corp.'s Wasatch Div. near Brigham City, Utah, had run fullduration full-thrust static firings on the SRB; high-speed sled runs in New Mexico had tested one phase of the SRB's parachute-recovery system (deployment of the pilot chute only) to see whether the nosecap of the system, when ejected, would clear the vehicle without becoming entangled; and the National Parachute Test Range at El Centro, Calif., had conducted airdrops of the entire parachute system. Other tests of SRB's electrical system and prelaunch checkout system were in progress. (*Marshall Star*, Mar 8/78, 4)

• NASA officials had unveiled the prototype of a parachute-cleaning machine strongly resembling a carwash operation, *Today* reported. If approved, the facility would become part of KSC's system to refurbish Space Shuttle parachutes. Each of the Shuttle's two solid-fuel rocket boosters jettisoned into the Atlantic Ocean shortly after launch would carry 4 nylon parachutes; both the boosters and the parachutes were designed for recovery and return to KSC for another launch. The carwash-type facility would rid the parachutes of salt water before washing and drying them in large machines. The current system had rinsed parachutes in the same water they were washed in; the new system would save some of the 20 000gal of water now used by the washing machine. In the carwash-type system, parachutes would move along an overhead rail, rinsed 3 times by water sent through high-pressure nozzles. The 3 main parachutes were 230ft long and 115ft in diameter, weighing 1600lb dry and 2000lb wet; the fourth parachute measured 145ft in length and 54ft in diameter, and would weigh 1200lb dry and 1400lb wet. On completion, the new system should require 2hr to wash and 2hr to dry each parachute. (*Today*, Mar 8/78, 8A)

• JSC announced plans for the week-long ninth annual lunar and planetary science conference focusing on the moon and other worlds, with the Lunar and Planetary Institute of Houston as co-host. Topics of presentations would include the formation of the solar system; new discoveries in moon rocks; the histories of planets; meteorites containing material from ancient stars; and comparative studies of Mars, Venus, and the earth. For the third consecutive year, the USSR would send a delegation. A number of topics would be emphasized: constraints on structures; composition and history of planetary interiors; characteristics and movement of material on lunar, planetary, and asteroid surfaces; characteristics and evolution of volcanic landforms; characterization and evolution of planetary crusts; nature and effect of impact processes; extraterrestrial materials; solar, interplanetary, and interstellar probes; and earliest history of the solar system. JSC had scheduled special sessions on industrial development of near-earth space, on origin of the solar system, on the future of planetary exploration, and on Mars and Mercury. (JSC Release 78-40)

• INTELSAT announced it had awarded the Yardney Electric Corp. of Pawcatuck, Conn., a $149 365 contract to develop an advanced nickel-hydrogen battery to demonstrate flight acceptance for use in future INTELSAT communications-satellite programs. The contract terms called for design, fabrication, testing, and delivery of two batteries by March 1979. INTELSAT had pioneered development of this technology, expected to improve energy density, increase cyclic capability, and prolong battery lifetime and reliability. (INTELSAT Release 78-6-I)

March 10: NASA announced receipt of the first 2 HiMAT (highly maneuverable aircraft technology) prototypes from the builder, Rockwell Intl., which developed the 0.44-scale version during a joint NASA-USAF multiphase competitive program. Flights of the subscale models at Dryden Flight Research Center would use the remotely piloted research vehicle (RPRV) technique developed there to test advanced aerodynamic technology for enhanced aircraft maneuverability. The RPRV offered a highly cost-effective means of flight testing advanced, high-risk technology without associated risks to test pilots. A pilot in a ground cockpit using telemetry, television, and radar air-launched unmanned craft from a carrier aircraft to fly them through maneuvers. (*DFRC X-Press*, Mar 10/78, 2; *AvWk*, Mar 20/78, 17)

• If the flight of *Soyuz 26* cosmonauts Yuri Romanenko and Georgy Grechko aboard *Salyut 6* space station continued to go well, they would establish a new space-duration record in the early morning of March 13, announced the Natl. Aeronautic Association, the organization maintaining official aviation and space records. The cosmonauts would surpass the previous record of 84 days set by U.S. astronauts Carr, Gibson, and Pogue aboard *Skylab 4* in 1973. The Soviets had not indicated when the mission would end. On Feb. 10 the *Soyuz 26* team equalled the previous USSR manned-spacecraft record of 63 days set by *Soyuz 18*. (NAA Release Mar 10/78)

March 13: NASA and ESA announced selection of experiments for the proposed 2-spacecraft solar-polar mission planned for launch in 1983 and designed for first-time observation of the sun from the unique perspective of its polar regions. The solar-polar mission would explore one of the remaining frontiers of the solar system: the third dimension of space, out of the plane of the orbits of the planets. All previous interplanetary space probes had flown in the orbits of the planets, essentially intersecting the sun's equatorial regions. Congress had not yet approved the project, but early selection of scientific participants and investigations would allow a prompt start on approval. NASA and ESA would each provide one spacecraft, and the combined scientific payload would be divided between U.S. and European investigators.

The Space Shuttle would launch both spacecraft simultaneously using an inertial upper-stage booster, directing them on a trajectory in the ecliptic plane (the plane containing all the planets) toward Jupiter. Swinging around Jupiter, the craft would use the gravity of that planet to move out of the ecliptic plane and back toward the sun in trajectories, one northbound and one southbound, essentially mirror images of each other. After passing over the north and south solar poles, the spacecrafts would swing through perihelion (the distance closest to the sun) in the ecliptic plane, pass respectively over the other solar poles, then fly back to the vicinity of Jupiter's orbit. The period from launch until shortly after the second pair of polar passages would be approximately 5yr.

The mission should provide important new knowledge about the solar wind, cosmic rays, and the 3-dimensional structure and evolution of the sun's corona (the outermost solar atmosphere), and increase understanding of solar phenomena that shape and control earth's space environment. In anticipation of a FY79 Congressional authorization of the mission, NASA's Jet Propulsion Laboratory, manager of the mission for NASA, had undertaken studies of the U.S. spacecraft, payload, and mission design concepts. More than 150 American and European scientists would participate in solar-polar investigations. (NASA Release 78-39; ESA Release Mar 13/78; JPL *Universe*, Mar 17/78, 1)

• NASA had mounted a "new effort" to make space applications the "administration's centerpiece," starting with a reorganization of applications management that could double that office's $283 million budget, *AvWk* reported. The new effort would include a major push for a "global information system" supported by all civilian earth-sensing spacecraft, to encompass all Landsat, Seasat, and environmental and meteorological satellite data, coordinated so that for any need an individual user could access all pertinent information available from space platforms. Theoretically, all space-derived information could some day come from a single distribution point.

NASA's effort would emphasize agency cooperation (NASA officials had met with representatives of the Depts. of Agriculture and Interior,

EPA, and AID); management changes (a plan awaiting approval would remove flight projects from the traditional applications branches; the project offices would instead develop user-oriented sensor systems to be passed along to an applications-systems division, which would then reconfigure the sensors into the proper spacecraft format); center changes (instead of the previous lead-center approach to managing flight projects, NASA Hq would formulate programs as the main user liaison); applied sciences (greater emphasis by space applications on the phenomena its spacecraft would measure); ground systems (speedup of work on the end-to-end data system); an operational comsat system (solution of years-old Landsat operating problems); and communications research (industry would foster the research, while government agencies would concentrate on creating and sustaining a real market for public service communications). (*Av Wk*, Mar 13/78, 67)

• For the year ending Dec. 31, 1977, ComSatCorp reported consolidated net income of $32 499 000 ($3.27 per share) compared with $38 271 000 ($3.83 per share) for the preceding year. ComSatCorp said the volume of its services through the INTELSAT system and the contribution to net income by its wholly-owned subsidiary, Comsat General Corp., had increased substantially in 1977. ComSatCorp's leases of INTELSAT fulltime half-circuits to its common-carrier customers had increased 19% between the end of 1976 and the end of 1977, and Comsat General's contribution to net income had increased from 16 cents per share in 1967 to 65 cents per share in 1977. These increases would have put 1977 net income ahead of 1976, except for the FCC proceeding challenging ComSatCorp rates for INTELSAT services.

The FCC's order to deduct from INTELSAT service revenue the amounts placed in escrow pending the outcome of the rate proceeding had adversely affected both 1977 and 1976 net income. The requirement had applied throughout 1977, but for only 6.5mo of 1976; the 1977 amount deducted from INTELSAT service revenue was therefore $37 779 000 more than that for 1976. ComSatCorp said its 1977 financial statements would reflect terms of a proposed settlement of the long-standing FCC rate proceeding. ComSatCorp management had agreed with representatives of the FCC general counsel and Common Carrier Bureau on the proposal, subject to approval by the FCC after an opportunity for public comment ending Mar. 30, 1978. Most of the settlement provisions related to current and future rates, and would affect operations results in 1978 and subsequent years. (COMSAT Release 78-8)

• A fundamental reappraisal by Western European governments of the overall European space effort had resulted in a demand by the West German government for more efficiency and better return on its investments in space, *Av Wk* reported. On hold during the debate were production of an initial batch of Arianes (European heavy launchers) and development of a new heavy telecommunications bus, as well as adoption of a broad new telecommunications package that would include the heavy bus. Meanwhile, determined to get more for its money, West Germany had begun exploring possible national or limited international programs outside the framework of ESA.

Industry sources said that West Germany, traditionally one of ESA's biggest contributors, was unhappy mainly because of repeated new beginnings of experimental communications-satellite programs as technological objectives changed. The Germans had wanted Europe to move directly into an operational comsat project after the successful launch and initial operations of the Franco-German Symphonie satellites. However, instead of building on this initiative, Western Europe (primarily through ESA) had taken up other experimental satellites like OTS (orbital test satellite), Marots (maritime OTS), and the H-Sat (experimental heavy telecommunications/multipurpose bus). West Germany, therefore, had begun thinking of developing independently (or more probably bilaterally) a direct-broadcast satellite that would be operationally feasible at home and commercially attractive abroad. (*AvWk*, Mar 13/78, 71)

March 14: Noise levels of Concorde departures at New York's John F. Kennedy Intl. Airport continued to be lower than those measured at Dulles Intl. Airport near Washington, D.C., according to the FAA's February monitoring report. On approach, average Feb. measurement was 98.7db at Cedarhurst on Long Island, compared with a cumulative average through Jan. of 100.4db. Vibration tests conducted by NASA during Feb. had shown the relationship between structural vibration and aircraft noise to be the same for Concorde as for other jet aircraft; the relatively higher levels of structural vibration measured during Concorde operations resulted from overall higher sound levels rather than from any unique characteristics of the Concorde. The FAA's monitoring equipment had recorded no sonic booms for the 103 Concorde operations during Feb. (FAA Release 23-78)

• JSC officials were puzzled after a weekend surprise, the *Houston Post* reported: 240 000 or so people came out to see the Shuttle orbiter Enterprise. With interest in the space program supposedly flagging, the question was why so many—some of whom walked for miles to get to the Space Shuttle orbiter—showed up to see an airplane-like craft just sitting atop a large airplane. The newspaper quoted a highway patrol trooper with 9yr of traffic-patrol duty: "Never in my life have I seen people come like that." Largest previous crowd had been the 135 000 who turned up at

Ellington Air Force Base for a 2-day Blue Angels show. Charles Biggs, JSC exhibits manager, said JSC had made no effort other than the usual press releases to attract a crowd, "no bands, no celebrities, no giveaways, just the orbiter sitting there." A tour guide commented: "What I think is that this is different than Apollo and going to the moon. I think the Shuttle is coming closer to the people. It is something they can relate to. They wanted to know when they can go on it." (*Houston Post,* Mar 14/78, 3A)

March 15: The *Marshall Star* reported that an MSFC researcher had been working to complete a weather experiment for the second Shuttle orbital flight test (OFT-2) scheduled for July 1979. His instrument, known as NOSL (nighttime-daylight optical survey of lightning), would obtain data on the many thunderstorms viewed by the Shuttle during a mission. The equipment consisted of a 16mm data-acquisition camera synchronized with a 2-channel cassette recorder, receiving the output of a photocell optical system serving as a sensor to pick up both visible and invisible lightning flashes and to record them as clicks on the tape.

Otha Vaughan, Jr., system developer, said he hoped to analyze the magnetic tapes containing cine photos and associated photo-optical data to get new insights into convective structure of cloud systems and lightning, the horizontal dimensions of lightning discharges, and possible electrical discharges that extend into the stratosphere. A successful experiment would permit adaptation of the data-acquisition and analysis technique to future meteorological satellites for identifying severe weather situations. (*Marshall Star,* Mar 15/78, 4)

• NASA announced it had judged the *UK 5 (Ariel 5)* mission successful in prelaunch objectives. This Explorer-class x-ray mission launched in 1974 had contributed significantly to x-ray astronomy and astrophysics. *Ariel 5*, a cooperative program between the U.S. and the U.K. Science Research Council, had discovered many new transient x-ray sources, although most of its studies were of steady x-ray sources. The second *Ariel 5* catalog had listed locations of 107 high-latitude x-ray sources, many of them extragalactic. The satellite had performed well, far beyond its 6mo design life, although gas for the attitude-control system had run out 2.5 yr after launch (18mo beyond prelaunch estimates). An onboard magnetic fine-torquing system had maintained attitude stability and allowed some spin-axis maneuvers. All 6 onboard instruments had performed as they did the first yr after launch and should continue to produce exceptional results. (MOR S-870-74-05 [postlaunch], Mar 15/78)

• NASA announced LeRC had awarded a $43 million contract to Detroit Diesel Allison Division of General Motors Corp. for support of the Dept. of Energy in development of ceramic materials, fabrication techniques, and components for new energy-efficient gas-turbine automobile engines. This extension through 1983 of an earlier contract would require use of special materials composed of elements available in abundance (such as silicon, nitrogen, and carbon) to be designed into components for gas-turbine engines; techniques to fabricate the components; and integration of the new components into advanced turbine-engine hardware. Allison Div. would test the components individually and in turbine engines, to verify the materials and the improvements needed to realize significant savings in fueling gas-turbine engines as well as to reduce costs of engine production. (NASA Release 78-43)

March 17: NASA announced it had selected four 2-man crews to begin training for Space Shuttle orbital flights. Those chosen were John Young, commander, with Robert Crippen, pilot; Joe Engle, commander, with Richard Truly, pilot; Fred Haise, commander, with Jack Lousma, pilot; and Vance Brand, commander, with Charles Fullerton, pilot. Young and Crippen would crew the first orbital flight test (OFT-1) scheduled for launch in spring 1979 from KSC, with Engle and Truly as backup.

NASA had planned 6 orbital flight tests increasing in complexity to check out the first U.S. reusable spacecraft. On the first 4 flights, the 75-ton orbiter would make an unpowered landing on a dry lakebed at DFRC; thereafter, it would return to a specially constructed runway at the KSC launch site. NASA had also considered an OFT mission to boost Skylab into a higher orbit; prime and backup crews to implement such a plan would be selected from those already named. (NASA Release 78-44; JSC Release 78-15; *DFRC X-Press,* Mar 24/78, 4; *JSC Roundup,* Mar 31/78, 1)

• Pioneer Venus, scheduled for launch May 20, would carry a recently completed JPL instrument to obtain new weather information on Venus, *JPL Universe* reported. VORTEX (Venus orbiter radiometric temperature experiment), a $4.5 million co-effort of JPL and Oxford Univ., England, would collect close-up data on winds, cloud structures, water vapor, and temperatures of Venus. Using technologies developed for satellite monitoring of earth weather, VORTEX would report 3-dimensional measurements of the Venus atmosphere. The complex instrument, an 8-channel radiometer combining 4 different subexperiments (three separate radiometers and a spectrometer) would weigh only 12 lb, partly because of the special hybrid microelectronics used. Orbiting as close as 100km above the cloudtops, it could "see" weather conditions at 7 different levels or layers. For a full Venus year of 243 days, scientists would process VORTEX data into weather maps and infrared images and compare meteorological conditions on Venus with those on earth. (*JPL Universe,* Mar 17/78, 2)

- NASA announced appointment of Neil Hutchinson, Charles Lewis, and Donald Puddy, all of JSC's Flight Operations Directorate, as flight directors at JSC for the first manned Shuttle-orbital flight, to plan and direct activities of the Mission Control Center during real-time Shuttle operations. They also would be responsible for integrating inputs from all elements of NASA, contractors, and the scientific communities. Hutchinson would manage all activities associated with the Shuttle-ascent phase; Lewis, the on-orbit phase; and Puddy, the orbiter reentry phase. All had served as flight directors during Apollo and Skylab missions. (JSC Release 78-16; *JSC Roundup,* Mar 31/78, 1)

- KSC's *Spaceport News* noted the 20th anniversary of the launch of *Vanguard 1,* a satellite which, together with *Explorer 1,* launched 6wk earlier, "invoked a national sigh of relief." Project Vanguard had begun in July 1955 with a White House announcement that the U.S. would launch a satellite as the U.S. contribution to the International Geophysical Year. The first attempt to launch a Vanguard, TV-3, ended when the spacecraft was destroyed Dec. 6, 1957, in an explosion on the pad. As one observer stated, "Following the explosion of TV-3, the Vanguard became a whipping boy for the hurt pride of the American people." Then came the success of *Explorer 1*. But the failures of the next Vanguard and next Explorer meant tremendous pressures on the Vanguard launch team. *Vanguard 1* had been launched from Cape Canaveral at 7:15am, March 17, 1958. The Vanguard team had in record time (2yr, 6mo, and 8da) developed from scratch a complete high-performance 3-stage launch vehicle, a highly accurate worldwide satellite-tracking system, and an adequate launch facility and range instrumentation. And, they had suceeded in getting the Vanguard into orbit during the International Geophysical Year. (*Spaceport News,* Mar 17/78, 4)

- Many significant technological advances of the past decade should be credited to NASA, said the *Lewis News,* quoting an article by Dr. Wojciech Rostafinski, an engineer in LeRC's Fluid System Components Division and frequent NASA spokesperson on Voice of America broadcasts, in the *Cleveland Plain Dealer.* The article continued: "The magnitude and importance of NASA-derived technology benefits have already been reported. Broadly, payoffs from technology transfer are direct and indirect. One direct benefit is satellites that report on weather, pollution, and crops. . . . An example of indirect benefits—the so-called spinoffs—is the electronic circuit miniaturization, so widely adopted today by terrestrial computer and calculator designers. Countless other innovations could be given for both categories.

"Now a question arises concerning the monetary return on the investment. In other words, why should we continue to spend large sums of money on 'obviously' abstract endeavors in space? . . . Let us check the figures. Since 1958, when NASA was created, in the span of a short (or long, depending on the 'yardstick' by which the time is measured) 20 years, NASA spent a total budget of $67.8 billion. This relatively lofty figure does not appear high at all when compared to other expenditures of the federal government. NASA appropriations have represented, on the average over the years, only 1.8% of the federal budget; in the last five years this has been down to 1%. Currently, NASA accounts for only 0.85% of the budget. Making a parallel with the budget of a family whose income is $20 000 a year, last year's NASA appropriation represents $170 per year set aside as investment. Not a large figure; a small one, actually, when compared to today's car prices, utility bills, or to the cost of eduction. . . .

"In 1971, Midwest Research Institute, studying the impact of technological activity, concluded that the $25 billion NASA spent in its first ten years had returned $52 billion through 1970 and by 1987 will have returned $181 billion. Five years later, Chase Econometrics also studied the economic impact of NASA R&D and concluded that an annual increase in NASA spending of $1 billion for the 1975-1984 decade would increase the GNP $23 billion by 1984; $23 billion for $10 billion during ten years . . . .

"Summing up, it becomes obvious and clear that it is much cheaper and wiser to seek technological progress through rational research, federally organized and performed by government and industry. The return on such investment is sound. The relatively modest sums invested in the national aerospace programs entrusted to NASA are well spent." (*Lewis News,* Mar 17/78, 1)

• Rockwell Intl. Corp. announced that the Air Force Flight Dynamics Laboratory had awarded its Los Angeles Division a $6.04 million contract to build test-aircraft structures using a revolutionary manufacturing process developed by the company and used on the Rockwell-built Space Shuttle. The AF Built-up Low-cost Advanced Titanium Structure (BLATS) program would use 2 Rockwell-developed processes (diffusion bonding and superplastic forming) in designing, fabricating, and testing a 1000lb section of fuselage for use on a future advanced fighter. The 46mo contract called for construction of a complex about 8ft × 10ft × 3ft, representing the center fuselage section where the wings attach, and a portion of the aft fuselage where the engines would be mounted, and including the necessary fuel tanks.

Rockwell engineers believed superplastic forming with concurrent diffusion bonding of titanium aircraft structure was the biggest breakthrough of the decade in advanced metals fabrication, because it might result (depending on the type of structure) in cost savings of from

50 to 70% and weight savings of from 30 to 50%, compared to conventional methods of building titanium parts. Besides the next generation of fighter aircraft, other applications of the process could be jet engines, spacecraft, and supersonic transports. (*Rockwell Release* LAD-8)

March 20: Ames Research Center announced the scheduled arrival this month of the XV-15, a tilt-rotor research aircraft, which would have wingtip-mounted turbine engines that turned 7.6m (25ft) prop rotors. The engine-prop rotor assembly could be tilted up for helicopter-type vertical takeoff and landing, or oriented forward in the normal position for conventional flight as an airplane. Bell Helicopter Textron, Fort Worth, Tex., had built the aircraft under a joint program of ARC and the U.S. Army's Research and Technology Laboratories at Moffett Field, Calif. The first XV-15 to arrive at ARC, modfied for remote control, would undergo 6wk of testing in ARC's 12.2×24.4m (40×80ft) wind tunnel. Flight test of the second aircraft would begin at the Bell facilities in Fort Worth after wind-tunnel tests. ARC would eventually house both aircraft for Army/NASA evaluation of the tilt-rotor concept, and for ARC research on terminal-area (airport) navigation and on vertical- and short-takeoff-and-landing. (NASA Release 78-45)

• The U.S. Army had devised a charged-particle beam weapon using negative hydrogen neutralized through an exchange cell as an antisatellite-kill mechanism, *AvWk* reported. The program had been code-named Sipapu, an American Indian word meaning sacred fire. Its application to ballistic-missile defense would be as an accelerator with small angular divergence; an antisatellite version with a range of several thousand km could destroy enemy vehicles without maneuvering to intercept. *AvWk* said that a smaller adaptation of the device, requiring lower power levels, could be orbited in 2 or 3yr, probably by NASA's Space Shuttle.

The Accelerator Div. at Los Alamos Scientific Laboratory had worked on Sipapu for the Army, with USAF participation; the Army's Ballistic Missile Defense Command would spend $4.3 million this year and $5 million in FY79 for charged-particle beam weapons, and $3 million in the coming yr for high-energy laser weapons. (*AvWk*, Mar 20/78, 13)

• The USAF had sent a destruct signal Mar. 19 to a Titan rocket carrying a pair of military comsats, the *W Post* reported, aborting an $80-million mission 8min after liftoff from Cape Canaveral. Air Force officials said a range-safety officer had sent the destruct signal when the rocket suddenly began slowing down and falling back to earth. Debris from the 31 000lb space mission, rocket, and satellites had fallen into the ocean, although the exact location was unknown. (*W Post*, Mar 20/78, B-9)

March 21: The *NYT* reported that the Soviet Union had built and bombers drop-tested a new manned space vehicle similar to the U.S. Space Shuttle. The delta-winged reusable vehicle had been dropped from a Soviet Tupolev TU-95 Bear bomber in atmospheric tests of the spacecraft's aerodynamic and handling qualities, like recent tests of the Space Shuttle. No known space testing of the delta-winged vehicle had occurred, although the Soviets might have begun orbital tests with a second new manned-flight craft with space-tug capabilities. Design characteristics of the test vehicle were apparently similar to those of both the U.S. Shuttle orbiter and the lifting-body research vehicles flight-tested earlier by the U.S. Air Force and NASA. The vehicle strongly resembled the USAF/Boeing X-20 DynaSoar concept, which had been viewed as a space station resupply vehicle before the program was canceled. The Soviet spacecraft was described as smaller than the 150 000lb NASA Shuttle orbiter, but larger than the Soyuz expendable manned transport. The *NYT* said the new vehicle indicated the USSR believed it had the technology to proceed with building reusable manned spacecraft. (*NYT,* Mar 21/78, 12)

March 22: LaRC announced it had begun a research program on problems of agricultural aircraft. Beginning in 1921, airplanes for agricultural use had required conversion from some other purpose; in 1950, the first aircraft was designed specifically for agricultural use. Agricultural aircraft had improved since then, but still relied on 1940s technology. The aerial-applications industry had recognized this, and asked NASA to study drift, swath guidance, liquid and dry material-distribution systems, and aircraft-handling characteristics causing pilot fatigue. The NASA study would also attempt to make the wake vortex of an airplane work for, rather than against, aerial applications. Aerodynamic concerns involved reducing drag for better fuel efficiency and developing appropriate high-lift concepts to improve takeoff, landing, and turning performance.

LaRC would conduct the study at its vortex-research facility, its full-scale wind tunnel, and in actual flight tests. The vortex-research facility would test small-scale models of the agricultural aircraft for interactions of the aircraft wake with dispersed spray and granular materials. The full-scale tunnel and flight tests would use an Ayres Thrush S2R-800, one of the largest agricultural aircraft. (LaRC Release 78-130)

• The *Marshall Star* reported that a NASA decision to boost the Skylab space station to a higher orbit could lead to reactivation and use of Skylab's onboard systems and instruments in a number of useful projects. MSFC had awarded parallel study contracts, each in the amount of $125 000, to Martin Marietta Corp. and McDonnell Douglas Astronautics Co. for independent reports on the possibilities and benefits of Skylab reuse. The 9-mo studies of the most effective ways to use Skylab in further missions, experiments, and demonstrations would concentrate on the experiments or equipment already aboard Skylab, for firsthand evaluation of how materials and equipment were affected by 10 or more yr in space; on crew quarters and other support, for Spacelab missions and experiments that could benefit from long duration and an additional energy supply; and on opportunities offered by the Skylab in itself, or in conjunction with other hardware elements, for new missions or experiments. For example, the relatively large facility (comparable to a 3-bedroom home) might provide a convenient platform for fabricating and constructing large space structures, or could aid in converting such structures into useful demonstrations or operational systems.

The ground station in Bermuda had begun operations Mar. 6 to 13 to reactivate Skylab, turning on command and telemetry systems and receiving data. Transfer of power from the airlock module to the Apollo telescope mount permitted reception of telemetry; the ATM solar arrays also provided power when turned toward the sun. Data received from the ATM computer indicated it was in good working order and had enough power. On Skylab's last working pass, controllers had used about 11min to check the status of the trickle charge to batteries 2, 4, and 5; the charge had been turned off on battery 7, which had remained fully charged and ready for the next mission. Controllers also had performed some systems checkoffs, Bermuda reporting the command receiver as strong throughout the pass. Plans were to reactivate Skylab in mid-April, using the TAC system to reduce atmospheric drag and prolong Skylab's orbital lifetime for several months. (*Marshall Star,* Mar 22/78, 1)

March 23: LaRC announced award to Micro Craft, Inc., Tullahoma, Tenn., of a contract to design and build precision models for wind-tunnel tests at the center. The small-scale models of controlled shape and finish, designed specifically for research and development projects, would consist of metals, moldable or formable materials, fiberglass-reinforced resins, woods, or combinations. The contract might also require specialized model-support equipment and other unique items needed for model tests. Micro Craft estimated the value of the 3-yr, cost-plus-fixed-fee contract at $2.8 million. (LaRC Release 78-14)

• NASA's Scientific and Technical Information Office announced publication of a 276-page summary, *The Search for Extraterrestrial Intelligence,* of the findings of a blue-ribbon group of 16 U.S. scientists on detecting possible radio signals from intelligent life in the universe. Based on the results of a series of SETI (search for extraterrestrial intelligence) workshops held during 1975 and 1976, the summary had 3 sections: Consensus, Colloquies, and Complementary Documents, covering subjects like preferred frequency bands, search strategies, and scanning devices used on radio telescopes. "Consensus" at the beginning of the book reviewed the conclusions reached by the SETI group: a serious search for extraterrestrial intelligence was both timely and feasible; a significant SETI program with potential for secondary benefits could be undertaken with only modest resources; large systems of great capability could be built; and SETI was intrinsically an international endeavor in which the U.S. could take a lead. NASA's proposed FY79 budget contained a request for $2 million for JPL to start the SETI program in Oct. 1978 and run it for 5yr. The funds, if approved, would be for an all-sky all-frequency search for radio signals from intelligent extraterrestrial life, using existing antennas of the Deep Space Network at Goldstone, Calif., and state-of-the-art hardware including a new very-wide-bandwidth supercooled preamplifier to be developed specifically for the effort. (ARC *Astrogram,* Mar 23/78, 1)

• NASA announced that the governors of Washington, Idaho, and Oregon had unanimously agreed to participate in a 3yr program demonstrating the use of NASA-supplied satellite data for natural-resources management in their states. Governors Dixy Lee Ray, John Evans, and Robert Straub, together with federal representative Pat Vaughan, constituted the Pacific Northwest Regional Commission. They voted $480 000 from commission funds to support first-yr activities beginning in Apr. 1979. The 3-yr plan would be a follow-up of the Land Resource Inventory Demonstration Project launched by the commission in 1975 during which more than 45 state and local agencies in the 3 states had used satellite data for surveys and inventories of various landcover types. Continuing the program would provide the state agencies doing planning and natural-resource management with an operational capability of extracting and using information from NASA's Landsat satellite system, and would aid them in using the more sophisticated data from Landsat-C after its launch. The Pacific Northwest program would join others across the U.S. sponsored by NASA's Office of Space and Terrestrial Applications demonstrating the use of space technology by state and local governments, businesses, and universities. (ARC *Astrogram,* Mar 23/78, 2)

March 24: DRFC reported that 5 of its employees had received NASA's Exceptional Service Medal for outstanding work on the Shuttle approach and landing test (ALT) program. The employees received the awards at JSC, where the Boeing 747 carrying the Shuttle orbiter had stopped on its way to MSFC.

Receiving medals were Fitzhugh Fulton, for "outstanding piloting skill and technical expertise as aircraft commander of the Shuttle Carrier Aircraft (SCA) during the Approach and Landing Test Program"; Tom

McMurtry, for achievements as first officer of the SCA "which contributed significantly to the accomplishment of all major flight objectives during the ALT program"; Vic Horton, for "personal dedication and technical expertise" including "design, development, testing, and installation of a unique crew egress system for the SCA"; Bill Andrews, for "an efficient and successful test team that was fully prepared to support the ALT flights"; and John McTigue, for effectiveness "in managing the overall support plan and in resolving technical, organizational, and resource problems as they developed." (DFRC *X-Press,* Mar 24/78, 2)

March 27: NASA and the Dept. of Energy announced they had signed a joint contract with Mechanical Technology, Inc., Latham, N.Y., to develop and test 3 generations of the Stirling engine for passenger cars. Produced earlier by United Stirling of Sweden, the baseline engine would be the hardware starting point for improving automotive fuel economy, performance, reliability, weight, and cost. The Stirling engine offered potential high fuel efficiency, flexibility in choice of fuels, inherently low noise and low pollution, and good driving characteristics. (NASA Release 78-48; DOE Release R-78-104)

• George Mueller, NASA's associate administrator for manned spaceflight 1963–1969, now chairman and president of Systems Development Corp., was named president-elect of the American Inst. of Aeronautics and Astronautics, *AvWk* reported. He would serve on the AIAA board of directors during 1978 and be inaugurated as president in 1979. (*AvWk,* Mar 27/78, 11)

• INTELSAT announced award of an international research contract valued at $75 380 to Cable & Wireless, Ltd., London, to determine the effects of rain on comsat radio signals. C&W would obtain annual statistics for rain-caused depolarization at 4GHz frequencies by measuring the relative amplitude of the 2 orthogonal circular-polarized components of a satellite signal. INTELSAT would use results to develop transmission and reception techniques that could double the capacity of its international telecommunications satellites. (INTELSAT Release 77-8-M)

March 28: KSC announced it had awarded the Univ. of Florida a $99 013 contract for development of a water-management system using satellite data and other NASA-developed technology. A Florida water-resources management information system could provide managers with a computerized bank of information to help meet increased demand for water; data from earth-resources satellites like NASA's Landsats and NOAA's Goes should reduce guesswork in water management.

Studies by the Univ. of Florida and the South Florida Water Management District had substantiated Florida's need for a comprehensive water-management system. The university study showed lack of water as the main problem facing Florida agriculture, and reported that water demand would equal the existing supply by about 1980. These studies, coupled with a 1972 mandate by the Florida legislature for a water-use and supply development plan, prompted state water-management districts to seek help from the university and KSC in solving their problems. (KSC Release 32-78)

March 29: The *Marshall Star* reported on a unique suspension system that would use air bags and cables to hang the Space Shuttle's orbiter Enterprise and its external tank in the test tower, for the first phase of 1978 ground tests scheduled to begin in Apr. The system would consist of a large overhead truss fitted with air bags and cables, installed like a crossbeam at the 65.8m (216ft) level between two walls of the tall test stand. Although the orbiter and external tank would weigh about 544 320kg (1.2 million lb), the suspension system would allow the freedom of movement needed to obtain test data. Suspension of the orbiter and tank in this first phase of testing would duplicate as nearly as possible the flight conditions of the Shuttle at high altitudes, after the solid rocket boosters drop off at 43.5km (27mi) 2min into the mission. Later tests of the entire Shuttle assembly would use a different suspension system.

The first Shuttle test phase would begin with the external tank temporarily fastened by knee braces to the stand, the orbiter lowered into place and mated as it would be for an actual flight. With the orbiter in place, engineers would install the suspension system overhead, attach cables from the air bags to the external tank, and remove the knee braces so that the cables would take the weight of the orbiter and external tank; lateral restraints would prevent sideways movement. The design of the truss, 25.3m (83ft) long and weighing 63 504kg (70 tons), would permit less than 2.5c (1in) deflection when fully loaded. Supporting members were of high-strength steel. (*Marshall Star,* Mar 29/78, 4)

March 30: Nature magazine quoted Aleksey Leonov, deputy head of the Soviet Space Training Centre, as saying that plans were final for the next 2 Eastern Europe cosmonauts—a Pole and an East German, each with a Soviet companion—to go into orbit "before the end of the year." Representatives of other socialist bloc countries (including Cuba and Mongolia) had arrived at the Gagarin Space Centre to train for flights "before 1983" furthering socialist cooperation in space. The scientific (as opposed to the symbolic) value of such participation would clearly vary from country to country, the article said: Cuba and Mongolia, for example, would contribute mainly by providing tracking facilities rather than new science. Czechoslovakia, however, had made scientific contributions; Czech microbiologists had provided special Chlorella strains, including mutations derived from chlorophyll, that for the first time demonstrated active growth, previously retarded by space conditions. Although Chlorella experiments had been performed on *Salyut 4* and at an earlier stage of the *Salyut 6* mission, the unique collection of strains produced in Czechoslovakia had exhibited considerable difference. (*Nature,* Mar 30/78, 394)

March 31: GSFC announced it would welcome a first group of guest astronomers Apr. 3 to its international ultraviolet explorer (IUE) satellite observatory, the first designed so that astronomers with little knowledge of spacecraft design and orbital operations could walk in, conduct observations, and receive data within 24hr. Astronomers would use NASA's *IUE,* placed in a modified synchronous earth orbit Jan. 26, to examine a wide range of celestial objects in the spectral ultraviolet region between 1150 and 3200A that included emissions of many common elements such as hydrogen, helium, carbon, nitrogen, and oxygen. Studies would range from the solar system to distant quasars, pulsars, and black holes.

NASA had selected nearly 200 astronomers from 17 nations (including the USSR) to conduct observations from GSFC's *IUE* installation and ESA's facility near Madrid, Spain. First guest astronomers would be Dr. Jeffrey Linsky, Univ. of Colo. specialist in cool stars, and Dr. Sara Head of GSFC, a specialist in hot stars. They would be succeeded by groups of 3 or 4 on a 7-day-a-wk basis throughout the yr. After 45 scientists had completed data reduction, the information gathered would indicate which aspects of astronomy should be emphasized during the spacecraft's remaining lifetime. (NASA Release 78-50)

• JPL announced that aerospace and aircraft design engineers from the U.S. and Europe would meet for the 12th Aerospace Mechanism Symposium (AMS) at ARC late in Apr. The symposium would be devoted to discussion of problems of design, development, and use of aerospace and aircraft mechanisms. Featured banquet speaker would be Dr. P.B.S. Lissaman, one of the designers of the Gossamer Condor, the ultralightweight and most successful manpowered aircraft built to date (*JPL Universe,* Mar 31/78, 1)

• The *JSC Roundup* reported on a 2-day conference, "Meeting Our Energy Needs Today and Tomorrow," with Dr. Peter Glaser, commonly thought of as the initiator of the solar-power satellite concept, as featured speaker. Glaser, designer of three instruments deployed on the lunar surface by the Apollo astronauts, had proposed a solar-power-satellite concept in 1968 before the days of concern with an energy shortage. He said the scientific community had slowly become aware that solar energy "shines brighter than ever as the major energy source."

Glaser summarized the advantages of collecting the solar energy constantly available in space, as opposed to disadvantages of any terrestrial energy source: absence of gravity and weather; avoidance of thermal pollution and waste; and the ability to direct a beam of collected energy to any location on earth, sunny or not, having usable land or not, close to the user or not. Solar-energy development would depend on development of the Space Shuttle, he noted, and was an idea becoming economically attractive and socially acceptable. Microwave and laser beams could transmit this energy to earth; directing a microwave beam accurately had been demonstrated. As for safety, Glaser pointed out that one cannot concentrate microwaves into a weapon system, whereas lasers could be lethal, so that all nations would have to agree that the laser method would never be used as a weapon. Glaser described the phased-array transmitting antenna already constructed on a 5-story-high project in the Aleutian Islands. He estimated 84% solar-power satellite efficiency in producing 30kw of power, as compared to current power plants with an efficiency range of about 40%. (*JSC Roundup,* Mar 31/78, 4)

• NASA announced that MSFC had developed a teleoperator-retrieval system (TRS) vehicle for first use in late 1979, either to boost the Skylab orbiting space laboratory to a higher orbit or to deorbit it in a remote ocean area. The decision to reboost or deorbit was expected early in 1979. Concept of the system evolved from teleoperator supporting research and development under way since the mid-1960s. Anticipated long-range usefulness of the TRS for payload survey, stabilization, retrieval, and delivery missions; its recovery and reuse capability; and its adaptability for the Skylab reboost/deorbit mission had led to its further development.

The TRS had a central-core propulsion system accommodating strap-on kits for additional propulsion. A 24-nozzle attitude-control system on the core provided 6 deg of freedom for control of the vehicle during rendezvous, docking, and initial orientation. Although thrusters originally used cold-gas propulsion, designers were studying use of low-cost hydrazine (hot gas). Preprogrammed instructions in the vehicle's core and computer, or a Shuttle crew member, would control guidance and attitude maneuvers. The TRS and orbiter structure were new designs; other systems would be fabricated almost completely from either off-the-shelf components or those being developed under NASA contracts. (NASA Release 78-49)

• Although to most people "aviation" would mean airliners and military aircraft, with a sprinkling of light planes, most of the aircraft in service and about two-thirds of the miles flown were in branches of flying known collectively as "general aviation," *Lewis News* reported, including private ownership for travel or sport; air taxi and commuter operations;

agricultural flying; prospecting and exploration; law enforcement; firefighting; air ambulance; and pilot training. About 230 000 aircraft of varied types did jobs in the Free World, more than 90% of them in the U.S. General aviation had served all 13 200 U.S. airports, compared to the airlines' 42%.

In 1977 general aviation had provided transportation for 100 million Americans, using less than 0.7 of 1% of all fuel used for transportation in the U.S. and less than 6% of fuel used by all aviation, while carrying a third of all intercity air passengers. The general aviation industry had provided jobs for more than a quarter of a million Americans in the manufacture, sales, and service of its products; grossed about $1.5 billion in new aircraft sales; and exported about 30% of its total production, contributing more than $500 million to the U.S. balance of trade. (*Lewis News,* Mar 31/78, 3)

During March: NASA announced arrival of the Space Shuttle orbiter, external tank, and solid-fuel rocket booster at MSFC for first assembly as a complete vehicle for ground-vibration tests. Using a facility originally designed for Saturn V moon-rocket testing, engineers would begin in early spring to evaluate the structural dynamics of the assembly and their effect on the Shuttle's system. The mated vertical ground-vibration tests would consist of "floating" the Shuttle in the tall test tower and applying vibrations with exciters powered by amplifiers similar to those found on home stereo sets; sensors on the skin would record characteristics of the vibrations passing from one area to another. The resulting data would verify system-design and mathematical models of the control-system's reaction to severe vibrations expected during launch and flight into orbit.

The first test-article configuration would include an orbiter and external tank to simulate the high-altitude portion of a mission after separation of the boosters. The second test configuration would consist of solid-fuel rocket boosters filled with inert propellants, stacked in the stand along with the orbiter and tank to simulate liftoff conditions. The third test configuration would be the same as the second, except that boosters would be empty, simulating a mission just before booster separation. Tests would continue through most of the year stopping only to change test configurations. (NASA Release 78-47)

• GSFC reported it had asked industry to build a computer that could process satellite images 10 to 100 times faster than present machines. The new computer would be based on 6yr of research in "massive parallelism," consisting of 16 384 processors working together, each processor a simple computer, with a single chip of silicon probably carrying 2 processors. Previous machines had not gone beyond a web of 64 simple computers. The new system would cross-correlate landmark images from earth-watching satellites 100 times faster than present

Space Shuttle orbiter Enterprise, assuming position in the dynamic test stand at MSFC where it would undergo vibration testing from May to November 1978. (NASA photo 78-H-222)

machines, making multispectral classifications (such as distinguishing wheatland from cornfields) and geometric corrections for comparison of pictures taken from different vantage points in one-tenth of the time previously necessary. (*Goddard News,* Mar 78, 3)

• Administrator Robert Frosch said NASA would analyze any chemical samples of UFOs received from credible sources but would not initiate any research programs, *Goddard News* reported. Frosch outlined NASA's position in a letter to the White House, in response to an inquiry from the director of science and technology policy. NASA was most often asked about the purported UFO sightings by astronauts and President Carter; NASA had satisfied itself that the astronauts had seen nothing abnormal in the space environment, and the bright object President Carter reported seeing in the sky when he was governor of Ga. had been classified as "unidentifiable" by the Natl. Investigations Committee on Aerial Phenomena. Some students of astronomy had suggested that what Carter saw was the planet Venus, at times much brighter than a first-magnitude star.

Although the USAF had routinely investigated reports of all UFO sightings until 1969, when it terminated its "Project Blue Book," it now limited its interest to sightings important to defense surveillance. The Air Force had concluded that never in almost 2 decades of investigating UFOs had it found "evidence of technological developments . . . beyond the range of present-day scientific knowledge." (*Goddard News,* Mar 78, 3)

• The Natl. Research Council reported that, at the request of NASA, a committee had reviewed potential uses of microwave sensors to supplement or replace visible-light and infrared detectors for earth-resource surveys from space. The committee had evaluated the potential of microwave sensors to penetrate cloudcover and possibly vegetation cover, to operate day and night, to detect soil moisture, to measure distance, and to aid in geological mapping; sensors specifically studied were synthetic-aperture radar and passive radiometers.

In a letter accompanying the report to NASA, committee chairman Arthur Anderson of IBM wrote: "The Committee concluded that an adequate experimental data base was available to support the initial development of an experimental radiometer sensor system (passive microwave sensor) for soil moisture, subsurface phenomena, and salinity measurements and a single-frequency single-polarization radar for geological explorations as shuttle experiments. The Committee did not feel that an adequate experimental data base was available to support the initial development of a multifrequency multipolarization radar for soil moisture measurements and vegetation or crop classification purposes." (NRC newsletter, Mar 78, 8)

- The Carter administration had submitted to Congress a $126 billion budget request for the Dept. of Defense, a 22% increase over last year, the Natl. Aeronautic Association newsletter reported. The $12 billion requested for procurement of 733 aircraft and helicopters was higher than last year, but the bill had asked for 34 fewer fixed-wing aircraft. Missile procurement dropped to $4 billion for 18 860 missiles, compared to 41 674 in FY78; military R&D was up to $12.7 billion. Higher personnel costs would consume more than half the budget. Civil defense at $100 million was the same as last year. The budget had forecast a 6.1% inflation factor, allowing for 2% growth over FY78. Except for the previously announced B-1 decision, the budget did not kill any major ongoing weapons-development programs. (NAA newsletter, Mar 78, 1; AFSC *Newsreview,* Mar 78, 2)

- The Natl. Academy of Sciences released a report of the ad hoc committee for review of the Space Shuttle main-engine development program. This review was requested by Senators Adlai E. Stevenson (D-Ill) and Harrison L. Schmitt (R-NM) in Dec. 1977 as an independent assessment by Natl. Research Council of the safety of the main engine during both orbital-flight and operational phases of the Shuttle program. The review was to be completed by the end of Feb. 1978 for consideration during FY 1979 NASA authorization hearings.

The committee reported that it saw "no reason to suggest that a safe and reliable main engine cannot be developed ultimately for the manned orbital flight tests and the later operational flights," but submitted a number of recommendations: first, since the flight schedule was not ready yet, the critical milestones should be reviewed later in 1978. Other recommendations included allowance of "adequate time to analyze root problems . . . rather than selecting technical 'fixes' to sure symptoms," as well as a complete tear-down inspection of the main engine after the first and sixth flights (which had not been envisioned in the schedule).

At a March 31 hearing before the Senate subcommittee on science, technology, and space, with Sen. Stevenson as chairman, Dr. Eugene E. Covert of the NRC described the ad hoc committee's activities and the reasons for its recommendations, expressing appreciation for the cooperation of NASA and its contractors during the review. Dr. Robert A. Frosch, NASA administrator, then summarized "the NASA reaction" to the report and its recommendations; Sen. Stevenson commended NASA's prompt and positive response, but proceeded with further questions on schedule and funding. In addition to NASA staff (including associate administrator John Yardley, MSFC engine project manager J.R. Thompson, Shuttle program director Dr. Mike Malkin, and deputy director Roy Day, and comptroller William Lilly), the Rocketdyne engine project manager Dominic Sanchini was on hand to explain administrative and technical problems encountered with the main engine.

The Senate committee report would contain all the testimony, the text of the ad hoc committee report, and further information on specific questions asked by Sen. Stevenson. (Text, hearing report 95-87, Sen Comm Cmte Mar 31/78; "Technical Status of the Space Shuttle Main Engine," Natl Res Cncl, March 78)

April

April 1: The *NYT* reported that the Soviet Union had apparently agreed to a U.S. proposal to begin talks on banning hunter-killer satellites, which the USSR had been developing for about a decade. A senior Western diplomat said the Kremlin had formally conveyed its willingness to negotiate but had not said when. Marshal Nikolai V. Ogarkov, chief of staff of USSR armed forces, had outlined the Soviet position when he met with 21 visiting U.S. congressmen in Moscow. The Carter Administration had revealed earlier that it had suggested a treaty after 3 successful USSR tests of a new satellite (known in military jargon as Asat) able to maneuver close to another satellite in orbit and explode. If perfected, such a weapon (which the U.S. did not have) could threaten satellites used to verify compliance with strategic arms agreements, to give early warning of missile attack, or for civilian and military communications links.

The Pentagon was unhappy about the administration's decision to negotiate because it had 2 U.S. hunter-killer satellites under development: one could be launched into an orbit similar to that of the satellite to be destroyed, and another more maneuverable model using infrared sensors could seek out targets over great distances. The administration apparently hoped Pentagon plans would induce the Soviets to begin negotiations in Apr. It was not clear whether a treaty would cover only hunter-killer satellites or would include ground-based antisatellite weapons such as powerful laser beams, which the Pentagon opposed banning. (*NYT*, April 1/78, 5)

April 2: The National Academy of Engineering elected 100 engineers to membership in the academy, including astronaut Neil Armstrong, the *NYT* reported. Election honored persons contributing to engineering theory and practice or pioneering in new or developing fields of technology. (*NYT*, April 2/78, 35; *Bridge*, Spring 78, 2)

April 3: NASA announced it had adjudged performance of the *Heao 1* satellite successful, as it had accomplished its primary and secondary objectives. *Heao 1* had demonstrated that, in addition to its operating mode, normal scanning, it could point at specific celestial objects for detailed investigations; it had performed more than 50 pointings. NASA had solved some minor problems with the stored-command programmer early in the mission. The experiment complement had continued to receive high-resolution data over an energy range of 150 to 10 million electron volts. Although the sensitivity of the large-area counter array instruments had degraded, several of the counters remained unimpaired because of their modular construction and were able to achieve overall objectives.

Heao 1 data clarified x-ray sources barely visible in older studies; the faintest sources seen were 10 to the 5th times weaker than the first x-ray star (ScoX-1) discovered 15yr ago. *Heao 1* had identified many distant active sources, such as quasars, Seyfert galaxies, and BL Lacerta objects. For example, *Heao 1* observed the BL Lacerta object Markarian-5011 spectra, which were extremely hard, establishing the lack of surrounding gas. *Heao 1* had discovered that the strongest active sources in the galaxy were emitting radiation into the MeV range as a result of stimulation by strong nonthermal emission mechanisms. A map of diffuse x-rays in the range 0.15 to 60KeV indicated a hot thermal plasma extending beyond earth's galaxy, and possibly early evolution of the universe. (NASA MOR S-832-77-01 [postlaunch] April 3/78)

• The Mar. 25 launch failure of a Martin Marietta Titan 3C carrying 2 DOD/TRW defense satellite-communications system (Dscs 2) spacecraft had left the U.S. with degraded defense communications, *AvWk* reported. The situation could worsen if one of the aging spacecraft in synchronous orbit should fail before another Dscs 2 launch scheduled in about 6mo. One Dscs 2 spacecraft serving Atlantic Ocean traffic, another serving Western Pacific communications, and a third serving Indian Ocean traffic had normally handled worldwide superhigh-frequency defense-communications requirements for the U.S. A *NATO 3B* spacecraft had been on loan to provide communications in the Eastern Pacific. One of the lost spacecraft would have orbited over the Western Pacific, and the spacecraft already there would have replaced the *Dscs 2* nearing the end of its usefulness above the Indian Ocean. The other spacecraft located over the Eastern Pacific would have freed the *NATO 3B* spacecraft for return to its primary command. Loss of the 2 spacecraft meant that the oldest *Dscs 2* might fail before being replaced, leaving the Indian Ocean areas with less coverage and placing an increased burden on the Atlantic spacecraft.

Two Dscs 2 spacecraft had been intended for launch in 6mo as in-orbit spares; now they would become primary replacements for the pair lost Mar. 25. A normal Dscs 2 system would require 4 operational spacecraft and 2 in-orbit spares. In view of the recent failures and 2 previous ones, future Dscs launches and orbital operations would have to be 100% successful to maintain the system using all 16 spacecraft that NASA had procured. (*AvWk*, April 3/78, 19)

• A Saturn IB rocket, twin of the one that launched Skylab and

Apollo/Soyuz test project (ASTP) astronauts into space, would begin a barge journey from KSC to Tokyo Apr. 5, KSC announced. The Saturn IB, part of the largest group of U.S. space hardware ever displayed overseas, would be only one of hundreds of artifacts borrowed from NASA centers across the U.S. for a Japanese space science exposition scheduled to begin July 16, ninth anniversary of the launch of *Apollo 11*. The U.S. display would include full-scale Mercury-Redstone, Mercury-Atlas, and Gemini-Titan rockets; a lunar rover; Mercury, Gemini, and Apollo spacecraft that had actually flown in space; a detailed mockup of the Viking lander; and an ATS-6 (applications technology satellite) communications satellite.

The Association for the Space Science Exposition, an umbrella organization for the Japan Science Society and the Japanese Maritime Science Promotion Foundation, had organized the exposition at an estimated cost of $21 million and would defray all costs of dismantling, loading, round-trip shipping, reloading, and reinstalling artifacts. The National Air and Space Museum of the Smithsonian Institution had coordinated the loan of the space artifacts. (KSC Release 38-78; *Marshall Star,* April 19/78, 2)

April 4: The *W. Star* "In Focus" column reported that NASA had awarded a grant to Dr. T. Stephen Cheston, associate dean of Georgetown University's graduate school, to inventory the literature produced by the social sciences on space activities and to plan for a scholarly journal, an academic forum for lawyers, economists, historians, and other specialists on ways to handle the new technology. The column quoted Cheston as saying they would have a lot to think about: "What we need is wisdom and a hell of a lot of it," he said. He commented on the many social scientists working on problems related to what is called "the industrialization of space." By establishing a scholarly journal, he had hoped to raise the "general quality of discussions of these issues because so many people are working in isolation now."

The column noted that the Space Shuttle had made possible the launching of satellites that could intrude into many aspects of human endeavor; Capt. Robert F. Freitag, NASA's deputy director of advanced programs in space transportation, said that by 1984 the Shuttle should be ready to begin construction of the first large satellites, those with a diameter of at least 100ft. Although NASA was weighing a number of options, according to Freitag, the technology "is for all intents and purposes here today. It's just a matter of nations, large corporations, and institutions deciding to go ahead and do it."

Although some disciplines had generated little space-related literature, lawyers had been at work for a dozen yr producing 4 treaties governing outer space, with several more in the works. J. Henry Glazer, legal counsel at Ames Research Center, had said that some of the proposed treaties were "needed yesterday," because "technology has rapidly overtaken the legal questions involved." Glazer had predicted that large comsats of the late 1980s would be the forerunners of much larger economic enterprises in space, such as an orbiting power-generating station about the size of an ocean liner that could convert solar energy into electricity and beam it to earth in the form of microwaves; or the idea in a book written by Dr. Brian O'Leary, Princeton astronomer and physicist who had helped develop the solar-power satellite, on pushing a small asteroid into high orbit around the earth and mining it for space-station materials. When asked about the legal and social ramifications of such a concept, O'Leary admitted there were quite a few "gaps: The social scientists are hopefully going to play catch-up ball here." (*W Star*, April 4/78, A1)

• Launch of *Cosmos 1000*, reported FBIS, had led Soviet scientists to review highlights of the series. The research program using Cosmos satellites had used several types of unified apparatus, and the USSR had always referred to them as a "series" irrespective of their scientific missions. The virtually mass-produced output of satellites had greatly assisted the high turnover speed of the series, which had reached the 1000 mark within 16yr and 2wk, more than 500 launches having come in the past 6yr.

The Cosmos series had begun Mar. 16, 1962, with the launch of an unnamed satellite (named *Cosmos 1* only 3wk later, when Tass reported launch of *Cosmos 2*). The USSR had seldom announced the purpose of Cosmos launches, a ploy that permitted concealment of their purposes as well as of their failures. Observers had classified *Cosmos 96*, for instance, as a Venus probe left or stranded in orbit; *Cosmos 146*, which had arrived in orbit measuring about 14m long and by next day was only 9m long, after emitting 2 capsules, was probably a Soyuz precursor. Georgy Narimanov of the Soviet Academy of Sciences Institute of Space Research said the USSR had used Cosmos satellites to test long-range communications systems, reentry installations, life-support systems, and assemblies for the Lunokhod moon-rover, as well as automatic-docking procedures. Lack of information had encouraged speculation about military applications, including hunter-killer tests.

USSR comments had stressed the practical aspects of the Cosmos program, such as experience accumulated during geophysical experiments for use on the "Meteor" spacecraft; basic research by *Cosmos 321* that had defined distribution of the magnetic field over 94% of the earth's surface; and Cosmos investigations of the upper atmosphere, the aurora, and the magnetosphere. (FBIS, Tass in English, April 4/78; Tass Intl Svc in Russian, April 1/78 [2 items])

April 6: Mechanical failures during tests of the Space Shuttle main engines had threatened to delay the whole program, although the first orbital launch had been scheduled for next year, *AvWk* and *Nature* reported. Turbine blades in the engine's high-pressure fuel pump had failed in 3 separate tests: twice the blades fractured or cracked, the third time a fire that followed the malfunction destroyed the evidence. Dr. Robert Frosch, NASA administrator, said that the setbacks had not lessened agency confidence in the basic engine design and that the Space Shuttle program should proceed on schedule; however, he admitted that chances of launching the first manned orbital flight in Mar. 1979 were "considerably less than 50-50."

A committee from the Natl. Research Council had begun in Jan. to assess the nature of engine failures and related problems, and to advise NASA and the U.S. Senate subcommittee on science, technology, and space on the engine-development program, according to the NRC *News Report*. The study centered on the safety and reliability of the engines and had included evaluation of engine design, materials, fabrication, and testing. The Senate subcommittee had requested a report of the study in time to use it in considering NASA's authorization request for FY79. (*Nature*, Apr 6/78, 482; *AvWk*, Apr 3/78, 18; NRC *News Report*, Apr 78, 1)

• The People's Republic of China had announced an 8yr crash plan to catch up with the rest of the world in scientific research, *Nature* reported. Vice-Premier Fang Yi told 6000 delegates to the All-Nations Science Conference in Peking that China had lagged 15 to 20yr behind the West in many branches of science and still more in others; the program would narrow the gap in some areas to 10yr by 1985, and would help in overtaking the rest of the world in all branches of science by the end of the century. The article said that China had planned to begin production of fast-mainframe computers and to establish a strong computer network and data base by 1985, with extra attention to peripherals, software, and associated areas of applied mathematics. PRC would devote its efforts to basic space physics, cosmic rays, and technical problems of remote sensing, and planned to launch a series of Skylabs and deep-space and communications satellites within 8yr. (*Nature*, April 6/78, 482)

April 7: American Indians of the Far West had used the Communications Technology Satellite, world's most powerful, in an unusual undertaking: tribes in Mont. and N.M. had linked up via the satellite with each other and federal officials in Washington, D.C., to exchange information and views on health, agriculture, education, and other tribal concerns. ARC said the *Cts* experimental satellite, operating on a new frequency and at power levels 10 to 20 times higher than current commercial comsats, had provided 2-way interactive video and audio communications between ground sites, allowing participants to view and confer with each other as though in the same room. Lewis Research Center had developed the transmitter (its key component) and would manage U.S. portions of the joint U.S.-Canadian satellite program, the *Lewis News* reported. Indian nations taking part in the broadcasts over 3days were the Crows and other tribes from the northwestern U.S., and the All-Indian Pueblo Council, Inc., and nearby tribes from the southwest. Government representatives included the assistant secretary of the U.S. Dept. of Agriculture, senators, and representatives, the governor of Montana, and the assistant secretary for Indian affairs in the U.S. Dept. of Interior. (ARC *Astrogram,* Apr 7/78, 2; *Lewis News,* Apr 14/78, 1)

• The Air Force had launched at 7:45pm the previous Thursday a secret payload aboard an Atlas-Agena rocket from Cape Canaveral, *Today* reported. The launch, not previously announced, resembled previous unannounced launches that had carried various spy satellites. The launch followed 3 consecutive unsuccessful military missions. In Feb., an experimental Trident fleet ballistic missile had failed after launch from a land pad at the Air Force station; in Mar., DOD officials had destroyed both a Titan III-C rocket carrying 2 Triple 7 defense communications satellites and a Poseidon fleet ballistic missile. (*Today*, April 7/78, 1A)

• NASA had not yet decided when to command the Skylab space station into a cartwheel maneuver that could extend its orbital life by 9 to 14mo, *Science* magazine reported. The maneuver could increase chances of a Space Shuttle mission either boosting Skylab or propelling it into controlled reentry. Major factor in the decision would be assessment of how long the station's control-moment gyros could sustain the maneuver before failing; without the cartwheel, NASA believed reentry would occur between May 1979 and July 1980. According to *Science,* however, Skylab's chances of remaining in the sky until NASA could get to it with a Space Shuttle were no better than 1 in 2. (*Science,* Apr 7/78, 28; *AvWk*, Apr 24/78, 15)

• The National Aeronautic Association announced the 1977 Collier trophy winners: U.S. Air Force Gen. Robert J. Dixon, and the Tactical Air Command headed by him since 1973. The NAA had awarded the Collier trophy annually for U.S. achievement in aeronautics and astronautics during the preceding yr; the selection committee had made the award for development and implementation in 1977 of the Red Flag combat-simulation flight-training program.

Red Flag, designed to enhance combat readiness of U.S. air crews, had been regarded as a significant contribution to national defense. Encounters in Red Flag, consisting of missions against simulated surface-to-air missiles, had been the most realistic undertaken in peacetime aerial-combat training. Gen. Dixon had established the program in 1977, conducting ten 4-wk exercises at Nellis Air Force Base, Nev.; more than 8200 air crews had trained during 1977, with 10 000 ground troops participating. (NAA newsletter Apr 7/78)

• The plane flown by Amelia Earhart when she disappeared would be the target of a search party headed by Vincent Loomis departing soon for the Marshall Islands, the *W Star* reported. Loomis, a USAF officer who had conducted aerial surveys to set air-navigation markers in advance of the H-bomb tests at Bikini atoll, said he had spotted plane wreckage but was in a hurry to finish his job. Some yrs later he had read a magazine article on the search for Earhart, and remembered the wreckage. Loomis said a simple navigational error could have put the Earhart plane over the Marshalls, north of her destination at Howland Island. Fred Goerner, who had combed the Pacific for clues to Earhart's fate, called the venture "a very, very, very long shot," especially since the area "has been surveyed and surveyed and every one of the islands has been visited and visited in recent years"; he said it was unlikely that evidence would have remained undiscovered. (*W Star*, April 7/78, C6)

• Flight testing of the Space Shuttle's microwave scanning-beam landing system (MSBLS) had begun at KSC, *Today* reported, with pilots and technicians using a specially equipped Jetstar aircraft operated by Dryden Flight Research Center to conduct the tests. The system-verification tests would be the initial checkout of the airborne MSBLS/Jetstar system and the ground stations with their laser systems to verify compatibility, and to check out the system processing test data. KSC had scheduled 5 test flights lasting a total of about 12hr. (*Today*, Apr 7/78, 12A; DFRC *X-Press*, Apr 7/78, 2)

April 10: NASA announced plans to launch orbital test satellite OTS-B, forerunner of an operational system, for ESA from Cape Canaveral no earlier than Apr. 27. This new comsat was 1 of 2 experimental models built by ESA to test satellite performance in orbit, with its operational descendants expected to provide satellite links in the 1980s for routing portions of intra-European telephone, telegraph, and telex traffic and providing TV relay services for Western Europe. Under an agreement between the U.S. and ESA, ESA would reimburse NASA for Delta launch vehicle, launch services, and other administrative costs totaling $17 million. NASA had launched *Ots 1* Sept. 13, 1977, on a Delta rocket from Cape Canaveral; the Delta first stage exploded 55sec. after liftoff. A burn-through of one of the solid-propellant strap-on Castor IV motors, subsequently modified, had caused the explosion. (NASA Release 78-53)

• KSC announced award to the International Hydrodynamics Co., Vancouver, British Columbia, of a contract for systems to recover expended Space Shuttle solid-fuel rocket-booster (SRB) casings. The $1 million contract would cover fabrication, testing, and delivery of the recovery systems with associated equipment and hardware. Recovery of SRB casings would mean sizeable savings to the Shuttle program; estimating 19 reuses of a casing, the savings realized by reuse (including cost of retrieval and refurbishment) would be $47 million. (KSC Release 37-78)

April 11: NASA announced that the solar-wind detector aboard Saturn-bound *Pioneer 11* had begun operating again, after lying dormant for 3yr despite scientists' repeated efforts to revive it. The solar-wind detector would apparently be ready for the 1979 encounter with Saturn for closeup pictures and other first-time measurements. Experimenters who completed an exhaustive checkout of the instrument were confident it was working properly. The solar-plasma analyzer, designed to map the flow of the million-mph wind continuously flowing out of the sun across the solar system, should provide important information about the interaction of solar-wind particles with Saturn and its rings. Investigators would compare these data with data gathered by the instrument 3yr ago when *Pioneer 11* swung around Jupiter.

The instrument had ceased operation shortly after the Jupiter encounter; in October 1977, after exhaustive analyses and a number of commands to the instrument (none of which worked), investigators decided to turn on the instrument's high-voltage power source to "thermally shock" the output circuits into operation. This had apparently worked, as 36days later NASA tracking stations reported the first transmission of data in nearly 3yr. Since then the instrument had responded to all radio commands, and appeared to be in condition for the Saturn encounter. (NASA Release 78-58)

April 12: Marshall Space Flight Center announced that the Bethpage, N.Y., plant of Grumman Aerospace Corp. had almost completed fabrication of an automatic device for forming continuous structural beams in space. MSFC had developed the "beam builder" to demonstrate techniques for fabricating large structures in space; the device would form triangular beams from flat rolls of very light materials, a complete section of the beam being easily supported with one hand. Housed in the cargo bay of the Space Shuttle in earth orbit, the machine would form continuous miles-long beams for building large structures such as satellite power systems and service platforms. (*Marshall Star*, Apr 12/78, 1)

- ESA announced that since the end of March its *Meteosat 1* had been transmitting daily meteorological data collected by special equipment aboard a French naval vessel in the North Atlantic. ESA had relayed the data to the European Space Operations Centre (ESOC) in West Germany. The satellite had transmitted messages every 3hr that included the ship's position, and parameters such as wind characteristics, sea states, and air and sea temperatures. The experiment, carried out jointly by the French meteorological service and ESA, had been planned to demonstrate the reliability of automatic communications between ships and the *Meteosat 1* satellite for regular production of meteorological reports. The experiment would last until July and would resume later with other European meteorological services. When operational, the system should include several hundred ships all over the world fitted with Meteosat equipment. (ESA Release, Apr 12/78)

• FBIS reported the answer of Konstantin Petrovich Feoktistov, USSR pilot-cosmonaut, when queried on the future of orbital stations. Feoktistov said they would develop as multipurpose scientific laboratories for extra-atmospheric astronomical observations, and for technological experiments to produce materials unobtainable under earth conditions. Orbital stations could seek out the most effective systems to develop methods for research and development and the design of scientific instruments and equipment. He went on to say: "Very broad prospects are opened up for manned stations if they are examined above all as the basis for industrial construction in space. Not everyone shares this view, but I personally believe that within the next few decades the development of technology will make it possible to actually set the task of creating power plants in space."

Feoktistov noted that such technology would mean putting millions of tons of diverse equipment into space to assemble enormous and complex structures. Such work could never be fully automated, he pointed out, and would require many people to control the equipment and carry out installation work. This would not happen quickly, he concluded, but it should not be relegated to the realm of fantasy. "After all, 17yr ago Yuri Gagarin's flight also seemed a fantastic achievement." (FBIS, Moscow Sotsialisticheskaya Industriya in Russian, Apr 12/78)

April 13: Langley Research Center announced it had awarded SDC Integrated Services, Inc., McLean, Va., a contract for support services to operate the digital-computer complex at the center. All segments of the center would use the computer complex in analytical studies, real-time simulations, and experimental data reductions. Services under the contract would include operation of the computing and data-handling equipment, operational and performance analysis, and keypunch and auxiliary equipment operation, in addition to research-data reduction such as programming, processing preparation, and data management. SDC had estimated the cost of the 2-yr base period of the contract at $5.7 million and a 1-yr option at $3 million; the cost-plus-award-fee contract also had two 1-yr unpriced options. (LaRC Release 78-17)

• Wallops Flight Center announced the first flight Apr. 10 in compound-helicopter configuration of its rotor-systems research aircraft (RSRA), a new research helicopter equipped with rotors, wings, and 2 jet engines. The compound configuration included the addition of 2 auxiliary-thrust engines and a 13.7m (45ft) wing. Powered by its 2 TF34 turbofan engines, the RSRA had used the combined lift of its wing and 5-blade S-61 rotor system to achieve an altitude of 610m (2000ft). By testing rotor systems to maneuver the craft at a wide range of altitudes and air speeds not possible in wind tunnels, RSRA would improve assessment of rotor performance by distinguishing inflight rotor characteristics from aircraft-system characteristics, which could not be done on the ground or in other aircraft. When fitted with fixed wings, the RSRA could test unproven rotor systems too small to lift it in normal flights.

Sikorsky Aircraft had built 2 RSRA vehicles under a program, jointly funded and managed by NASA and the U.S. Army, to develop technology to increase rotor-aircraft speed, performance, reliability, and safety, and to reduce helicopter noise, vibration, and maintenance. Sikorsky, under contract to LaRC, would test both RSRA vehicles for approximately 80hr flight time before delivering them to NASA and the Army for flight research at NASA's ARC. Sam White, NASA RSRA project manager, said, "The RSRA will help eliminate the costly practice of having to develop or extensively modify helicopter aircraft in order to conduct flight investigations for each promising new rotor concept. The RSRA is readily adaptable to accommodate new rotors and provide precise measurements of a variety of rotors under repeatable test conditions." (NASA Release 78-59; LaRC Release 78-16)

April 14: NASA announced successful launch of a FltSatCom-A Feb. 9 on an Atlas-Centaur from ETR's launch complex 36 at 1617:01hr EDT. Transfer-orbit parameters were 35 968.5km apogee, 167.20km perigee, and 26.4° inclination. NASA had fired the apogee-kick motor on Feb. 11 to inject the spacecraft into the desired synchronous orbit. All spacecraft systems were operating normally. (MOR M-491-202-78-01 [postlaunch] Apr 14/78)

• NASA announced that the Jupiter-bound *Voyager 2* was having problems with both its radio receivers that prevented Jet Propulsion Laboratory engineers from sending commands to the spacecraft. Early spacecraft data had indicated failure of one receiver, and the backup receiver might have had trouble receiving commands from earth; the spacecraft had continued sending information to earth. The Voyager had been designed to switch radio receivers if it had not received earth command for 7 days; the receiver that appeared to have failed had been the one on-line, and in 7 days the spacecraft computer should automatically switch to the backup receiver. Engineers would then attempt to reestablish communications by sending commands to the spacecraft. (NASA Release 78-56)

The *Voyager 2* spacecraft had accepted a command, the *W Star* reported, ending a week-long radio failure that had threatened its mission to Jupiter and Saturn. To protect the mission, engineers at JPL had begun reprogramming the spacecraft's onboard computer "so it will still do its science when it passes Jupiter and send back the results whether we can talk to it or not," said JPL spokesman Frank Bristow. Earlier in Apr., the spacecraft had automatically switched from the primary to secondary receiver because Voyager had not received an uplink from tracking stations for 7 days; the commands had not been sent because controllers had been preoccupied with scan-platform problems on *Voyager 1*. Controllers were reviewing their activities to pinpoint why they had overlooked an uplink, to ensure the lapse would not recur during the mission. (*W Star*, Apr 14/78, A-7; *AvWk*, Apr 17/78, 20)

• Scientists at JPL had assisted UCLA doctors in treating tumors with heat therapy, or hyperthermia, JPL *Universe* reported. While other researchers were exploring various ways of treating cancer with heat, UCLA neurosurgeon Robert Rand and veterinary radiologist Harold Snow had developed the concept of using a magnetic field to focus intense heat on specific diseased organs without damaging surrounding tissue or other parts of the body; To get the complex equipment required to test the theory, the doctors had enlisted the help of JPL scientists Dr. David Elliott and Glenn Haskins.

Dr. Rand had reasoned that, if heat would aid destruction of cancer cells, applying heat to an organ injected with magnetic material might more effectively deaden the tumor and any peripheral cancer cells. He asked Elliott and Haskins to design a magnetic heating system and build the equipment to see whether isolated organs could be heated. The JPL scientists had created a field coil to surround a patient's body, generating a 0.1tesla (1,000gauss) magnetic field at a frequency of 5000Hz to heat a magnetized organ to temperatures up to 55C for approximately 5min. Initial tests on rabbits had elevated the temperature of an isolated organ as anticipated; further testing on dogs would determine if the method worked on larger subjects. (*JPL Universe*, Apr 14/78, 2)

• The NASA 928 aircraft, Johnson Space Center's high-altitude WB-57F operated by its Aircraft Operations Division for DOE, had made 2 flights for the National Center for Atmospheric Research (NCAR), *JSC*

Roundup reported, to support research on the transfer of pollutants from the troposphere to the stratosphere in the vicinity of jet streams. The plane had taken air samples at altitudes between 35 000 and 60 000ft and measured the samples with onboard sensors: the stratospheric Aitken-nuclei detector, and the ozone monitor. The former would determine concentration and size of Aitken nuclei, considered tracers for tropospheric air, and the ozone monitor would determine the concentration of ozone, a tracer of stratospheric air; Nuclei-detector data superimposed on ozone-monitor data would define pollutant transfer. NCAR would analyze the data and reimburse JSC for the cost of the flights.

NASA-928 aircraft, equipped with various sensors, were used primarily to determine concentrations of radioactive particulates and gases and other trace constituents at several altitudes in the upper atmosphere (40 000 to 65 000ft). (*JSC Roundup,* Apr 14/78, 1)

• JPL's Edward Divita would be featured speaker at a meeting on nuclear waste disposal in Los Angeles Apr. 18, sponsored by the Institute of Electrical and Electronics Engineers, Nuclear and Plasma Sciences Society. Divita, a professional nuclear engineer who had performed several radiation studies of nuclear and space systems, analyzing and assessing the various disposal options, would speak on "Technical Status of Salt Bed, Sea Bed, and Space Waste Disposal." (*JPL Universe,* Apr 14/78, 1)

April 17: NASA announced it had accepted the review board's final report on failure of the Delta launch vehicle carrying ESA's orbital test satellite OTS-A that exploded after liftoff from Cape Canaveral Sept. 13, 1977. The review board had concluded that the failure resulted from a defect in the propellant of the solid-fuel strap-on Castor IV motor No. 1: either incomplete mixing and curing of propellant ingredients, or introduction of a contaminant, most likely water.

The board's findings had led to changing the design of the Castor IV solid-fuel rocket motor by putting additional insulation between the propellant and motor case to protect against such defects. NASA had also revised procedures for mixing and curing the propellant, and had improved the contractor and government test, inspection, and manufacturing surveillance procedures. New Castor IV motors incorporating the corrective actions had been manufactured and requalified. A second orbital test satellite, OTS-B, to replace OTS-A had been scheduled for launch May 4 from Cape Canaveral. (NASA Release 78-61)

• NASA had been considering FY81 and 82 missions to send a probe toward Halley's comet and land it on Tempel 2 comet, and one that would return a sample of Martian soil to earth, *AvWk* reported. Return of a Mars sample—to be studied first aboard the Space Shuttle in earth orbit—had become the key mission sought by scientists planning Mars strategy for NASA, who viewed a sample mission as a direct follow-on to the Viking mission in progress. NASA would have to make its decisions this summer on specific missions for FY82.

NASA might request the comet mission for FY81, in view of the determination by NASA planners and scientists to send a spacecraft near Halley's comet, even though a plan to rendezvous with the comet had disappeared when NASA was unable to obtain a Halley mission in FY79. The Halley/Tempel 2 comet mission would consist of a 1985 Space Shuttle launch of a vehicle using solar-electric propulsion to encounter Halley's comet in Nov. 1985 about 93 million mi from earth, eject a probe that would dive into the head or nucleus area of the comet, and return atmospheric data. The mother spacecraft would continue a high-speed flyby of the comet, either passing through the comet's tail or flying a safer trajectory in front of it. (*AvWk*, Apr 17/78, 14)

• NASA had under consideration for FY83 a supersonic-technology demonstrator, *AvWk* reported: an arrow-wing design in the 75 000-lb class suggested by a Boeing concept, although the company was not pursuing development of the vehicle. Boeing had proposed building such an aircraft to obtain cost information and to validate aerodynamics and manufacturing concepts for other U.S. aircraft manufacturers thinking of building operational supersonic transports. If NASA decided to pursue the project with DOD, agencies would probably seek participation by Lockheed, McDonnell Douglas, and Boeing. (*AvWk*, Apr 17/78, 31)

• NASA had cancelled the Lockheed/JPL SeaSat ocean-research spacecraft launch planned for May 17, *AvWk* reported, because of a failure in its synthetic-aperture radar downlink transmitter. Lockheed had detected the failure during thermal-vacuum testing of the spacecraft. Launch would be postponed until early June; the postponement would delay launch of the RCA/Goddard Tiros-N weather satellite to early Aug., because both missions would go on General Dynamics's Altas boosters that must use the same pad at Vandenberg AFB. Analysis of the SeaSat failure had revealed damage to a ceramic circuit board during manufacture; all other SeaSat systems functioned normally in the test. (*AvWk*, Apr 17/78, 27)

• *AvWk* reported that the Soviet Union had labeled the U.S. Space Shuttle a killer-satellite development, and U.S. officials planning talks with the USSR had been debating how to refer to the Space Shuttle in these discussions. NASA had been concerned about the effect on the Space Shuttle program of antisatellite negotiations, although U.S. negotiators had stated they had no intention of trading away Space Shuttle capability for USSR concessions. Informal negotiations on limitation of antisatellite-weapons tests had been set for May; U.S. officials said that NASA must have expected the Space Shuttle to be part of antisatellite discussions. As the Soviets had not cited the Space Shuttle by name, but rather as a "manned system of dual use," U.S. officials had not decided whether the concessions sought by the USSR would be on overall Shuttle capability, on individual Shuttle capabilities, or on individual potentially antisatellite Shuttle missions. (*AvWk*, Apr 17/78, 17)

April 18: NASA announced award of a $5 million contract to Vought Corporation, Dallas, Tex., for 5 Scout launch vehicles and associated hardware, such as separation systems and heat shields. Vought had been the Scout project's prime contractor since 1959. The first Scout, designed to place a 130-lb payload into a 300-nautical-mi circular earth orbit, had been launched in 1960. The new Scout could launch a 404lb payload from Wallops Island, Va.; Vandenberg Air Force Base, Calif.; or San Marco offshore at Kenya, Africa, into that orbit. The latest configuration of Scout vehicles was known as Phase 8; phases 1 through 7 had produced 107 vehicles, 96 of which had been launched. (NASA Release 78-20)

April 19: NASA announced plans to conduct the first full design-limit test of the parachute-recovery system for Space Shuttle solid-fuel rocket boosters. A B-52 aircraft flying at an altitude of 21 000ft would drop a 48 000lb dummy booster over the National Parachute Test Range at El Centro, Calif. Three main parachutes plus the pilot and drogue parachutes would deploy to slow the rate of descent. Primary objective of the test would be to simulate at a high dynamic pressure the loads that the main cluster of parachutes would encounter in actual use. MSFC would be responsible for the solid-fuel rocket boosters; Martin Marietta Corp. had been under contract to develop the recovery system; and DFRC would be responsible for B-52 operations. (*Marshall Star*, Apr 19/78, 3; DFRC Release 10-78)

• The USAF's Space and Missile Systems Organization (SAMSO) announced award of a contract to Boeing Aerospace Company, Seattle, Wash., for full-scale development and initial production of an inertial upper-stage (IUS) vehicle system, consisting of a 3-axis-stabilized vehicle of multiple solid-propellant stages with an avionics section including a reaction-control system. Although the USAF had primary responsibility for development of the IUS, NASA would participate in development because it also would use the IUS. The basic concept, a 2-stage vehicle for use with the Space Shuttle and on the Titan for some USAF missions, had one 20 000lb solid-propellant motor and one 6000lb solid-propellant motor that would deliver a 5000lb spacecraft to geosynchronous orbit.

NASA also planned two other vehicle configurations, one to carry large spacecraft on planetary missions and another for deep-space missions such as Galileo, a probe of the planet Jupiter. The Space Shuttle would deliver all NASA IUS vehicles to low earth orbit for use in pushing spacecraft to higher orbits or to earth-escape trajectories. The first NASA IUS flight would be a mission planned for summer 1980 to launch a tracking and data-relay satellite (TDRS); first use of a 3-stage IUS would be the Galileo mission planned for launch Jan. 1982. (*Marshall Star,* Apr 19/78, 1)

April 20: NASA announced that Dr. Eugene Emme, NASA historian since 1959 and developer of the NASA history program of professional and general publications recording the agency's development, would retire May 1 to devote full time to lectures and writing on aerospace science and technology history. Prof. Thomas Hughes, chairman of the NASA History Advisory Committee, said that Emme was "not only present at the creation of what is virtually a new discipline—space history—but he also played a central role in establishing sound institutional support for that discipline."

Emme had been founding chairman of the history committee of the Natl. Rocket Club (later the Natl. Space Club) and had served on the history committees of the American Inst. of Aeronautics and Astronautics and the American Astronautical Society. He had been a corresponding member of the Intl. Academy of Astronautics, a fellow of the American Association for Advancement of Science, the British Interplanetary Society, and the AAS; and he was an associate fellow of AIAA. (NASA Release 78-63)

April 21: NASA announced it had invited scientists to submit proposals for life sciences experiments to be carried aboard the Space Shuttle and Spacelab in the early 1980s. The experiments would study physical effects of spaceflight on humans and other living systems, and would test equipment and procedures. These experiments would be performed in the pressurized cabin of the Space Shuttle orbiter, in a pressurized Spacelab module, or (if requiring continuous exposure to the space environment) attached to an unpressurized Spacelab pallet. Proposed investigations would have to meet one or more of the following life sciences objectives: investigate and understand physiological, performance, and biochemical changes observed in humans who had flown in space; identify and investigate significant biological phenomena that might occur during or after exposure to the space environment; and/or test and demonstrate under operational conditions the equipment and procedures that would be needed by the NASA life sciences program. (NASA Release 78-64)

• DFRC announced it had tested on one of the center's YF-12 aircraft a dual-mode landing-gear system that used variable stiffness landing gear to reduce structural fatigue and dampen vibrations for future large aircraft ground operations. Three different landing-gear configurations tested at various speeds and weights over rough portions of the runways and taxiways resulted from changing loads on both the main and nose landing gear. (*DFRC X-Press*, Apr 21/78, 2)

April 24: AvWk reported that the Senate Commerce subcommittee on science and space had become the second congressional group to add another $4 million for a fifth Space Shuttle orbiter to NASA's FY79 budget, challenging President Carter's decision to defer purchase. This equalled the amount provided earlier by the House Science and Technology Committee. Overall, the Senate subcommittee had increased NASA's $4.371-billion budget request by $13 million, but failed to agree with some earlier House changes that had given NASA a $15.5-million increase. Among Senate subcommittee changes were $3 million for advanced manned-spaceflight programs, $4 million less than the House; language added to stress the importance of continuing lunar exploration; $2 million of $4 million deleted by OMB, restored for Spacelab experiments; and $3 million added for large space-structures work, making a total of $111.3 million for this item. (*AvWk* Apr 24/78, 27)

• The International Civil Aviation Organization's All Weather Operations Div. announced it had selected the U.S./Australian scanning-beam microwave-landing system at its meeting in Montreal, following West Germany's withdrawal of its MLS system from the competition. The final 39 to 24 vote (with 8 abstentions) had been secret to ensure that delegates could freely express their preferences.

France had proposed the secret ballot, strongly supported by the United Kingdom, Ireland, and Sweden, all of whom favored the Doppler MLS, as well as others. Asked whether the U.K. planned to continue the battle in ICAO's Air Navigation Commission, which must act on the decision rendered in Montreal, a representative said the U.K. had accepted the decision of delegates from 71 countries.

The meeting had been remarkably free of charges of improper efforts to win delegate votes. Asked if they had been subjected to any political arm-twisting or offers of political plums, delegates from Third World countries acknowledged that they had "intensive technical briefings" but nothing more. In view of the secret ballot, an observer commented that "if any quid pro quos were offered during this bitter battle, the offerer has no way of knowing now whether he got full value in return." (*AvWk* Apr 24/78, 23)

• The ESA Council had given the go-ahead to produce an initial batch of 5 operational Ariane launch vehicles, and had approved interim funding of approximately $23.8 million for preliminary production-related expenses, *AvWk* reported. Although individual participating countries must approve the expenditures, Council approval had been sufficient to begin production. The first operational launcher should be available by the end of 1980 to launch ESA's Exosat scientific satellite in early 1981. Cost of the 5-launcher package had been estimated at approximately $182.6 million, with an additional 10% proposed as a contingency reserve. Total budget would amount to about $200 million. At West Germany's insistence, the individual satellite programs that would use the Arianes would fund the launcher costs.

Program managers at ESA had sought approval to produce a first batch of 6 Arianes; however, policy of governments involved was to fund production of launchers only for identified markets. Once production buildup was completed, production rate of the Ariane was expected to reach 2 vehicles per year; program officials predicted this could increase rather quickly to 4, or possibly 5, a year to meet market demand. (*AvWk* Apr 24/78, 96)

April 25: NASA announced plans to launch an orbital test satellite (OTS) May 4 on a Delta rocket from Cape Canaveral; KSC would launch the experimental comsat for ESA. (KSC Release 45-78)

• The House of Representatives, in a Committee of the Whole House, by a vote of 345 to 54 passed H.R. 11404, authorizing appropriations to NASA for research and development, construction of facilities, research and program management, and other purposes. Alterations in the bill resulted in a net increase of $43.7 million to the budget NASA had submitted. Included were a $4-million increase to maintain the option for a fifth Space Shuttle orbiter; a $5-million decrease for expendable launch vehicles, slated to be phased out; a $3-million increase for space structures, power conversion, and microwave technology; a $3-million increase in energy-technology applications for solar-satellite power; a $10-million increase for applications research and technology development; a $1.5-million increase for industrial applications; a $4-million increase for technology applications (bioengineering); a $4-million increase for Spacelab systems implementation; and a $4-million increase for a Stereosat development program.

In debate recorded in the *Congressional Record*, Rep. Olin Teague (D-Tex.) of the House Committee on Science and Technology noted that

NASA had scheduled 25 launches in 1978, 11 on Delta (launch vehicles) and 8 on Atlas Centaurs, with NASA personnel supporting 3 Atlas-F launches from Vandenberg Air Force Base: "Most of the launches in 1977 emphasized the use of space for the direct benefit of people on earth—such as communications, geodetic, environmental, navigation, meteorological, and earth resources. In 1978 plans call for missions virtually equally divided between emphasis on these applications and launches of spacecraft for primarily scientific and exploratory research.

"Fifteen launches will be for paying customers other than NASA. They include the European Space Agency, ComSatCorp., the U.S. Navy, Japan, the National Oceanic and Atmospheric Administration, the United Kingdom, and Canada.

"The launch schedule for 1978 could not have been conceived of in 1958 when our space program began. In 1998 people will look back to 1978 and likely say the same thing. That will only occur if we support a strong continuing space program." (*CR*, Apr 25/78, 3164-3166)

April 26: The *Marshall Star* reported that 6 American and 2 British scientists would compete for the 2 experimenter seats on the second flight of the Shuttle-borne Spacelab experiments in 1981. The experimenters, called payload specialists, would be employed by the payload sponsor and would differ from NASA's pilots and mission specialists in being selected and trained for a specific mission. Four selected from the group of 8 would train to operate the mission's 13 scientific investigations; NASA would choose 2 of the 4 to fly on the 7- to 12-day mission.

The investigators working group (IWG), scientists representing 13 experiment teams for the U.S. and the United Kingdom charged by NASA with directing Spacelab scientific investigations, had selected the candidates for payload specialist. MSFC would manage the training programs for all 4 payload specialists. The Spacelab 2 finalists would be the second group of candidates for flight assignments; 10 Spacelab 1 payload-specialist finalists had been selected in Dec. 1977 to compete for 2 flight positions.

Finalists for the Spacelab 2 payload-specialist positions were Loren Acton, Lockheed's Palo Alto research laboratory; John-David Bartoe, Naval Research Laboratory; John Harvey, Kitt Peak National Observatory; Bruce Patchett, Astrophysics Research Division of Appleton Laboratory; N. Paul Patterson, Ball Brothers Research Corporation; Dianne Prinz, NRL; George Simon, Sacramento Peak Observatory; and Keith Strong, Mullard Space Science Laboratory. (*Marshall Star*, Apr 26/78, 4)

• MSFC announced it had completed acoustic testing of NASA's second high-energy astronomy observatory HEAO-B, first in a series of environmental tests on the satellite before its launch scheduled for Nov. HEAO's prime contractor, TRW Systems, Inc., Redondo Beach, Calif., performed the test that simulated sound pressure generated during launch. HEAO-B, second of 3 large unmanned orbital observatories designed to study x-rays, gamma rays, and cosmic rays emitted by stars and star-like objects throughout the universe, had been moved into TRW's thermal-vacuum chamber for further testing in June. (*Marshall Star*, Apr 26/78, 2)

- MSFC announced it had established a project office to develop an atmospheric cloud physics laboratory (ACPL) as a cargo for Spacelab in the early 1980s, so that scientists could for the first time study cloud physics without gravitational disturbances. Comparisons between earth-based observations had frequently not correlated with occurrences in actual clouds. The ACPL facility would permit observation of the processes of freezing, thawing, collision, electric charging, and temperature changes, photographing them as frequently and for as long as necessary. NASA had planned to send up the first ACPL mission on Spacelab 3, tentatively scheduled for June 1981. (*Marshall Star*, Apr 26/78, 3)

April 27: NASA announced successful launch Apr. 26 from Vandenberg AFB on a Scout of the first spacecraft built to measure variations in the earth's temperature, the heat-capacity mapping mission (HCMM). The experimental satellite had assumed a circular sun-synchronous 620km (385mi) orbit to measure areas of the earth's surface for maximum temperatures, then to measure minimum temperatures of those same areas about 11hr later. A 2-channel scanning radiometer would acquire images in the visible and near-infrared spectrum during the day, and in the thermal-infrared spectrum both day and night. Data resolution of the 2 channels would be about 0.5 × 0.5km (1800ft) as the spacecraft covered a 700km (435mi)-wide swath. Project officials said the position of the 300lb satellite in orbit was "almost perfect. If all goes well, we could take a picture of the earth in the visible spectrum as early as tomorrow."

HCMM was first in a series of low-cost modular spacecraft built for the Applications Explorer Missions (AEM)—small, experimental spacecraft in special orbits to satisfy mission-unique experimental data-acquisition requirements. Scientists designed the mission to determine the feasibility of using day-night thermal-infrared remote sensor-derived data to detect various rock types and mineral resources; monitor surface soil-moisture changes; measure plant-canopy temperatures at frequent intervals to detect transpiration of water and plant stress; measure urban heat islands; map surface-temperature changes on land and water bodies; and analyze snow fields to predict water runoff. HCMM data, correlated with data from other satellites (especially Landsat) and from ground observations, could aid in detecting time variations of temperatures on the earth's surface. Goddard Space Flight Center had been responsible for design, integration, and testing of the satellite and for data processing. Boeing Aerospace Co. had built the base module, International Telephone and Telegraph the instrument. (NASA Release 78-60; MOR E-651-78-01 [prelaunch] Apr 19/78; *NYT*, Apr 27/78)

April 28: JPL reported it would participate in tracking and data acquisition as well as in navigation during the upcoming Pioneer Venus mission. Hughes Aircraft Co.'s Space and Communications Group had designed 2 spacecraft and 4 probe vessels for the mission, and General Electric Co.'s Systems Division would provide a probe-entry system. R. B. Miller would be tracking and data systems manager at JPL, with William Kirhofer responsible for the navigation phase.

First would be an orbiter carried in the nose cone of an Atlas-Centaur scheduled to go aloft from KSC May 20; the multiprobe vessel would be launched Aug. 7, also on an Atlas-Centaur. The flight to Venus would be relatively short, about 6mo for the orbiter and about 5mo for the multiprobe. The orbiter would study the atmosphere and other characteristics of Venus for 243 days.

In the multiprobe phase a large entry probe would make a detailed sounding of the lower atmosphere and clouds of Venus, and 3 smaller probes would measure widely separated areas of the planet's atmosphere. To conserve energy, JPL trackers and navigators would turn on the battery-powered radio transmitters of the probes only about 20min before entry into the atmosphere; ground controllers would not be able to check frequencies between launch in Aug. and shortly before entry in Dec. The orbiter-and-multiprobe mission would carry about 30 instruments for gathering scientific data on Venus. (JPL *Universe*, Apr 28/78, 1)

• NASA had completed an ignition test Apr. 21 on 3 Space Shuttle main engines at the National Space and Technology Laboratory at Bay St. Louis, Miss., the JPL and KSC newsletters reported. The firing planned for 2.5sec had actually lasted only 1sec, but NASA officials said the test had met most of the objectives. A low-temperature reading in the fuel-pressure pumps caused premature cutoff. NASA and Rockwell Intl. engineers would decide whether to repeat the test before moving to the next phase of the program. (*JSC Roundup*, Apr 28/78, 1; *Marshall Star*, Apr 26/78, 2)

During April: NASA reported it had increased emphasis on general-aviation research, addressing problems and concerns of the industry, which had used results from recent programs in designing new aircraft. Objectives of NASA's general-aviation research had been to improve safety and efficiency, reduce environmental impact of general-aviation aircraft, and ensure a base of new technology adequate to support continued growth in the light-airplane utility. Aerodynamic research had emphasized airfoil development, resulting in significant improvements in performance in all speed ranges. Avionics research had aimed at reducing the complexity and cost of aircraft interaction with traffic-control systems improving navigation, guidance, stabilization, and systems management.

In generating information needed to design low-cost advanced avionics systems applicable to general aviation in the 1980s and beyond, NASA had integrated results of its studies into specifications for final systems design, fabrication, and installation on a twin-engine general-aviation aircraft for flight evaluation in 1979 and 1980. NASA had also defined technical problems that had limited accuracy and efficiency in aerial application of agricultural materials. The NASA program also had emphasized safety measures, including an automatic pilot-advisory system (APSA), structural crashworthiness, stall/spin research, and investigation of alternative fuels such as automotive gasoline for aircraft. Reducing the environmental impact of noise and exhaust emissions generated by light aircraft had also been an objective of NASA's aeronautical program, which would eventually assist industry and regulatory agencies in meeting future noise and pollution-reduction goals. (NASA Release 78-41)

• NASA announced that the Delta vehicle scheduled to launch the Japanese broadcasting satellite for experimental purposes (BSE) from KSC Apr. 7 had been qualified for flight, following an investigation and tests resulting from detection of a leak in the first stage of the Delta used in the successful launch of NASA's *Landsat 3* from Vandenberg AFB Mar. 5. The pogo accumulator that suppressed pogo-stick-like vibration in the fuel system during flight had been the most probable cause of the leak. NASA had installed a new pogo accumulator, previously subjected to screening and test, in the Delta for BSE launch. (NASA Release 78-55)

• NASA's role in basic and applied research under its charter—the Space Act of 1958—had created a paradox, NASA Administrator Dr. Robert Frosch said in a speech to GSFC employees, the *Goddard News* reported. NASA had been in the business of producing change, but the really interesting changes had not been predictable from past experience. "The interesting things happen because somebody not only changes the idea of how to solve the problem but questions the basic problem," Frosch said. "Thus the R&D problem faced by the applications part of our agency is that of trying to do work that not only satisfies the perceived mission questions, but will change the nature of the mission, or change the nature of the problem." Current NASA assignments included defining its R&D

posture, deciding the proper balance of basic and applied research, selecting what problems to solve, and attaining the proper balance between in-house and contractor effort.

Frosch discussed the proper role of NASA in future operational systems. He saw no great difficulty in NASA's taking on certain operational jobs with judicious interpretation of the Space Act. The question remained of NASA's operating national systems in a given part of the U.S. government. NASA's R&D effort in applications satellites was leading, Frosch said, "into an evolving global information system. But if . . . it is to become useful, then it will have to be an operating system. I think we are going to play two roles in such a situation: one of these is to do the R&D to improve such a system, and the other is to take the leadership role of making sure that the valuable research gets turned into an operational system which delivers data in a reasonable way."

Frosch concluded that, although NASA does not want to be the ultimate operator of such a system, it might have to do so in the early stages: that is, NASA would be an initial operator with the intent of transferring the system to another operator, either another government agency or a contractor. "Temptations posed by continuing large and everyday operations might lead us into difficulty in remaining a lively R&D organization." (*Goddard News*, Apr 78, 1)

• NASA's struggle to save 3 astronauts trapped in a foundering spaceship during the flight of *Apollo 13* had begun the most harrowing 6 days in NASA's history, *NASA Activities* reported in observing the 8th anniversary of the flight. The astronauts had heard a loud bang; when Fred Haise, Jr., scanned the instrument panel, he saw that one of the main electrical systems was deteriorating. The words Jack Swigert spoke to mission control long haunted NASA: "Houston, we've got a problem."

At the time of the explosion, *Apollo 13* traveling at 2100mph was 205 000mi from earth and more than 50 000mi from the moon, with a dead mothership (Odyssey) that included the main propulsion engine. After moving the astronauts into the lunar module, NASA focused on getting the Apollo on course for home while conserving limited supplies of power, oxygen, and water. Not only had NASA employees worked round the clock, but also specialists and computers from North American Rockwell, the contractor; Grumman Aerospace Corp., manufacturer of the LM; TRW Systems, builder of the descent propulsion engine; and many others. NASA had decided to drop the service module and put the command module Aquarius into the correct attitude for earth-atmosphere reentry; when the latter entered earth's gravity 216 277mi from home, it needed another course correction. The third descent-engine burn was successful, putting Aquarius into the proper reentry corridor.

As the spacecraft drew closer to splashdown, one of the astronauts described to ground control what he was seeing: "There's one whole side of that spacecraft missing . . . right by the high-gain antenna, the whole panel is blown out, almost from the base to the engine . . . it's really a mess . . . man, that's just unbelievable—looks like a lot of debris is just hanging out of the side near the S-band antenna." The astronauts had returned safely to earth 142hr, 54min, 41sec after launch. As Jim Lovell said, "We do not realize what we have on earth until we leave it." (*NASA Actv*, Apr 78, 14)

• NASA announced 3 key personnel changes during April: Dr. William Schneider would become associate administrator for space tracking and data systems, responsible for planning, development, and operation of global tracking networks, facilities, and systems for communications and data acquisition and processing for all NASA spaceflight programs. Dr. Schneider had joined NASA and the Gemini program in June 1963, after 2yr as director of space systems at ITT's Federal Laboratories. He had been mission director for 7 of the 10 manned Gemini missions, and had received NASA's Exceptional Service Medal for his service as deputy director of the Gemini program. He had also been mission director and program deputy director for the Apollo mission, and had received NASA's Distinguished Service Medal for his contribution to the success of the Apollo 8 mission.

At the end of Apr., Clarence Syvertson had become director of ARC, where he had served as acting director and deputy director. Syvertson had begun with the NACA at ARC in 1948 as a research scientist and assistant branch chief, then chief of the Hypersonic Wind Tunnel Branch, and later was director of the Mission Analysis Division. In 1966 he had become ARC's director of astronautics, a position he held until appointed deputy director. Syvertson had received NASA's Exceptional Service Medal for leadership of the joint DOT-NASA Civil Aviation R & D Policy Study, and in 1976 had been named a fellow of the American Institute of Aeronautics and Astronautics.

R. D. "Duff" Ginter, assistant associate administrator for energy programs in the Office of Aeronautics and Space Technology, had announced he would retire in June after 28 years of government service with NASA and the U.S. Navy. Ginter's contributions to space solar power, photovoltaic developments, wind energy systems, and automotive propulsion had placed him in the forefront of advanced technology applications. Ginter had begun his career with NASA in 1960 as chief of the Scout vehicle program. He later held several positions including director of the Centaur program, for which he received NASA's Exceptional Service Medal; director of the Technology Applications Division; director of the Energy Systems Division; and assistant administrator for energy programs. (NASA anno April 3, 10, 26/78; NASA Release 78-57, 78-65, 78-52; ARC Release 78-16)

• The Air Force Systems Command noted that Atlas launches planned at Vandenberg AFB as well as at Cape Canaveral in 1978 would use one of the oldest booster systems still working: first use of Atlas as a launch vehicle was in 1958. Atlas launches planned from Space Launch Complex 3 would support the USAF NavStar global-positioning system and space-test program; 2 NASA-sponsored satellites, Sea Satellite-A (SeaSat) and Tiros-N, would also use Atlas vehicles. (AFSC *Newsreview,* Apr 78, 3)

• The Air Force Flight Test Center at Edwards AFB, Calif., was target site for a Navy Tomahawk cruise missile successfully launched from a submarine, the USS Barb, a major milestone in the cruise-missile development program, AFSC *Newreview* reported. The test off the coast of southern Calif. was the Tomahawk's 24th, and its first launch from a submarine. The missile flew a fully guided land-attack test flight terminating at AFFTC. Objectives accomplished in the test included underwater ignition of Tomahawk's solid-propellant boost motor, and a midair engine start of the missile's turbofan engine for cruise flight. An onboard inertial-navigation set guided the missile to the Pacific coastline; once over land, the missile's terrain-contour matching system guided it to a recovery area. The flight ended with deployment of the test missile's parachute over a remote section of the test center; the missile, recovered by Navy personnel with AFFTC assistance, would be reused in future tests. (AFSC *Newsreview,* April 78, 11)

• Decisions not yet made by the U.S. government would determine the U.S. aerospace industry's fate for years to come, said Aerospace Industries Association president, Karl Harr, Jr., in the AIA's *Aerospace Review and Forecast, 1977.* Aerospace-industry sales in 1977 had increased appreciably, if not dramatically, in comparison with the previous year. Profit as a percentage of sales had climbed half a percentage point but remained below the average for all U.S. manufacturing. The industry had anticipated similar sales gains in 1978 and expected space activity to continue at approximately the same rate. Military sales should top those of 1977.

The aerospace industry had pressed for elimination of barriers to "free and fair trade" in the General Agreement on Tariff and Trade (GATT) negotiations; foreign-government subsidy of their manufacturers had made them highly competitive. The U.S. government's attempts to control export of "strategically critical" technology could severely impact sales abroad. Other issues of concern to the aerospace industry were capital formation and dealings with the government such as renegotiation, overregulation, expanding paperwork, and government competition with industry. Harr said that negative decisions could lead to "forfeiture of the pre-eminent position the U.S. has long held in high technology." (*Aerospace,* Spring 78, 3)

• In 1909 the Aero Club of Great Britain had established prizes of $125 each for the first four 250-yd flights of a heavier-than-air craft and three prizes of $250 for a one-mile flight in a closed circle, the NAA newsletter observed, in noting the progress of aviation. The Royal Aeronautical Society had just announced the $200 000 Kremer prize for the first crossing of the English Channel west to east by a manpowered aircraft. The society also had offered a $20 000 prize to the first Britisher who could duplicate in a manpowered aircraft the course flown by the U.S. Gossamer Condor, a one-mile figure 8. (NAA newsletter, Apr 78, 1)

May

May 1: Langley Research Center announced it had presented to Thomas Blackstock its first major award for an energy-saving proposal. LaRC Director Donald Hearth had presented Blackstock with a check for $1245, largest monetary award ever given at the center for a suggestion. Blackstock's idea was to upgrade the high-pressure air-supply system in the continuous-flow hypersonic tunnel so that it would run in an intermittent short-duration-flow blowdown test rather than in the longer duration continuous-flow mode. The tunnel, which represented LaRC's only large-scale Mach 10 aerothermodynamic-test capability, had been important to the Space Shuttle testing and would aid in developing low-cost advanced space-transportation vehicles. Blackstock's suggestion could save $30 000 the first yr without compromising the tunnel's research capability; LaRC's wind tunnels had consumed one-third of the electrical energy purchased by the center. (LaRC Release 78-24; *LaRC Researcher,* May 5/78, 1)

• In its report on U.S. R&D spending for 1979, the National Science Foundation pointed out that doubling of the nation's R&D expenditures since 1969 had been almost entirely attributable to inflation. However, the overall U.S. R&D effort had experienced real growth every year since 1975, mainly from increased emphasis on finding alternative energy sources; it had not provided major stimulus to economic growth or productivity in the short term. The federal government, currently supporting more than half the nation's R&D effort, had over the last decade emphasized civilian programs rather than defense and space. Largest federal R&D spending increases for 1978 and 1979 were expected in the energy and health areas.

During the 1970s, the U.S. had allocated proportionately less of the national R&D effort to basic research. Factors influencing this trend included the federal government's deemphasizing basic research in defense and space programs during the early to mid-1970s; the industrial sector's seeking quicker payoffs from R&D efforts aimed at a specific product or process, as opposed to projects with no specific commercial application; and universities' increased emphasis on applied research over basic research since 1970. (*NSF Highlights,* May 1/78, 1)

• *Av Wk* reported that the House of Representatives had authorized a $4.415 billion NASA budget that included $4 million for an option to purchase a fifth Space Shuttle orbiter, which President Carter had tried to defer. The House had not changed the authorization bill submitted by its committee on science and technology. House members had defeated an amendment by Rep. Ted Weiss (D-N.Y.) to delete funds for advanced supersonic-transport research, but had passed an amendment requiring the NASA administrator to report to Congress by Dec. 31, 1979, regarding NASA policy on conflicts of interest, standards of conduct, and financial disclosure. The House version of the NASA budget had represented a $43.7 million increase over the Carter budget.

The Senate Committee on Commerce, Science, and Transportation had approved a $4.388 billion NASA FY79 authorization that included $4 million for the fifth orbiter and an additional $7 million for manned-flight advance planning. Former Apollo astronaut Sen. Harrison Schmitt (R-N.M.) had helped convince the committee that NASA needed the $7 million increase to maintain a viable advance-planning program for manned spaceflight. Total Senate Commerce Committee authorization was $17 million higher than NASA's original $4.371 billion request to Congress. (*AvWk*, May 1/78, 22)

• Latest word on *Progress 1* refueling of the *Salyut 6* propulsion system in orbit indicated that redesign of the space station had been more extensive than the USSR had previously revealed, *Av Wk* reported. The Soviets had made a fundamental change in the station's rocket engine and fuel system by switching to pressure-fed from turbine-driven main engines, and adopting a common hypergolic propellant for both the main propulsion and reaction-control system engines.

Before *Salyut 6,* the Soyuz and Salyut propulsion systems had been the same: in the reaction-control system, hydrogen peroxide thrusters had provided attitude adjustments and hydrogen peroxide had also driven turbines pumping hypergolic oxidizer and fuel to the main propulsion system. Fuels in the main propulsion system were a form of nitric acid and of hydrazine. This arrangement had required 3 sets of propulsion lines and tanks: 1 for the hydrogen peroxide, 1 for the main-engine oxidizer, and 1 for the main-engine fuel.

Changing the *Salyut 6* had allowed both the main engines and the reaction-control thrusters to use propellants from the same tanks, permitting much simpler and more effective orbital refueling than if it had to include hydrogen peroxide. The change had required design of an entirely new internal-tanking and propellant-line fuel system, and development of the pressure-fed engines and the hypergolic-thruster system. The aft section of the station containing the main engines and a concentration of reaction-control-system engines could now accommodate docking at that end by a second Soyuz transport vehicle. The Salyut's standard forward port was able to accept both Soyuz and Progress-tanker spacecraft. (*AvWk*, May 1/78, 1)

May 2: NASA announced signing of an agreement with the National Eye Institute (NEI), Bethesda, Md., for laboratory and clinical tests of a new surgical instrument to remove hard cataracts. Lewis Research Center had developed the instrument jointly with Cleveland ophthalmologist Dr. William McGannon. The NASA-NEI program would include refinement of surgical techniques using the cataract instrument, and design and implementation of clinical trials.

The new instrument consisted of a surgical handpiece, a regulated flow system for infusing fluid, and a peristaltic outflow pump with necessary controls. The handpiece would use a high-speed air-turbine drive to power an auger-type cutter capable of both end and side cutting, rotating at speeds between 300 000 and 270 000rpm. The control system had a flow-sensing pressure regulator to maintain intraocular pressure under widely varying flow conditions. Use of the rotary cutting instrument would require only a small opening in the cornea, and should apply to the full range of cataract hardness. As part of the joint agreement, NASA and NEI announced that interested companies might apply for a nonexclusive patent license to manufacture and market the flow-sensing pressure regulator, which might be applicable to other types of eye surgery. (NASA Release 78-66; *Lewis News,* May 12/78, 1)

• NASA announced that Frank Nola, engineer at Marshall Space Flight Center, while working on a NASA solar heating and cooling project, had invented an inexpensive yet revolutionary device called a power-factor controller that could save energy used by billions of electric motors throughout the U.S. The small device would continuously determine the load on the motor to which it was attached by sensing shifts between voltage and current flowing to the motor. When it sensed a light load, it would reduce the voltage to the motor to the minimum required, with accompanying drop in current flow, minimizing power going to heat loss. A typical washing machine using a power-factor controller would drop its power usage from 160watts to about 55watts on an idling motor. Savings would be less on a loaded motor but, as most domestic motors never needed the full 125volts available, the power-factor controller could save up to 8% of the power used by a motor under heavy load. Energy savings for larger motors would be even more dramatic. NASA said patent licensing and technical information for Nola's invention would be available from the MSFC chief counsel. (NASA Release 78-67)

May 4: NASA reported that 8mo after launch and 8mo before beginning close Jupiter observations, *Voyager 1* had traveled nearly 555 million km (348 million miles) from earth and was then at a distance from the sun of about 3.1AU. Its velocity relative to the sun was about 19.9km (12.4mi) per sec.; time needed for one-way communications was 30min 45sec. On May 3, the Voyager had performed a midcourse correction of its Jupiter-bound trajectory, as part of a set of commands relayed to the craft Apr. 26.

NASA had constructed a model to investigate the sticking of *Voyager 1*'s scan platform. One possible explanation for the sticking was that a small piece of plastic from the scan-actuator locking mechanism might have lodged in the actuator's gear. Before moving the scan-platform on the Voyager into the position where it had previously stuck, NASA would try to reproduce the failure in laboratory tests and determine its effects. (NASA Mission Status Bulletin, May 4/78)

May 5: NASA announced appointment of Samuel Keller as deputy associate administrator for space and terrestrial applications (OSTA), succeeding Leonard Jaffe, who had been appointed special assistant to NASA's chief engineer. In his new position, Keller would be responsible for general management and direction of OSTA programs. He had held numerous management positions at Goddard Space Flight Center before coming to NASA Hq as assistant administrator for personnel programs; had worked in industry and government; and had been a pilot in the U.S. Air Force from 1956 to 1959. (NASA anno May 5/78)

• Dryden Flight Research Center reported it had aborted the first design-limit test of the parachute recovery system for the Space Shuttle's solid-fuel rocket boosters (SRB), when the 48 000lb dummy booster scheduled for airdrop from a B-52 flying over the National Parachute Test Range would not separate from its aircraft. After several unsuccessful attempts, pilots returned the B-52 to DFRC without incident and rescheduled the flight for the week of May 5. (*DFRC X-Press,* May 5/78, 2)

• The *W. Post* reported that a subcommittee of the House Appropriations Committee had cut NASA's budget, endangering 3 of the agency's most ambitious plans for the next 5yr: to search for life beyond earth, to keep the Skylab space station from falling back to earth, and to fly a rare joint mission with West Germany around the north and south poles of the sun.

The subcommittee had cut $1.4 million from NASA's request for $2 million to begin a program called SETI (search for extraterrestrial intelligence), in effect killing the program. A NASA official said the funds would not be enough "to start design work on the antennas we were going to use to listen for extraterrestrial signals." The subcommittee also denied NASA $20.5 million to use for Space Shuttle astronauts to fire the abandoned Skylab space station into a higher and safer orbit. NASA considered the most critical cut to be $5 million from the $13 million sought to start in 1983 the solar-polar orbiter mission that would have used two spacecraft, one built by the U.S. and the other by West Germany. The West Germans had warned that any U.S. delay meant that Germany would drop out of the mission. (*W Post,* May 5/78, A13)

• LaRC announced it would sponsor during May a program on "Women in Aviation," with slides and tapes highlighting women's contributions to aviation. An international organization of women interested in aviation, The 99s, which was founded in the 1930s and whose original members included Amelia Earhart and Anne Morrow Lindbergh, provided the program to LaRC. (LaRC Release 78-27)

May 8: LaRC announced that Donald Hearth, center director, had received an honorary Doctor of Science degree during commencement exercises May 7 at George Washington Univ. in Washington, D.C. As guest speaker, Hearth discussed Technology and the Future, noting, "There will continue to be a strong tendency to characterize technology as either the savior of society's problems or the villain in the drama of life. Technology is not magic," he continued, "but it can help solve some of society's problems. We want machines, but not the kind that run us. We want technology, but we want to be in charge of it." The address included a short history of technology, and projections of possible uses of technology in the near future. (LaRC Release 78-29)

• LaRC announced it would hold a conference on safety, occupational medicine, and environmental health May 15-18, with about 60 representatives from all NASA centers attending. Welcoming remarks by Fred Bowen, LaRC technical manager for management operations, would cover audiometric testing, types of hearing loss, rehabilitative techniques, noise protection equipment, and computerized treadmill data. LaRC Director Donald Hearth's remarks on the second day would include discussion of workmen's compensation, disability, continuation of pay status, and LaRC's occupational medicine and environmental health and safety programs. The May 17 agenda would cover space suit testing, new approaches in workplace health standards, fire protection, Space Shuttle safety challenges, the kinetic method of lifting, and system safety in wind-turbine generators. On May 18 participants would discuss nuclear safety in space systems, Space Shuttle occupant safety, and new firefighting equipment. (LaRC Release 78-28)

• The long controversy in the International Civil Aviation Organization over choice of a microwave landing system — which had ended in a 39-24 victory for the U.S.-backed time-reference scanning-beam (TRSB) system — had taught the group important lessons, *Av Wk* reported. First: the large turnout of Third World countries indicated that future proposers of new systems must focus their appeals to such countries. Roy

Cox, director of telecommunications for the British Civil Aviation Authority, described the problem: "Those . . . in the meeting faced the impossible task of studying and acting upon some 2,000 pages . . . on complex operational and technical issues and the time allotted did not permit adequate study by those states which had not previously participated in the many years of preparation." Second: although the ICAO had recognized the complexity of evaluating competing systems on the basis of extensive avionics expertise and background, 5 of the 10 specialists appointed to the All Weather Operations Panel had represented countries favoring particular microwave landing systems. Although selected as individuals, not as official representatives of their governments, the 5 had consistently favored the entry sponsored by their government. Hindsight would recommend including specialists from many countries on such an evaluation panel, to ensure that members predisposed to one of the choices would represent a minority, rather than a majority, of the panel. (*AvWk*, May 8/78, 44)

• Walter Morgan, ComSatCorp senior staff scientist, said at a conference sponsored by the American Institute of Aeronautics and Astronautics that a network of high-capacity comsats could be in orbit by the early 1990s, increasing transmission capacity and reducing the need for large ground antennas, *Av Wk* reported. Called orbital antenna "farms," the 14 000lb clusters of communications satellite antennas and sensors could replace many smaller individual satellites in geostationary orbit. Five antenna farms linked together in circular fashion would form a major cluster, which in turn could make up 1 part of a network of 5 separate clusters serving the Americas, Europe, the Pacific, Africa and the Middle East, and Asia. Such antenna farms would reduce the number of individual satellites and emit higher powered radio frequencies, permitting numerous earth stations to have smaller antennas with attendant lower costs and greater mobility.

Basis of an antenna farm cluster would be a structure about 40ft long and 14ft in diameter, housing electronic modules to control mission-payload equipment. Although current technology was capable of handling orbital antenna farms, two possible technological problems would be in-orbit construction of the platforms and in-orbit servicing of the equipment. Disadvantages would include compromises on optimum orbital locations for various missions, radio-frequency interference between payloads grouped on a common platform, and the effect of a single system's failure on other antenna farm systems. Although ComSatCorp had not taken an official position on the concept, Morgan predicted that an antenna farm could be operational in the early 1990s, and that a smaller experimental farm might be operating earlier. (*AvWk*, May 8/78, 56)

May 9: ESA announced that on May 12 the Villafranca del Castillo ground station near Madrid would be inaugurated by H.M. Juan Carlos, king of Spain. ESA had designed the station, built under an Aug. 2, 1974, agreement between ESA and Spain, to monitor and control the *IUE* satellite launched in Jan. 1978, OTS-2 to be launched the first half of May, and Marots to be launched in 1980. The station had antennas measuring 15m for *IUE,* 12m for Marots, and 3m for OTS-2. ESA also would use the station beginning the end of 1978 to control the U.S. meteorological satellite Goes. Like other ESA ground stations, Villafranca would be linked with the European Space Operations Centre (ESOC) at Darmstadt in West Germany, which housed control and computer facilities for all European satellites. (ESA Release May 9/78)

May 10: MSFC announced that its weather scientists working with colleagues from Texas A&M had encountered practically every kind of severe weather—thunderstorms, tornadoes, torrential rains, hail, and snow—within the study of the zone, roughly the south central states, during the 7th atmospheric variability experiment (AVE-7). As a 3-day weather pattern had moved across the test area, 24 rawinsonde stations operated by NOAA and one each by MSFC and Texas A&M had acquired data. The experimenters had also received for comparison a weather picture from a Goes satellite every 15min and reports from weather radars throughout the area every hour.

In the AVE-7 experiment, rawinsonde stations every 3hr released balloons carrying instruments to record temperature, humidity, atmospheric pressure, and wind velocity and direction from ground level to an altitude of about 25km (15.5mi). The balloons, normally released about every 12hr, on a 3-hr schedule had allowed researchers to spot small-scale events that could dissipate quickly. The balloon data, combined with satellite pictures at 15-min intervals, had permitted computerized weather forecasts by showing scientists how to spot approaching bad weather much earlier; comparison of ground-based with satellite data would help identify the causes of severe weather. With more such experiments using space technology, scientists could eventually accumulate enough information to more accurately predict severe weather. (*Marshall Star,* May 10/78, 1)

• MSFC announced it had awarded contracts totaling $42 751 900 for scientific investigations in support of the Space Telescope project to five educational institutions: Research Foundation of the State Univ. of New York, Albany, Dr. J. J. Caldwell, Space Telescope interdisciplinary scientist, $126 000; Beloit College, Beloit, Wis., Dr. D. J. Schroeder, telescope scientist, $69 000; Institute for Advanced Study, Princeton, N.J., Dr. J. N. Bahcall, interdisciplinary scientist, $257 700; Johns

Hopkins Univ., Baltimore, Md., Prof. W. G. Fastie, telescope scientist, $151 300; Univ. of Texas, Austin, Dr. William Jefferys, astrometry science team leader, $1 862 500; and Dr. K. L. Lambert, interdisciplinary scientist, $285 400.

The Space Telescope, planned for launch into earth orbit in the early 1980s, would permit scientists to see 7 times farther into deep space than before. MSFC had been lead center for the Space Telescope; GSFC had been responsible for scientific instruments and for planning telescope operations. (*Marshall Star,* May 10/78, 2)

• To put space processes and space products within public reach as early as possible, NASA had sought private firms to do materials processing in space, MSFC reported. The center had named a special task team to work exclusively with firms interested in pursuing space-made or space-derived processes and products; the team, headed by Richard Brown, would be the point of contact for information and assistance to firms with ideas for new or improved products that might be processed in space. "We intend to develop simplified working arrangements," Brown said, "and make it as easy as possible for commercial firms to try out new ideas for materials processing in the space environment."

Promising results from earlier space experiments, coupled with the opportunity offered by the Space Shuttle to work routinely in space, could make material processing commercially attractive. According to Brown, gravity reduced to a millionth of that on earth could make "interesting things" happen: elimination of the forces of buoyancy, sedimentation, and movement in fluids caused by heat could permit uniform mixing or separation of materials that would be impossible on earth. Families of new materials might be derived from materials processed in space.

To reduce industry's initial capital outlays usually needed for joint ventures, NASA had planned to build an inventory of general purpose processing hardware for lease to any U.S. firm, institution, or individual that would conduct materials experiments or demonstrations consistent with NASA's objective of public benefit. Brown had invited anyone interested in the program to ask MSFC's task team for more information. (*Marshall Star,* May 10/78, 4)

May 12: NASA announced launch of ESA's orbital test satellite 2 *Ots 2* at 6:59pm EDT, May 11, from Cape Canaveral toward a stationary orbit 22 300mi over Gabon in West Africa. A Delta vehicle had put the satellite into synchronous-transfer orbit with apogee of 36 042km, perigee of 185km, and inclination of 27.4°. An apogee boost motor fired 37hr after launch at fifth apogee had circularized the orbit so that *Ots 2* could drift to its final location above the equator at 10°E. *Ots 2* would remain for 6 or 7yr relaying telecasts among nations of the European Broadcast Union and carrying as many as 7000 telephone calls at a time.

This new comsat was one of two experimental models built by ESA to test performance in orbit; their descendants, European operational comsats, would serve in the 1980s as satellite links for intra-European telephone, telegraph, and telex traffic and would provide western Europe with television relay.

The first launch of an Ots from Cape Canaveral Sept. 13, 1977, had failed when the Delta first stage exploded 55sec after liftoff. The probable cause (burn-through of a solid-propellant strap-on Castor IV motor at the bottom of the first-stage Thor booster) had been avoided for further flights by increasing the thickness of insulation between the solid propellant and the exterior motor case.

Industries in Belgium, Denmark, France, Germany, Italy, Netherlands, Sweden, Switzerland, the United Kingdom, and Spain had built the spacecraft under ESA supervision; prime contractor was British Aerospace Dynamics Corp. Under an agreement with the U.S., ESA would reimburse NASA for the cost of the Delta launch vehicle, launch services, and other administrative costs totaling $417 million. GSFC had provided technical direction of the Delta project; Kennedy Space Center's expendable vehicles directorate was responsible to GSFC for launch operations. (MOR M-492-210-78-02 [prelaunch] Apr 27/78; NASA Release 78-53; *W Post,* May 12/78, B-6; ESA Release Apr 7, 12/78; May 8, 17/78)

• Jet Propulsion Laboratory announced it had agreed to participate during May in an experiment called Avefria (Spanish for lapwing) to perturb the atmosphere so that scientists could study ionospheric irregularities and their effects on radio communications. The experiment would create a high-altitude barium-ion cloud in the ionosphere (an atmospheric layer 25 to 250mi above earth in which radio signals typically travel for long distances) by rocket-borne release of barium-metal vapor. Optical-observation and radio-communications stations in Nev. would make scientific measurements of the cloud and its interaction with the ionosphere. Major observation point would be the Tonopah (Nev.) Test Range; JPL would operate a second observation station at its Table Mountain astronomy site near Wrightwood, Calif. (*JPL Universe,* May 12/78, 1)

• NASA announced Lockheed Missiles & Space Co. biotechnologists and engineers would use frogs on the Space Shuttle as research subjects to study inner-ear balance as the cause of space sickness that had afflicted some astronauts. Under a 6mo contract with Ames Research Center, Lockheed researchers would first study the behavior in space of the frog's otolith nerve bundle, which closely resembled man's. Microelectronic sensors inserted in the frogs would periodically feed data into recorders on the ground and in space. Preflight ground-based tests using a centrifuge would furnish baseline data to compare with information obtained in weightlessness. (*JSC Roundup,* May 12/78, 4; *NASA Act,* May 78, 10)

May 15: NASA announced it would participate in a joint American-Soviet particle-intercalibration (JASPIC) project to compare techniques used by both countries to measure intensity of energetic electrons and protons entering the lower ionosphere. At a 1975 meeting of the International Union of Geodesy and Geophysics in Grenoble, France, U.S. and Soviet scientists had agreed that the first question to be resolved was instrument credibility. Joint measurement of energetic electron and proton intensity at the same time and same place would be a first step in reconciling the conflicting results reported in the scientific literature. A Soviet research ship, the Professor Vize, off the Va. coast near Wallops Flight Center would serve as the Soviet launch platform.

The researchers would study the role of the particles in creating ionization in the lower ionosphere at night at mid-altitudes. They would compare data from 4 sounding-rocket launches from Wallops Island and 5 Soviet MR-12 rocket launches from the ship. One of the Wallops-launched rockets would release a chemical cloud possibly visible to east coast residents. Each organization would be responsible for its own countdown and launch operations. (NASA Release 78-71)

• LeRC announced it had signed a memorandum of understanding with the Papago tribe of Ariz. and the U.S. Public Health Service to provide the Papago Indian village of Schuchuli, Ariz., with electric power from the sun by Sept. 1978. This would be funded by DOE as a first step toward construction of a 3kw totally solar electric village power system, part of DOE's Natl. Photovoltaic Conversion Program. The new system would provide Schuchuli's 96 residents with sufficient electricity to power a community refrigerator, freezer, washing machine, sewing machine, water pump, and lights for the village's 16 homes, church, and feast house. Under the cooperative agreement, NASA would furnish all materials and technical assistance necessary to install the equipment and supporting facilities. The Papago Construction Co., an arm of the Papago tribal council, would erect the solar arrays and install all power-system equipment. (NASA Release 78-70; *Lewis News,* May 26/78, 1; DOE Release ET-0044/2)

• *Av Wk* reported that a long deadlock on airline regulatory reform had been broken when the House Transportation and Public Works aviation subcommittee approved a bill consigning automatic route entry to further study before planning a permanent program. The White House had wanted a strengthened bill from a joint House-Senate committee. The Senate-passed bill had allowed an airline one new route in the first 2yr after the bill became law in 1979, and two new routes in each of the following 3yr. The House bill approved by voice vote had allowed a carrier one new route in the first calendar yr after enactment, as well as designation of one market, disallowing automatic entry to a competitor.

The bill would have required the CAB to complete a study on success of automatic-entry provisions by June 30, 1980; if that study should find the experiment successful, Congress would pass additional legislation for airline deregulation on a more permanent basis. A compromise bill pushed by the Carter Administration would terminate the CAB, continue the small-communities subsidies, and give U.S. international carriers fill-up rights to carry domestic traffic on all international routes, among other provisions. (*AvWk*, May 15/78, 26)

• Congress had remained divided over continuing the separate development by the Army and the USAF of two battlefield nuclear weapons for a similar purpose: delivery of quick-reaction nuclear strikes to free for other missions the dual-capability NATO aircraft now committed to nuclear alert, *Av Wk* reported. The Army had been working with Martin Marietta on the Pershing 2 mobile tactical nuclear ballistic missile; the USAF had been working with General Dynamics on the Tomahawk ground-launched cruise missile.

The House Armed Services Committee had asked the Pentagon to make an early decision on one of the two weapons for the European theater. The committee would have to take into consideration transferring the Pershing 2 program to the USAF and halting the ground-launched cruise-missile production program; adding $10 million to the $10.1 million requested for the Pershing 2 in recommending program transfer to the USAF; the questionable survivability of the ground-launched cruise missile, and the capability of Pershing 2 to penetrate 1980's Soviet air defenses; and development studies by the USAF of at least two ballistic missiles, at the time it was requesting ground-launched cruise-missile production funds. The Senate Armed Services Committee had recommended continuing the Pershing 2 under the Army program office, and providing full funding for the USAF ground-launched cruise missile. The House and Senate now had to seek a compromise on the two systems during a conference scheduled for late May or early June. (*AvWk*, May 15/78, 18)

• ComSatCorp had agreed to a 48% reduction in its rates for international satellite communications, following FAA approval of a proposed settlement of its 13-yr-old rate case, *Av Wk* reported. ComSatCorp management and FCC staff had agreed on settlement last Feb. As U.S. representative to the 102-nation International Telecommunications Satellite Organization (INTELSAT), ComSatCorp had leased capacity on INTELSAT circuits to other U.S. carriers such as American

Telephone & Telegraph, ITT World Communications, RCA Global Communications, and Western Union International. The FCC now had to decide how these carriers would pass rate reductions along to their customers.

As part of the settlement, ComSatCorp would relinquish all claim to an escrow account totaling more than $100 million that it had set up in mid-1976 pending judicial review of an FCC Dec. 1975 decision in the rate case. The amount in escrow had been set aside for possible refund of overcharges since mid-1976, and the FCC also had to decide how refunds would be made. Under the settlement, ComSatCorp's overall allowed rate of return would be 11.48%, compared to 10.8% under the FCC's 1975 decision. (*AvWk,* May 15/78, 22)

May 16: NASA announced it had awarded Pratt & Whitney Aircraft Group of United Technologies Corp. an $80.4 million contract for advanced technology to improve fuel efficiency in future turbofan engines. Options in the contract could raise its total value to $94.4 million. This effort, a cooperative government-industry technology program to reduce fuel consumption in future turbofan engines by 10 to 15%, while also reducing direct operating costs, emissions, and noise levels, would constitute the major part of the energy efficient engine project, an element of NASA's aircraft energy efficiency (ACEE) program. At the end of the 5-yr technology contract, Pratt & Whitney should be able to begin development of advanced commercial turbofan engines for airplanes flying in the late 1980s and early 1990s. NASA had signed a similar technology-development contract with General Electric. (NASA Release 78-74)

May 17: MSFC announced it had begun preparations for stress tests of the Space Shuttle orbiter at Lockheed-Calif.'s 430ton (390 metric tons) steel test rig at Palmdale, Calif. The 6-story "reaction frame" rig had more than 350 hydraulic jacks exerting loads on the orbiter to simulate those to be encountered at launch, during spaceflight, and on reentry into earth atmosphere. Lockheed would perform the tests under contract to Rockwell Intl., builder of the orbiter. (*Marshall Star,* May 17/78, 2)

• Johnson Space Center announced it would sponsor a 2-day conference with the Univ. of Texas health science center on ways to use spaceflight in biomedical research. The first session, entitled Space: A Challenge for the Life Sciences, would have as speakers the JSC director, Christopher Kraft, Jr.; NASA's director for life sciences, David Winter, M.D.; JSC director of space and life sciences Richard Johnston; and Univ. of Tex. health science center resident Truman Blocker. The conference would conclude with a panel discussion on spaceflight applications. (JSC Release 78-19)

• The *W Star* reported that spokesmen for four government agencies using portable nuclear-power generators could not respond immediately to charges that the federal government lacked adequate control over the use or whereabouts of such devices. Charges came from members of the House Commerce Committee who said the public had not been informed of potential hazards to health, safety, and the environment posed by the devices or how they were used.

Committee members had sent President Carter a letter asking for a public accounting of portable nuclear devices used by the U.S. around the world. Of major concern was a device called SNAP (system for nuclear auxiliary power) generating from any of three radioactive and highly toxic fuels—cesium, strontium, or plutonium—heat that the device could transform into electricity for satellites or other remote power needs. Designed in the early 1960s primarily as a power source for space satellites, SNAPs had later been applied to terrestrial uses including a CIA project in which a SNAP, left on a Himalayan mountain to monitor China's missile testing, was apparently buried by an avalanche.

After a 1964 mishap when an Air Force spy satellite powered by a SNAP reentered and burned over the Indian Ocean, spreading radioactive fallout over a wide area, DOD had ordered manufacture of SNAPs in case-hardened containers that would remain intact if a satellite carrying one should fall from orbit. In 1968, a rocket launching another spy satellite from Vandenberg AFB had misfired and dropped two SNAP generators into the Pacific just off the southern Calif. coast; a 6mo search had recovered the SNAPs. (*W Star*, May 17/78, A-18)

May 18: NASA had officially changed the date of the first Space Shuttle orbital flight from Mar. to June of 1979 because of engine problems, the *NY Times* reported. NASA said the June date would be tentative until additional tests confirmed that the Shuttle's three main engines were free of problems. There had been signs the engine situation was improving. Launch date had been important because of NASA's plan to use the Shuttle to rescue Skylab from an uncontrolled fall to earth next year, possibly by early summer. (*NYT*, May 18/78, B-10)

• The *NYT* reported a leading Soviet astrophysicist's forecast that the next 250yr would see construction of a vast "artificial biosphere" in outer space capable of supporting 10 billion people, more than twice the world's present population. Iosif Shklovsky had visualized a system of space colonies with total surface area hundreds of thousands of times greater than earth's, that could capture and use huge amounts of solar energy. Looking further into the future, Shklovsky had also predicted that 2540yr from now mankind would have fully colonized the solar system.

His article in the journal *Social Sciences* emphasized U.S. plans to use large satellites to beam solar energy to earth as microwaves. Prof. Gerald O'Neill of Princeton Univ., head of a group of physicists and engineers designing solar satellites, had estimated that a space colony accommodating 100 000 or more people could be built by the year 2025. Skhlovsky had agreed that expansion into space would be inevitable because of mounting pressures of population and environment on earth's resources. Only the colonizing of space could offer a long-term solution, Shklovsky contended, claiming mathematical proof that adopting a global-balance strategy of limited growth could delay, but not avert, a world crisis. Shklovsky added that construction of large-scale colonies would require raw materials from the moon, asteroids, and other planets.

Although other Soviet scientists had stressed the importance of the search for extraterrestrial life, Shklovsky said that chances of finding other intelligent life were extremely remote. The distance to the nearest extraterrestrial civilization (one whose signals could be reaching the earth now) had been calculated at 3300 to 9800 light yr, or 19 800 trillion to 58 800 trillion mi; however, the absence of life, at least intelligent life, in earth's region of the universe should not discourage, but rather encourage, the conquest of space, he said. (*NYT,* May 18/78, B-10)

May 19: NASA announced plans to launch the geostationary operational environmental satellite GOES-C, third in a series of operational spacecraft funded by NOAA to meet DOC requirements, no earlier than May 25, 1978, from the ETR on a 3-stage Delta. In May 1974 and Feb. 1975, NASA had launched two operational prototype meteorological spacecraft, *Sms 1* and *Sms 2*, developed with NASA funding and still in orbit; it had launched the first two operational spacecraft funded by NOAA, *Goes 1* and *Goes 2*, in Oct. and June 1977, respectively.

NASA had planned to use a 3-stage inertially-guided Delta launch vehicle to inject GOES-C into a transfer orbit where its apogee boost motor would put it in a synchronous orbit. A GOES-C on station at 135°W would augment existing satellite coverage of earth weather and become a part of the Global Atmospheric Research Program (GARP). (MOR E-612-78-01 [prelaunch] May 19/78)

• NASA announced that Kenneth Chapman, associate administrator for external relations since Jan. 1978, had resigned from NASA to join Vought Aircraft Co., Dallas, Tex., as vice president for market research and analysis. Chapman had assumed his NASA position in Jan. 1978. Before joining NASA, Chapman had been director of nuclear material safety and safeguards at the Nuclear Regulatory Commission. He was a retired Air Force major general. (NASA Release 78-78; NASA anno May 18/78)

May 21: FBIS said Tass had commented on the upcoming U.S. launch of

Pioneer Venus-A, noting that the orbiter's radar resolution would permit it to transmit only an approximate map of Venus that would show only very big objects. Tass had pointed out that the mission would not include soft landing or transmission of photographs from the surface of Venus as the Soviet *Venera 9* and *Venera 10* had, because Pioneer Venus-B (the lander) would not survive landing and all transmission of information would stop. (FBIS, Tass in English, May 21/78)

May 22: NASA reported it had successfully conducted the second of two chemical-cloud experiments from WFC at dawn on May 22; it had launched the first at 5:29am EDT May 20. The launches had been designed to measure turbulence transport in the stratosphere by observing chemical vapor trails at an altitude of between 15 to 50km (9 to 31mi). Two-stage Nike sounding rockets had released vapors of titanium dioxide particles in spurts at 10-sec intervals, forming a series of dashes similar to an aircraft contrail; ground cameras had recorded the dispersion of the vapor trails. (WFC Release 78-4, 78-6; NASA Release 78-73)

• French and West German space officials had told a House committee that increasing costs would limit use of the U.S. Space Shuttle and the European Spacelab, *Av Wk* reported. A NASA official had admitted that Spacelab could have been built less expensively under a single nation's management, but denied that Space Shuttle operational costs had increased prohibitively.

Wolfgang Finke, director of West Germany's Dept. of Space and Transportation, told the House Committee on Science and Technology that not only had the price of conventional launches increased steadily, but also "the new space transportation system will not revolutionize the cost of space transportation. On the contrary, the estimates for the fare of a Shuttle roundtrip and the fringe costs going with it went up considerably, putting a real brake on our more ambitious plans to utilize the new system." As for Spacelab, Finke said the results of a West German materials-processing experiment would need a series of experiments for confirmation, with a cost level "well above the range of traditional materials research costs."

Hubert Curien, president of the French space agency, said that "anticipated cost of Spacelab experiments now reaches such heights that their funding raises very serious problems for national budgets and constitutes a very severe limitation." Finke noted in conclusion that administrative costs of international cooperation had increased costs of the overall project. A NASA official agreed that if Spacelab had been built in the U.S. under one management, total cost would have been less. (*Av-Wk,* May 22/78, 28)

• *Av Wk* reported that U.S. Air Force ground controllers had begun to bring the second NavStar global-positioning system satellite up to full operational status, after its successful launch on an Atlas F May 13 from Vandenberg AFB. The 955lb spacecraft had gone into an 85 × 10 900nm elliptical orbit, and its solid-fuel rocket motor had fired 2days later, putting it in a circular orbit; thrusters would shift it to its final station. Modifications of the rubidium-vapor atomic clocks had delayed launch of *NavStar 2* for a day; when ground tests indicated a malfunction of the NavStar frequency-divider circuit, the USAF had ordered modifications of all three clocks on *NavStar 2* to prevent a similar occurrence in orbit. The USAF planned to launch a third NavStar from Vandenberg in Aug. (*AvWk,* May 22/78, 24)

• *Av Wk* reported that the House of Representatives had committed the U.S. to membership in Inmarsat, the $200 million 40-nation international maritime-satellite organization, and had designated ComSatCorp as the U.S. representative. The Senate Commerce Communications Subcommittee had completed hearings and was expected to approve similar legislation shortly. A protracted controversy had gone on within the government and among international telecommunications carriers over the entity or combination of entities that should represent the U.S. in an organization to own and operate a global satellite-communications system for ships. Other nations had not encountered this problem, because their communications entities were all government agencies anyway.

Two documents establishing Inmarsat had opened for signature Sept. 3, 1976: a statement of organization policies and structure, and an operating agreement calling for initial $200 million capitalization. The U.S. and other participating governments would sign the former; ComSatCorp, under the House-passed legislation, would sign the latter. The 40 nations planning to join would divide investment initially according to anticipated usage, later according to actual usage. ComSatCorp would finance $34 million, the U.S.'s initial 17% of the capitalization. Nations that would account for 95% of the initial investment had to sign by a July 3, 1979, deadline, or the Inmarsat documents would be void. As the documents would take effect 60days after, by Sept. 3, 1979, Inmarsat would be either operative or dead. (*AvWk,* May 22/78, 21)

• U.S. share of the 102-nation INTELSAT organization had dropped from its 61% starting point in 1964 (when 19 nations had joined to establish a global communications consortium) to 25.1%, *Av Wk* reported. Charges that the U.S. had dominated INTELSAT had caused friction in the organization since its inception. The 1964 interim agreements had ensured U.S. majority ownership of not less than 50.6%; expanded membership and traffic patterns had decreased U.S. control as well as the friction. Other ownership shares under the new allocation were those of the United Kingdom, 10.8%; France, 5.98%; Brazil, 4.23%; and Japan, 3.68%. The 1964 interim consortium had been capitalized at $200 million; the 1971 agreements had boosted the figure to $500 million, including members' net capital contributions and the outstanding contractual capital commitments. The amount had reached $900 million in 1976 to finance Intelsat V satellites. (*AvWk*, May 22/78, 56)

May 24: MSFC reported it had successfully conducted a first major test firing of the three Space Shuttle main engines during a 15sec run at the Natl. Space Technology Laboratories in Bay St. Louis. MSFC had already successfully fired one of the modified engines for 520sec at 100% thrust, more power than needed for the first Shuttle mission. Test results would aid in establishing countdown procedures. More than 300 contractor and NASA employees had worked on the test series, described by program officials as the most complex large propulsion-system evaluation undertaken in the U.S. space program. Test budget for firings and facilities was $52 million.

Main propulsion-test articles were the Martin Marietta external tank containing propellant for firing tests, just as it would during Shuttle missions; the Rockwell Intl. orbiter simulator; three Rocketdyne main engines; and the Rockwell Intl. shuttle avionics test set, replacing the Shuttle-orbiter computers during tests. Key Shuttle launch and flight parameters to be verified by propulsion-test data were the cryogenic boil-off rate of external-tank propellants on the launch pad; rate of external-tank self-pressurization; engine-prestart requirements, such as pressure and temperature milestones; starting, throttling, and cutting off the engine, using external-tank propellants with the engines operating in close proximity; engine gimbaling; control-system response; loads placed by the engines on the orbiter; and launch acoustic data. Flight-dynamics engineers had been especially interested in verifying control-system models with the main propulsion-test data.

Test 1, scheduled for 2.35sec, had been terminated at 1sec because of an instrument problem. Test 2, scheduled for 15sec with the engines at 70% thrust, would assess propellant-flow paths from the tank through the orbiter and into the engines under thrust conditions. Test 3, planned for 15sec at 90% thrust level, would reverify overall propellant-system operation, and effect of the engine plumes on the test stand's flame deflector. Test 4, planned for 40sec at 90% thrust, would define propellant flow and flame-deflector hot spots. Firings 5 through 7, lasting 500 to 600sec, would duplicate Shuttle flight profiles; firings 8 through 10 would run for long durations with flight-rated 77:1 nozzles; and firings 9 through 13 would have the 35:1 nozzles reaffixed for engine throttling, to demonstrate the effect on flight profiles.

Although the main engine had taken the blame for Shuttle launch-date slip, the external tank and solid-fuel rocket booster tests also had failed to support a March launch. (*Marshall Star,* May 24/78, 1, May 17/78, 1; *JSC Roundup,* May 12/78, 1; *AvWk,* May 22/78, 55; May 29/78, 49)

• NASA announced that, as interrogations from tracking stations in Mar. and Apr. had successfully made contact with Skylab, the agency would begin in early June to uplink from JSC mission control a series of Skylab attitude-change procedures to put a new program into Skylab's onboard computer. The attitude-control system would be used to maneuver the spacecraft into the desired position and operate the control-moment gyros to keep it there. The procedures establishing control of Skylab, to be completed by mid-June, might extend its orbital lifetime by 6 to 12mo. The attitude change, if successful (and if the gyros continued to operate), could delay Skylab's reentry into the atmosphere until late 1979 or mid-1980. The maneuvers would decrease atmospheric drag on the spacecraft by aligning it with the flight path; once in the desired attitude, Skylab could survive from day to day without further maneuvers or adjustments. (JSC Release 78-20; *JSC Roundup,* May 26/78, 1; *Marshall Star,* May 24/78, 1)

• MSFC reported that a patent had been issued for a reel mechanism and associated control system invented by MSFC employees Ralph Kissel and Charles Rupp for the Shuttle tethered satellite system (TSS). NASA had considered the TSS as a means of enhancing Space Shuttle performance of scientific investigations and operational activities in space. The system would use a flexible tether line to deploy and retrieve subsatellites at distances as great as 100km from the orbiter. NASA had judged the mechanism and control-system invention would contribute significantly to overall feasibility of the TSS concept. (*Marshall Star,* May 24/78, 4)

May 25: NASA announced that *Heao 1,* first of a series of observatories designed to study celestial x-rays and gamma rays, was officially a success. Primary objective of *Heao 1* was to obtain highly detailed experimental data on astrophysical phenomena by measuring sizes and locations of x-ray sources and determining their intensity and time variation; according to Dr. Noel Hinners, NASA associate administrator for space science, the objective had been accomplished. NASA had authorized an extension of *Heao 1*'s mission to a second scan of the celestial sphere, and would broaden scientific participation through a guest-investigator program. NASA planned to launch in Nov. 1978 a second high-energy astronomy observatory designed for detailed study of interesting sources pinpointed by *Heao 1*; a third would be launched in 1979. (NASA Release 78-80)

May 26: NASA reported it had halted MSFC preparations for vibration tests of the Shuttle orbiter Enterprise May 12, when the upper dome of the prototype liquid oxygen tank buckled inward while being filled with water. Pressurizing the tank to 1lb later removed the wrinkles. The tank implosion would delay Enterprise vibration testing, but the program office would have to review delay of the overall Space Shuttle program. (*JSC Roundup,* May 26/78, 1; *Marshall Star,* May 10/78, 1)

• NASA announced it had postponed to at least June 24 the scheduled June 10 launch of SeaSat-A because of precautionary changes and tests on the Atlas F launch vehicle. Temperature increases in the aft sections of recently launched Atlas vehicles had necessitated the changes and resulting tests.

NASA would use SeaSat-A, first satellite designed to study the world's oceans, to see if microwave instruments scanning from space could provide scientific data useful to oceanographers, meteorologists, and commercial users of the seas. The spacecraft would send information on sea-surface winds and temperatures, currents, wave heights, ice conditions, ocean topography, and coastal storm activity.

NASA would attempt to demonstrate the feasibility of an operational multiple-satellite network to monitor oceans on a continuous near-real-time basis; twice daily, the system could provide ships at sea with detailed charts updated to show latest weather conditions, sea states, and hazards. Long-range use of the system could influence ship design, port development, and selection of sites for offshore facilities such as power plants. Other potential users of SeaSat data would include commercial fishermen, oil exploration firms, the Weather Service, pollution-control agencies, and the Coast Guard and Navy. (NASA Release 78-77; *JPL Universe,* May 26/78, 1)

• LeRC reported that the communications technology satellite *Cts,* world's most powerful comsat, had made possible quick diagnosis of burn and multiple-injury victims of a simulated catastrophe at the Baltimore/Washington International Airport. The Maryland Institute for Emergency Medical Services had conducted a disaster exercise called Emergency Management of an Airport Catastrophe as part of a 2-day seminar for physicians, interns, residents, nurses, and other health care professionals. The "disaster" (a simulated collision between a commercial airliner and a ground aviation fuel tanker) had tested airport and statewide emergency medical plans to aid the "injured," including about 180 crash victims. The scenario at the site included a breakdown in local communications, and an overload of local emergency medical facilities. A joint program between the U.S. and Canada, *Cts* operated on a new frequency at power levels 10 to 20 times higher than those of other satellites; the higher broadcast power had allowed use of much smaller and far less expensive ground-receiving equipment. (*Lewis News,* May 26/78, 2)

• JPL announced that more than 100 scientists from universities and corporations in the U.S. and Europe had attended its solar-probe science workshop at CalTech May 22-23 to discuss the proposed closest encounter with the sun. The proposed mission, being studied at JPL under NASA's Office of Space Sciences, would launch a uniquely designed spacecraft in 1985 into an orbit 2 100 000km (1.3 million mi) above the sun's surface, arriving in 1989. Scientists had theorized that the sun probe would yield valuable new insights into solar physics, providing data on solar wind, solar particles, the solar magnetic field, and interplanetary dust. The workshop had reviewed new technologies needed for a spacecraft orbiting so close to the sun. JPL had begun studies of a new heat shield design, a high-powered telecommunications system with dual X- and X-band frequencies, new selenide-isotope power generators, and a drag-free spacecraft whose small rocket motors could counteract opposing external forces to keep the craft in correct solar orbit. (*JPL Universe,* May 26/78, 1)

May 29: Av Wk reported that President Carter had signed a policy directive establishing a unified policy for all U.S. military and civilian space programs that would broaden the civilian role in both military and civilian areas. Civilian space programs had not previously benefited from military R&D because of security restrictions; the new policy would reduce restrictions on technology transfer from the military to the civilian sector. Civilian space projects and users of civilian space technology (such as the Dept. of Agriculture) that had not had access to data from military programs (such as imagery from U.S. photoreconnaissance spacecraft) would be able to use in the civilian sector data gathered by military spacecraft. Space programs on weather and climate, land use, and earth resources could now use military data where applicable.

The directive had named presidential science adviser Frank Press to head a U.S. space program coordinating committee of users and suppliers of space data. The committee would oversee development and coordination of both military and civilian programs. The policy also attempted to identify and reduce stresses among the four major U.S. space programs: intelligence-community programs using spacecraft such as Lockheed's "Big Bird" high-resolution reconnaissance satellite; DOD space programs using spacecraft such as the defense satellite communications system (Dscs) vehicles; civilian/government space programs using vehicles such as those flown by NASA and NOAA; and civilian/private-program spacecraft users such as commercial communications-satellite operators, and possibly future commercial remote-sensing companies. (*AvWk*, May 29/78, 23)

• *Av Wk* reported that the Soviet Union had launched its 15th killer-satellite test since 1968, less than a mo before the scheduled start of anti-satellite-limitation talks with the U.S. The USSR had launched on May 19 an interceptor, *Cosmos 1009,* toward a target, *Cosmos 967,* launched Dec. 13, 1977, which had also been target for the USSR's 14th unsuccessful killer-satellite test Dec. 21, 1977. (*AvWk*, May 29/78, 23)

May 30-June 1: NASA launched *Pioneer Venus 1* from Cape Canaveral at 9:13am EDT May 20 atop an Atlas Centaur rocket. It would launch a Venus lander toward the planet 3mo later, about Aug. 7. *Pioneer Venus 1,* an orbiter, should reach its destination by Dec. 7, 1978; *Pioneer Venus 2,* a multiprobe spacecraft, would arrive 5days later after splitting into a bus and 4 atmospheric-entry probes 13 million km (8 million miles) and 20days out from the planet. Most engineering systems and the interplanetary experiments had signaled normal operation. "Performance so far has been extremely good," said Marshall Johnson, Venus-orbiter flight director at ARC. "Of course, we have 300 million miles and 6mo to go," he added. At 9am PDT June 1 the spacecraft was 4 800 000km (2 890 000mi) from earth and had traveled toward Venus at 16 580kph (10 296mph). Controllers would make the first midcourse correction June 8, changing velocity to 12.6kph (7.8mph) to aim the Pioneer at the exact point for Venus orbit insertion.

Pioneer Venus 1, a spin-stabilized spacecraft with its spin axis perpendicular to the ecliptic plane, consisted of a cylindrical thrust tube serving to transmit acceleration loads into the spacecraft; a 97.5in-diameter circular equipment shelf for mounting electronic units and scientific instruments; a 100in-diameter cylindrical substrate 48in long for mounting the solar-cell array; and a despun mast assembly supporting three antennas, including a high-gain antenna. A single-piece fitting that would remain on the Centaur at spacecraft separation would be the mechanical/structural interface with the launch vehicle. Payload was 12 scientific instruments with a total weight of 100lb, using average power of 58watts and peak power of 85watts. The instruments could handle 75 redundant discrete commands and 6 redundant quantitative commands, as well as 17 serial-digital, 24 bilevel, and 31 analog telemetry channels. Three of the instruments would use pyrotechnic firing pulses for various functions. The magnetometer had been put at the tip of a 15.7ft beam to isolate it from spacecraft magnetic fields. The *Pioneer Venus 1* launch vehicle was an Atlas SLV-3D Centaur D-14 131ft tall, weighing about 328 640lb at liftoff. The nosecover protecting the spacecraft, a conical/cylindrical fiberglass fairing about 29ft high attached to a 56in-high aluminum split barrel, had been designed for jettison early in the Centaur burn.

NASA had launched *Pioneer Venus 1* into an ascent trajectory varying from 3° to 18° south of due east during the first 15days; the 7mo flight to Venus would take the spacecraft more than halfway around the sun (through about 200°), covering about 480 million km (300 million mi). The orbiter would take 3mo longer for the trip than the multiprobe, and would have a slower speed upon arrival at Venus, requiring less power from the orbit-insertion motor. For the first 82 days after launch, *Pioneer Venus 1* would fly outside the earth's orbit. In Aug. it would return inside earth's orbit and, during the last 4mo of the journey, would cross 42 million km (26 million mi) on a long curving trajectory between the orbits of the earth and Venus. The orbiter's flight path would resemble that of the Venus multiprobe, to be launched a few days after the orbiter crossed back inside earth's orbit.

NASA had designed the flights to answer questions such as: Why had two planets with about the same mass, formed of similar materials and situated at comparable distances from the sun, evolved different atmospheres? Why had the surface of Venus baked in searing heat, while earth had climate friendly to life? Answers would aid in understanding the evolution of Venus and earth atmospheres and the forces affecting earth's weather. The Pioneer Venus flights, first devoted to studying another planet's atmosphere and weather on a global scale, would use the largest number of vehicles devoted to such studies and would make measurements at the greatest number of locations. The flights would record characteristics of Venus's upper atmosphere and ionosphere, as well as of the lower atmosphere, and interactions of these regions with the solar wind (the continuous stream of ions and electrons flowing outward from the sun) and with the solar magnetic and electric fields. Circling the planet for at least 80mo, the Pioneer Venus orbiter would make the longest observations of Venus to date and would be the first U.S. spacecraft to orbit the planet.

A NASA-contractor team under KSC's expendable vehicles directorate had been responsible for preparing and launching *Pioneer Venus I.* Hughes Aircraft Co. had built the spacecraft and radar mapper. (Ames Release 78-21; MOR S-825-78-01 [prelaunch] May 15/78; [prelaunch summary] May 11/78; NASA Release 78-68; NASA press brief May 9/78; *Marshall Star,* May 17/78, 1; May 24/78, 1; *Lewis News,* May 26/78, 1; *JSC Roundup.* May 26/78, 1; *DFRC X-Press,* May 19/78, 2; *C Trib,* May 21/78, 1-3; *W Post,* May 21/78, A6; *NYT,* May 22/78, A10; *W Star,* May 21/78, A-2)

May 31: MSFC reported that 17 scientists had signed contracts to develop materials-processing experiments, first to be conducted in the weightless environment of the Space Shuttle and Spacelab and partially based on results of investigations during the Apollo, Skylab, Apollo-Soyuz, and SPAR (space processing applications rocket) programs. Among the products of interest to the scientists were improved medicines; electrolyte materials for smaller, higher capacity batteries; and larger, more perfect crystals for electronic applications. The $12 million program, covering a 5yr initial period, should produce significant scientific results for developing specific useful materials and products in space. NASA had aimed the programs at science and technology in both research and manufacturing activities leading to privately funded materials-processing in space.

Five of the 17 scientists would have a chance to fly their experiments on an early orbital test of the Shuttle in a materials-experiments assembly (MEA) package that would not only produce valuable scientific data but would also define further experiments and hardware needed before the first extended flight on Spacelab or Space Shuttle. A Shuttle satellite-deployment mission would fly four of the MEA experiments again, along with five others; the fifth would fly again on the third Spacelab mission devoted almost exclusively to materials processing. Four more of the scientists would have experiments on this mission. (*Marshall Star*, May 31/78, 1)

• NASA reported it had amended a contract with Rocketdyne Division of Rockwell Intl. Corp. to procure 9 Space Shuttle main engines, in addition to the 7 already on order. The amendment had authorized Rocketdyne to manufacture and test the 9 engines during a 180-day period in which prices would be negotiated. Estimated cost of the amendment had been $250 million. The 9 engines would support the Space Shuttle program through the first 4 orbiters approved for production; the present contract had covered engines for the first Space Shuttle flight scheduled for 1979 and had included 3 ground-test engines and 1 spare. (*Marshall Star*, May 31/78, 3)

• MSFC announced it had successfully completed at the National Parachute Range May 23 an airdrop test of the parachute system designed to recover Space Shuttle's reusable solid-fuel rocket boosters after launch. This was the fourth drop test in which a dummy booster had been carried aloft beneath the right wing of a B-52 aircraft; the recovery system's three main parachutes had all deployed, and all had functioned as expected. The dummy booster had landed virtually undamaged. NASA engineers planned two more airdrops before certifying the system for flight. (*Marshall Star*, May 31/78, 2)

During May: The Natl. Aeronautic Assn. announced that a 15-member U.S. aerobatic team would travel in Aug. to Ceske Budejovice, Czechoslovakia, to compete in the ninth World Aerobatic Championships sanctioned by the Intl. Aeronautic Federation (FAI). The 1978 U.S. team would be all male, as no woman had qualified in competitions to determine team members. Three-time U.S. national champion Leo Loudenslager, an American Airlines pilot, said, "We have high expectations. We have even better trained pilots and more sophisticated aircraft than we did in 1976, and we are more knowledgeable about what to expect." The 1976 U.S. team had lost to the USSR the titles it had captured in France in 1972: top team, best male, and best female. Loudenslager would fly his experimental modified version of the 200hp Stephens Akro, the Laser 200. Henry Haigh might fly the modified Pitts Special biplane in which he had competed in 1976, basically the same design flown by the remainder of the team. The five aircraft for the competition, disassembled Aug. 8 at Kennedy Intl. Airport for transport to Frankfort, would be reassembled and flown to Peine near Braunschweig in West Germany for 2wk of intensive practice, then to Budejovice for the competition.

The championship event would mark nearly two decades of U.S. participation in FAI-sanctioned precision aerobatic competitions, first held in 1960 in Czechoslovakia. The event also would commemorate two-thirds of a century since the first recorded aerobatic flight: Lt. Peter Nesterov of the Imperial Russian Air Service was credited with performing in 1913 the first inside loop in a Nieuport monoplane. NAA, the aero club representing the U.S., had sanctioned as the official team representing the U.S. at the world championships the group selected by the Aerobatic Club of America, which would also be the only team competing in Czechoslovakia without government financing. (NAA newsletter, May 78, 1)

• The USAF reported it had begun initial screening of sites in 10 states for the MX intercontinental ballistic-missile system. DOD officials said screening would identify and evaluate potential MX deployment sites in the continental U.S; the USAF would not make a selection, but rather evaluate candidate sites that could serve MX needs. Initial screening earlier in 1978 had identified 12 geologically suitable sites in 15 states; current studies would focus on Arizona, California, Colorado, Kansas, Nebraska, Nevada, New Mexico, Oklahoma, Texas, and Utah. Final decision would require many additional studies, including environmental impact analyses. Deployment of an MX system could begin several yr after completion of its design and evaluation of environmental impacts. DOD officials had not decided whether to recommend ultimate deployment of the MX, which would depend on the outcome of development and analyses now under way and on approval by Congress. (AFSC *Newsreview,* May 78, 4)

• Aerospace engineers at AEDC reported they had precisely measured damage from a single particle of rain, snow, or dust—or one from a meteorite—striking the surface of a missile or spacecraft at speeds up to 7000mph. Materials better able to protect missiles or spacecraft would result from this research and testing. The precision measurement used a 2-stage launch "gun" to propel small projectiles (3.2in long, 1.625in diameter) through a track formed by four guide rails inside the range's sealed tank, to simulate altitudes up to 46 000ft. The projectile, tipped with a sample of nosecone material, passed through the track and was photographed as it collided with single particles of water, glass, or nylon.

Previous tests had sent projectiles into clouds of small particles; however, engineers had arranged the single-particle impacts to get a more precise understanding of nosecone erosion and its effect on the accuracy of missile and spacecraft trajectories. Uneven or excessive erosion could cause a missile or spacecraft to miss an intended impact point or landing site. SAMSO had conducted the tests on a track 62ft long and 1.625in in diameter, with a 110ft recovery tube. (AFSC *Newsreview,* May 78, 5)

• The Dayton Air Fair had accelerated efforts to complete construction of a Wright B Flyer replica by the Apr. 30 fair date. Operation Tailwing, organized to ensure completion of the project, had had more than 400 participants in the project during the past 2yr. As plans for the original plane no longer existed, the builders had used measurements of one of the two original remaining Wright B craft, replacing wood with aluminum tubing and steel. Computer studies had verified that propellers turning at 1000rpm with a 210hp motor could maintain the original 17-to-1 power/weight ratio. Modern safety standards had required modification of the airfoil design and gas tank. (NAA newsletter, May 78, 6)

June

June 1: NASA announced it had appointed two U.S. scientists to an international group of five seeking to become payload specialists on the first Spacelab mission scheduled for late 1980. One American and one European would be selected to fly on the laboratory to operate the science instruments. The investigators, working group (IWG), composed of scientists representing all investigators, had selected Dr. Michael Lampton of Berkeley, Calif., space physicist at the Univ. of Calif., and Byron Lichtenberg of Natrick, Mass., vestibular researcher at the Mass. Inst. of Technology. The three payload specialists not chosen to fly would be backup specialists and would particpate in ground-based mission activities at Johnson Space Center.

In a similar screening program, ESA had selected its payload specialists from among thousands of applicants from its member states. The Spacelab 1 mission would investigate stratosphere and upper atmosphere physics, materials processing, space plasma physics, life sciences, astronomy, solar physics, earth observations, and space technology. Marshall Space Flight Center had been responsible for payload-specialists training as part of its overall management responsibility for the Spacelab mission; ESA's Spacelab Payload Integration and Coordination in Europe (SPICE) had managed training in Europe. (NASA Release 78-76; *JSC Roundup,* June 1/78, 4; *DRFC X-Press,* June 1/78, 3)

• ESA announced it had organized a tour of several Mediterranean countries to demonstrate use of data from the Meteosat weather satellite. Tour objectives were to display Meteosat-system capabilities exhibited in the documents or images produced by the spacecraft; and to show potential users (meteorologists and others working on the study and management of earth's natural resources) how to use the system. The system included a secondary data-user's station (SDUS) mounted on a trailer towed by a vehicle containing a 2.5m-diameter receiving antenna and all equipment needed to receive analog signals in the WeFax (weather-facsimile) format. Demonstrations had begun in Cairo and would continue in Athens, Tunis, and Algiers. ESA had planned similar demonstrations in other countries covered by Meteosat. (ESA Release June 1/78)

June 2: Kennedy Space Center announced it had awarded Planning

Research Corporation (PRC), McLean, Va., a $23 626 661 contract extension for engineering support services. Under the 12-mo cost-plus-fixed-fee extension, PRC would continue through May 19, 1979, its design engineering of Space Shuttle equipment and other tasks for KSC's design-engineering directorate, including design of new and modified ground-support equipment and minor facilities-design support for all KSC programs. The Spaceport and Cape Canaveral in Fla., Vandenberg AFB, and Dryden Flight Research Center had received support under the contract. The extension brought the total contract amount since the original award in May 1974 to $97 306 220. (KSC Release 53-78)

June 5: AvWk reported U.S. avionics manufacturers were sure that a NavStar receiver suitable for general aviation use, with features comparable to those of a VOR omnirange distance-measuring equipment and area-navigation computer, would be available for under $5000 by the late 1980s. AEL, Inc., which had designed a microwave landing receiver that could sell for less than $1500 through normal distribution channels, said the price of a NavStar receiver in 1978 dollars could eventually be as low as $2000. The role of NavStar (DOD's satellite-based navigation system) in civilian aviation would depend on availability of inexpensive receivers at least as accurate as the present VOR/DME navaids. The potential market for civilian NavStar receivers had aroused considerable interest among civil-avionics manufacturers, resulting in Langley Research Center's receiving a number of bids in a recent competition for a technology/design study of a low-cost NavStar receiver. Magnavox Government and Industrial Electronics Co. had won the contract. (*AvWk*, June 5/78, 91)

June 6: Although the U.S. had planned to begin talks with the USSR June 7 on killer-satellite control, the Carter Administration had not clarified its own stance toward antisatellite systems, *The Washington Star* reported. The administration had asked the Soviets what they wanted to consider in the 10 days of Helsinki talks described as exploratory, partly as a guide to its own approach. The Kremlin had greeted coolly an earlier Carter proposal to ban antisatellite weapons, which it had been testing since 1967. As Defense Secretary Harold Brown said in Oct. 1977, the USSR already had missiles that could knock down U.S. reconnaissance satellites. The U.S. had studied killer-satellite possibilities for yr without moving to develop them; when the USSR intensified its testing 2yr ago, U.S. interest increased, and officials opposed entering into an agreement that might hinder U.S. efforts to catch up with the Soviets.

On March 30, 1977, Carter announced that the U.S. and the USSR would establish a study group to "develop an agreement whereby we might forgo the development of a capability of destroying satellite observations vehicles, so that we can have an assured way to watch the Soviets [and] they can have an assured way of watching us from satellites" (the principal means of detecting noncompliance with strategic arms-control limitations). Yet, according to Brown's statement, Oct. 1977 was already too late for both sides to give up killer-satellite development. The Navy had been interested in killer-satellite capability because the USSR had begun to use ocean-surveillance satellites to locate American ships. The USSR had also tried to use satellites to detect submerged submarines, making missile-carrying submarines vulnerable and offering a threat to the undersea leg of the U.S. triad of nuclear deterrence. (*W Star*, June 6/78, A-4)

• The British Interplanetary Society's journal *Spaceflight* reported that NASA had developed a lunar- and planetary-sciences teaching aid, using actual samples of lunar material encased in a clear plastic disk, to educate both the earth-science student through the Aerospace Education Program and the museum visitor through the NASA exhibits program. The hand-held aid would permit looking at, or viewing through a microscope, both sides of a sample. Under an 8yr program for lunar- and planetary-science education, colleges had been able to borrow thin-section microscope slides of lunar material along with a teaching manual, and museums and fairs could display several dozen prepared lunar samples in cases. But until now, NASA had directed no program toward the secondary school earth-science student.

The new school-use program which included a film on lunar science, the sample disk, workbook material, slides, and an audio cassette) would require considerable interaction between teacher and class. Student reaction to the museum program, which included a shorter sound-slide presentation and the disk, had been highly favorable. NASA expected to test museum-visitor reaction later in 1978 and to use about 2/3lb of lunar materials to make 100 lunar-sample disks available for use in both programs. Apollo astronauts had brought back 843.5lb of lunar material during lunar explorations in the late 1960s and early 1970s. (*Sf*, June 6/78, 218)

• In recognition of the energy shortage and the need to conserve fuel, the Natl. Aeronautic Association announced it had established a new national aviation-record category for competition among U.S. airlines, Efficiency on a Commercial Air Route. NAA had administered for a number of yr the category Speed on a Commercial Air Route, which had proved popular with airlines and airline pilots. This category would continue, in conjunction with the new category, to measure efficiency over a route by the amount of fuel used according to the mileage and number of passengers carried. The measurement would result in an efficiency-index number, and the highest number over a given route would hold that particular record. To maintain equity, five categories had been based on maximum gross aircraft weight, and one open category would include all aircraft. (NAA newsletter, June 6/78)

• *Spaceflight* reported that British aerospace engineers had completed a study that might lead to construction in space of a huge power station capable of generating an electrical output equal to 5 to 10 power stations on earth. Britain's BAe Dynamics Group, the European company most experienced in assembling space-solar arrays, had done a 6mo study under ESA contract of solar arrays to provide extra power for the European Spacelab and space platforms, as well as for a space power station. A pilot-unit space platform designed to operate in a low-earth or geostationary orbit would supply 1 to 2megawatts, as precursor of a power station in geostationary orbit with a power output of 5 to 10 gigawatts (5000 to 10 000megawatts). Europe had already demonstrated it was capable of significant contributions to large space power-plant technological development. (*Sf*, June 6/78, 219)

June 7: The *Marshall Star* announced it had begun ground-vibration tests of the Space Shuttle, using a computerized shaker system turned on by engineers in instrumentation trailers located near the test stand. The shaker system, the Shuttle modal test and analysis system (SMTAAS), had applied vibration cycles and force inputs and had acquired response information from the vehicle suspended inside a tall stand and canted 9° from vertical. In upcoming months engineers would use SMTAAS to "tune in" to the vehicle's various vibration modes, comparing responses to mathematical predictions. Results of the tests would allow engineers to verify predictions of Shuttle reaction to the much more severe vibrations expected during launch. (*Marshall Star*, June 7/78, 1; *JSC Roundup*, June 9/78, 1; *DFRC X-Press*, June 16/78, 2; *AvWk*, June 19/78, 75)

• MSFC announced plans to begin thermal-vacuum testing of NASA's second high-energy astronomy observatory, HEAO-B, as part of a series of environmental tests on the satellite before scheduled launch in Nov. TRW Systems, Inc., HEAO prime contractor, would conduct the 2wk-long test to simulate thermal-vacuum conditions in space. HEAO-B had already undergone a number of environmental tests, including acoustic and mechanical vibration tests. (*Marshall Star*, June 7/78, 1)

• A House subcommittee on space science and applications would hold panel discussions later in June on opportunities for international cooperation in space, the U.S. House Committee on Science and Technology announced. Moderated by Dr. Jerry Grey of the American

Inst. of Aeronautics and Astronautics, the discussions would follow subcommittee hearings scheduled May 16, 17, and 18 on international space programs. Panelists reviewing the hearings would make recommendations on international cooperation for publication with the hearings. Rep. Don Fuqua (D-Fla), subcommittee chairman, said he favored use of the space environment "not only for our own nation but for the benefit of all mankind. All reasonable efforts should be made in the international community to assure that worldwide understanding and mutual commitments are developed to provide sound basis for these future efforts." (U.S. House Committee on Sci and Tech Release, June 7/78)

June 9: NASA Deputy Administrator Dr. Alan Lovelace, addressing members of LeRC's launch directorate, had termed the launch-vehicle team "probably the best in the world. The record speaks for itself," the *Lewis News* reported. Lovelace had visited LeRC to discuss the role of launch vehicles in NASA's overall plans and to be briefed on other Lewis work. The expendable vehicles would make 13 launches through 1980, when the Space Transportation System (STS) would begin carrying payloads into space. Members of the STS team responsible for launching the Space Shuttle who would work with the LeRC launch-vehicle group "can learn team development, discipline, and mental rigor from the exemplary record of the Lewis team," Lovelace said. (*Lewis News,* June 9/78, 1)

• In its "Profile of a Mission," *Lewis News* reported that a new era of opportunity in space had become available. The Space Transportation System (the Space Shuttle) would open the door to activities ranging from experiments in basic research on physical and chemical process, to using the weightlessness and vacuum of space as an environment for manufacturing products. LeRC's Space Experiments Integration Office (SEIO) would help principal investigators proposing experiments that needed a space environment, by preventing duplication of effort and standardizing paperwork required for project approval.

SEIO had management responsibility for one facility and four experiments, three of which were to fly on the long-duration exposure facility (LDEF) managed at LaRC:

—The advanced photovoltaic experiment (APEX) on the LDEF consisted of three related approaches to investigating the solar spectrum and the effect of exposure to space on advanced photovoltaics and materials. For the first time, space-calibrated solar cells would be available as laboratory standards for assessing performance of new solar cells.

—The space power experiment (SPEX) on the LDEF would evaluate the use of commercially available components for a space-power system. With three LDEF trays carrying commercially available terrestrial solar cells and a fourth carrying commercial power electronics and batteries, successful operation of these components in space could significantly reduce costs of future space-power systems.

—In the ion-beam texturing and surface-coating experiment on the LDEF, which would expose such surface samples to launch stress and near-earth space, ground tests before and after exposure would record the optical properties of the surface of each sample and compare durability of ion-beam-textured surfaces with that of state-of-the-art surfaces. If the ion-beam-textured surfaces exhibited higher durability, use of thermal-control coatings for spacecraft would increase.

—The cryogenic fluid-management experiment (CFME), a payload attached to a Spacelab pallet, would obtain long-duration low-gravity engineering data on systems for storage/acquisition/supply of liquified gas (subcritical) to verify the possibility of successfully operating such a system in space.

. The zero-gravity combustion facility (OGCF), a versatile reusable assembly that would permit a Spacelab experimenter to study combustion free of earth's buoyancy forces, would carry out tests ranging from rapid burning to long-term smoldering of a variety of liquid, solid, and gaseous fuels. (*Lewis News,* June 9/78, 3)

June 11: The *Washington Post* reported that the USAF had launched on June 10 from Cape Canaveral a Titan 3C rocket carrying a classified satellite cargo under unusually tight security; the rocket had operated in a normal Titan flight pattern. A USAF statement said only that a USAF industry team had participated, but not whether or not the launch had been successful. The *Post* said the security measures—with guard dogs and USAF combat personnel supplementing the usual guard force—had been heavier than for any previous launch at the Cape. Workers at the launch pad did not know what the payload was, and received passes only to the specific areas of their assigned duties. A Titan 3C, which could lift up to 10tons into space, had been used several times to launch military-intelligence satellites; it could also lift a series of satellites and drop each in various orbits. The most recent classified launch from the Cape had been on April 6, making this the shortest interval between classified missions in more than 11yr. (*W Post,* June 11/78, A9)

June 12: Av Wk reported that engineers suspected a hot-gas system leak in the sustainer engine of the USAF/General Dynamics Atlas F launch vehicle as the cause of overtemperatures registered during recent DOD NavStar launches from Vandenberg AFB. Overtemperatures in the boat-tail area of the Atlas F (recorded in telemetry from the launches of *NavStar 1* in Feb. and *NavStar 2* in May) had forced postponement for at least 14days of SeaSat-A's launch from Vandenberg. Rockwell Intl.'s Rocketdyne Div. had test-fired the engine to determine whether a leak in the sustainer engine's hot gas system was responsible, and to pinpoint the location of the leak. They also had hot-fired sustainer-engine gas generators as part of the investigation. General Dynamics had monitored aerodynamic flow around the Atlas F during flight for hot gas from the exhaust plume impinging on the boattail.

Temperatures in the boattail had climbed earlier than expected during the NavStar launches; in one launch, that area had become hot enough to melt insulation on copper wires, causing failure of some onboard instruments and intermittent operation of others. NASA had increased insulation on some wire bundles in SeaSat-A's Atlas F; more insulation around other rocket's components might also be added. Additional instruments installed in the launch vehicle should obtain more details on temperature and pressure conditions in the boattail area during launch, program officials noted. They had not decided whether to postpone launch of the third NavStar satellite, planned for Aug. (*AvWk*, June 12/78, 18)

- *Av Wk* reported MSFC's intent to demonstrate within 5yr the assembly of a large space structure in low-earth orbit, to exploit Space Shuttle capability. MSFC had proposed assembling a 10 × 30m (33 × 98ft) structure in space to demonstrate fabrication and assembly techniques, followed closely by a Space Shuttle mission to demonstrate deployment of a large antenna 50 to 200m (164 to 656ft) in diameter. The USAF Space and Missile System Organization (SAMSO) had also expressed interest in flying a Shuttle-based antenna-deployment test before 1985.

Although most U.S. aerospace companies had been working on structures in space, Grumman Aerospace Corp. had been in the forefront in large space-structures work; MSFC had awarded Grumman four of six contracts for large space structures development, and JSC had awarded it at least one significant contract. MSFC's approach had been to define tools and procedures needed, and to evaluate whether such operations would be possible and to what degree. Most building-block study for large space structures had assumed the use of open-truss aluminum beams joined by diagonal and cross braces, with beam sections 1m (3.2ft) wide and length varying according to mission. Grumman had just begun to demonstrate a beam-building machine [see Apr. 12]. Both NASA and Grumman had been scheduled to evaluate the beams for deviations from specifications such as straightness or basic strength, and for individual subsystems such as the automatic welders used to affix the cross sections. In Sept. Grumman would transfer the beam-builder to MSFC for tests of beam manufacturing in a vacuum, to simulate space conditions. (*AvWk*, June 12/78, 49)

June 13: NASA announced it had appointed Robert Allnutt associate deputy administrator. Allnutt, currently acting assistant general counsel for legislation at DOE, had been deputy assistant administrator for the Energy Research and Development Administration, responsible for program areas including procurement, personnel, labor relations, and construction. Following a career both inside and outside the executive branch, Allnutt in 1970 had been appointed associate general counsel to the congressionally-established Commission on Government Procurement to improve governmentwide policies, procedures, and legislation. In 1973 Allnutt had been appointed staff director and counsel of the Senate Committee on Aeronautical and Space Sciences, with jurisdiction over federal aerospace research and development activities including all NASA programs.

NASA had named Isaac Gillam, IV, director of DFRC, where he had been acting director since the departure of David Scott in Nov. 1977; he had been deputy director there since Aug. 1977. Gillam, an associate fellow of the American Inst. of Aeronautics and Astronautics and a senior member of the American Astronautical Society, had joined NASA Hq in 1963 as a resource-management specialist and had gone to DFRC in 1976 as director of Space Shuttle operations. Before joining NASA, Gillam had served in the USAF.

Lt. Gen. Duward Crow (USAF-ret.), assistant to the deputy administrator of NASA, had announced plans to leave NASA Sept. 1, 1978. Gen. Crow had begun NASA service in 1974 as assistant administrator for DOD and interagency affairs. In Oct. 1975 he had become associate deputy administrator of NASA, assuming his present duties following a reorganization in Nov. 1977. Crow had graduated in 1941 from the U.S. Military Academy and had served in the China-Burma-India theatre of operations. He had become comptroller of the USAF in 1969 and was assistant vice chief of staff from Oct. 1973 to July 1974 when he retired. He had received NASA's Distinguished Service Medal for his work on the NASA launch vehicles program. (NASA anno June 13/78)

• KSC announced it had awarded Management Services, Inc., Huntsville, Ala. a $1 774 404 contract extension to operate the Spaceport's component-refurbishment and chemical-analysis laboratories. The 1-yr extension, a cost-plus-award-fee contract set-aside for small business, had brought the total value of the original contract to $4 536 339. The basic contract, awarded in 1976 for 3yr with annual renewals, had been negotiated competitively. (KSC Release 58-78)

June 14: NASA announced plans to launch Comstar-C on an Atlas Centaur launch vehicle for Comsat General, a subsidiary of ComSatCorp, on a fully reimbursable basis. NASA had successfully launched *Comstar 1A* and *Comstar B* on May 13 and July 22, 1976, respectively; a fourth satellite being built would serve as a spare. KSC and its Expendable Vehicles Directorate would be responsible for preparing and launching the Atlas Centaur vehicles; a joint government-industry team would supervise all launch-vehicle and pad operations during countdown. The ETR, launch-vehicle contractor General Dynamics, and NASA would supply all personnel and equipment required to handle assembly, prelaunch checkout, and launch.

The Comstar spacecrafts used by American Telephone and Telegraph Co. as part of its nationwide communications network had a capacity of more than 14 000 high-quality 2-way voice circuits. In geostationary orbit at 22 300mi altitude, they could provide reliable communications to the 48 contiguous states, Alaska, Hawaii, and Puerto Rico. Each Comstar carried centimeter-wave beacons to test higher frequency propagation at 19 and 28GHz, measuring the adverse effects of rainfall on signals at these frequencies, and leading to possible use of higher frequencies to future commercial comsat systems. Under an agreement with Comsat General, AT&T would lease the entire capacity of the three in-orbit Comstars; AT&T, which was building its own earth-station network, would integrate satellite communications with its terrestrial facilities to enhance its domestic U.S. services. (MOR M-491-201-78-01 [prelaunch summary] June 14/78; *Spaceport News,* June 23/78, 1)

• MSFC reported that an invention by MSFC engineer Frank Nola to decrease energy consumed by electric motors [see May 2] had aroused the most public interest of any invention by the center. Accounts of the invention in *Chemical and Engineering News* and *Industry Week* drew more than 600 inquiries from readers. An item in *Kiplinger Letter* had produced about 1430 reader responses. George Porter, patent counsel in MSFC's Office of Chief Counsel, said his office had received three applications for licenses to manufacture the power-factor controller. MSFC's Technology Utilization Office also had received almost 300 letters and telephone calls. The invention had been tested in a textile mill in Alexander City, Ala., using 3700 industrial sewing machines operated by half-horsepower 3-phase motors. One of the motors equipped with the power-factor controller had recorded a power saving of 33% over an identical motor without it. (*Marshall Star,* June 14/78, 2)

June 15: NASA announced that its scientists had apparently discovered a way to account for the formation on earth 4 billion yr ago of nucleic acids, one of two essential "building blocks of life." The discovery had followed earlier work by the same investigators (Dr. James Lawless, team leader at Ames Research Center; Dr. Edward Edelson, National Research Council fellow; and Lewis Manring, student at the Univ. of Santa Clara) that revealed a mechanism to explain the formation of the other critical building block, protein. Taken together, the findings answered a years-old question about the chemical evolution of life on earth.

Theorists had wondered how the building blocks of life, randomly scattered on the shores of primitive oceans, could continuously congregate and organize over millions of yr in concentrations high enough to produce living organisms. The explanation was found in substances common on the shores of primitive oceans: metal clays. Dr. Lawless had mixed low-concentration solutions of DNA-forming nucleotides with commonplace metal clays, and found that most clays attracted them. One type of metal clay containing zinc had preferentially attracted all six of the building blocks of DNA and RNA nucleotides. Especially significant was that zinc clay had attracted 97% of the nucleotide 5-prime adenosine monophosphate (AMP), most common DNA building block in living systems and essential precursor to ATP, the basic energy molecule present in every life form. Zinc also had been a constituent of the enzyme DNA polymerase, which linked DNA building blocks (nucleotides) in living cells; enzymes had acted as super catalysts, drastically speeding many life processes.

Dr. Lawless had done another experiment in which zinc clay had preferentially attracted the "5-prime" life-form over the "2-prime" and "3-prime" forms, the 5-prime form being the only one found in living organisms. The result had suggested a mechanism for incorporating this life-specific building block into the first DNA-like material.

Scientists had previously applied electric discharges or other energy release to ammonia, methane, and water vapor to produce small quantities of basic life molecules; however, until Lawless's discovery, they could not explain the behavior of life-building blocks in primordial water. Lawless's group would also try to show metal clays linking nucleotides into polynucleotides, the next step toward forming a DNA-like molecule. (NASA Release 78-85)

June 16: Spacecraft controllers had completed a maneuver to move the planet-arrival point of *Pioneer Venus 1* from the southern hemisphere to an orbital-injection point 216mi above the northern hemisphere, *DFRC X-Press* reported. The course change also had slowed the spacecraft so that it could use the sun's gravity to increase its speed. NASA expected that the flight-path change would allow *Pioneer Venus 1* to enter its planned long-oval orbit tilted 75° to the planet's equator, with the closest approach 180mi from Venus, greatest distance 41 000mi. *Pioneer Venus 1* had been scheduled to reach Venus in Dec. *(DFRC X-Press,* June 16/78, 4; *ARC Astrogram,* June 1/78, 1)

• For 4yr, the LaRC Space Systems Div. and Flight Dynamics and Control Div. had analyzed the Space Shuttle orbiter's entry guidance and control system at the request of JSC, the *Langley Researcher* reported.

The system had been designed to control the orbiter from deorbit until landing, regulating both the aerodynamic surfaces and a reaction control system (RCS) composed of small rocket thrusters such as those on Apollo spacecraft, so that orbiter control had turned out to be a hybrid between aircraft and spacecraft control. Onboard computers could direct the entire orbiter entry without pilot input, or a pilot could take over control if necessary.

A primary tool in the LaRC analysis had been a reentry flight dynamics simulator (RFDS) that permitted evaluation of the onboard-control system in both the automatic and the manual modes. It had also aided in developing alternate control systems to be used by the pilot if the main control system could not handle a situation. The simulator cockpit, though not an exact representation of the orbiter cockpit, did contain all the instruments and controls necessary to simulate flight: cathode-ray tubes (CRTs) showed displays identical to those on board, changing guidance information automatically as the orbiter entered various flight regimes. Push buttons and toggle switches would permit the pilot to take over any of the control functions.

LaRC investigators had found that the system was dangerously sensitive to possible sensed angle-of-attack errors in the automatic mode at hypersonic speed. Shuttle astronaut Col. Henry Hartsfield had been unable to maintain vehicle control in the manual mode. This and other problems had led to major system redesign that had been tested at LaRC and found to solve the sensed angle-of-attack problem. Investigators had then begun work on other system problems. (*Langley Researcher,* June 16/78, 1)

• Under a cooperative agreement between NASA and the U.S. Navy, a Navy experimental plane would undergo in-place tethered-flight testing at LaRC's Impact Dynamics Research Facility, the *Langley Researcher* reported. The Navy's XFV-12A plane built by Rockwell Intl.'s Columbus Aircraft Div. was the only one of its kind, a prototype V/STOL aircraft designed to demonstrate the thrust-augmented wing concept in supersonic flight. The tests at LaRC had tethered the XFV-12A to the gantry by a single cable from a powerful winch mounted 210ft above ground. Static restraint tests kept the aircraft in place at various heights by ground cables attached to the nose and main landing gear, as its engines operated at full power. Using results of the static tests, LaRC had put the XFV-12A through dynamic tests, permitting greater freedom of motion, as the aircraft was not restrained by ground cables. The upper tether cable automatically reeled in as the aircraft maneuvered within the gantry, and a shock absorber on the winch cable guarded the airplane against loads greater than 2g. (*Langley Researcher*, June 16/78, 6)

• ESA announced that the full-scale propellant mockup of the European launcher Ariane (47m high, maximum diameter 3.8m) had left the integration site near Paris for its launch base at Kourou in French Guiana on the South American coast. Two Seine barges had transported three pressurized containers (one for each stage) to Le Havre, where they were loaded on a freighter bound for Cayenne. Tests of the mockup would begin in Aug. when the launcher would be erected on the pad for the first time.

ESA had planned 3mo of tests to check general conditions for launcher assembly and vehicle compatibility with ground facilities, as well as fueling and draining facilities and systems both on the ground and on the vehicle. Centre National d'Études Spatiales (CNES) would handle transport, assembly, and test operations. Aerospatiale, the system integrator, would evaluate dynamic and thermal behavior of the launcher under ambient climatic conditions in tests simulating vibrations like those at liftoff. ESA had planned four qualification launches between June 1979 and Oct. 1980 (ESA Release June 16/78)

June 19: As the climax of a 4-mo attempt to stabilize and trim the orbital position of Skylab, a NASA team of engineers and controllers from JSC, MSFC, IBM, and three tracking stations (Bermuda; Madrid; and Goldstone, Calif.) had successfully put the space station in an orbital position they hoped would prolong its lifetime in space, JSC announced. Maneuvers in June had put the space station in an attitude intended to reduce atmospheric drag at its orbital height of 389km (242 statute mi). Assuming that the gyros continued to function, NASA hoped the new attitude would extend the orbital lifetime of Skylab by 6 to 12mo to late 1979 or early 1980, giving NASA time to arrange an early Space Shuttle test flight to reboost Skylab into higher orbit, or deorbit it into a remote ocean area.

Maneuvers had begun in Mar. when engineers and flight controllers at the Bermuda tracking station checked out Skylab systems and brought its batteries to a fully charged state. On June 8, flight controllers had turned on the dormant control-moment gyros in the Skylab attitude-control system. (The other part of the system, the thruster attitude-control system, would expel nitrogen gas through nozzles to move Skylab into various attitudes in its orbit.) A computer with various sensors to indicate position had controlled both systems. Although one gyro that failed during the last Skylab mission in 1974 could not be used, the other two gyros were activated and worked as expected.

On June 9, controllers had commanded Skylab into a solar inertial attitude in which the spacecraft's solar cells always faced the sun. On June 11, they had maneuvered it into the desired "end-on velocity-vector" attitude, docking port forward and long axis parallel to the ground along the flight path. Skylab had remained in that position while engineers continued to monitor it, periodically transmitting minor corrections to the onboard computer controlling Skylab's position. (JSC Release 78-25; *Marshall Star,* June 7/78, 1; June 14/78, 1; *JSC Roundup,* June 9/78, 1; June 23/78, 1; *DFRC X-Press,* June 2/78, 2; June 30/78, 2; *W Post,* June 12/78, A-19))

• MSFC had designed a geostationary-orbit platform that could perform the functions of more than a dozen domestic communications and weather satellites serving North America, *AvWk* reported. As early as 1986, NASA could use the Space Shuttle to assemble the platform, first U.S. large space structure in synchronous orbit. NASA would assemble the platform over the course of three Space Shuttle missions, using technology under design at MSFC and JSC. Analysis had predicted that over a platform's 15-yr lifetime it could save 50% of the cost of operating communications and weather systems now scheduled to be placed in synchronous orbit during that period.

Institutional rather than technical barriers had delayed completion of the platform project, but MSFC had reached agreement among common carriers and hardware manufacturers on the savings possible by using the platform. One concern of NASA planners was that grouping communications systems on a single platform would make them more vulnerable to Soviet attack; however, MSFC planned to issue an RFP of about $500 000 for major systems studies in FY79 and FY80 that would constitute Phase A evaluation of typical platform requirements; phase B would require about $1 million for FY80 and FY81 for system design concepts. (*AvWk,* June 19/78, 67)

June 21: Preliminary results from a 6-yr experiment at the San Andreas fault in Calif. indicated that earth motion along the fault had accumulated at a rate much faster than expected, equivalent to 6 to 7m over the past 70yr, *Goddard News* reported. Measurements from tracking sites on both sides of the fault in the San Andreas Fault Experiment (SAFE) to determine gross plate motion across the fault, and to demonstrate practical application of satellites and ground-based lasers in tracking relative motion of the earth's upper plates, had produced satellite tracking data accurate within a few centimeters. Gross motion across the fault had provided a measure of strain energy as it accumulated, eventually to be released as an earthquake. Goddard Spaceflight Center's pulsed laser ranging systems had measured earth motion by determining the range to a satellite, measuring the time required for a short pulse of intense light to travel from the laser to the satellite and back, and repeating the measurements once per second throughout each satellite pass. GSFC had performed the SAFE experiment in cooperation with a broad cross section of the scientific community. (*Goddard News,* June 21/78, 1)

- GSFC announced that, in a cooperative program with NOAA, its scientists had demonstrated the benefits of using short-interval full-resolution satellite images to monitor tropical cyclones. The agencies had undertaken the program to determine optimum resolution and image frequency for selecting winds to study, in forecasting tropical-cyclone intensity. Experiments using the *Sms 2* to study hurricane Eloise and cyclone Caroline, and *Goes 1* for tropical storms Belle and Holly, had taken scan images of the storms at varying resolutions and at a variety of time intervals. GSFC using its atmospheric and oceanographic information processing system (AOIPS) to examine the experimental data, had found that rapid-scan full-resolution visible images could best disclose the wind vectors. Scans taken every half or quarter hr had been inadequate, because many clouds of the type and size best suited for tracking did not persist or maintain their shape for as long as 30min. GSFC scientists also had found that the full-resolution visible images could improve cirrus tracking and could increase the number of trackable elements available from coarser-resolution infrared images. (*Goddard News,* June 21/78, 4)

- MSFC reported that the American Inst. of Aeronautics and Astronautics had selected William Rice, deputy manager of Space Shuttle solid-fuel rockets at MSFC, to receive the Wyld propulsion award for "outstanding contributions to development of solid propellant rocket motors and superior leadership of aerospace technology programs." AIAA had presented the award each year to reward work in development or application of rocket-propulsion systems. Rice would receive a medal and certificate at an awards luncheon during the AIAA/SAE Joint Propulsion Conference. (*Marshall Star,* June 21/78, 1)

June 22: Nature magazine reported that NASA planned to draw more academic scientists into its R&D activities. NASA Administrator Robert Frosch had issued a policy statement that, in the future, academic scientists would conduct "a substantial proportion" of basic research in all agency disciplines. Non-NASA scientists would take a greater part in all NASA basic research, from conception and planning through programming and execution to interpretation of data and publication of research results. The statement said NASA would increase peer evaluation of research projects in order to guarantee high quality. NASA critics had charged that agency research, particularly in areas such as space applications, traditionally proceeded in-house with little outside evaluation.

The new policy, partly in response to President Carter's directive on increased government support of basic science, had aimed at counteracting the criticisms. After NASA's divisions had prepared a list of proposed activities, NASA would publish a 5-yr plan based on these proposals and conforming to the new policy. The Office of Space and Terrestrial Applications, for instance, whose programs at the outset had been essentially technical exercises to test and develop means of data acquisition, had realized its concern with science relatively late, so that scientists had had to do the best they could with the data available. Under the new system, NASA would ensure better working arrangements between scientists and instrument designers in deciding on the most useful data a satellite could collect, from both a scientific and a technical point of view. NASA stressed that greater use of academic scientists would improve quality control at the same time it would help to develop specific fields of scientific knowledge. (*Nature,* June 22/78, 586)

• ESA announced plans to launch its scientific satellite Geos-2 in mid-July from ETR/Cape Canaveral on a Delta. This satellite would carry out the mission originally planned for *Geos 1,* which a malfunctioning Delta injected into a transfer orbit too low to permit attaining its scheduled geostationary orbit. The European Space Operations Centre (ESOC) at Darmstadt, W. Germany, had fired *Geos 1*'s apogee boost motor to inject it into a 12-hr elliptic orbit; after a yr in this orbit, results showed *Geos 1* had made a significant contribution to the International Magnetospheric Study (IMS).

Main mission objective of Geos-2 would be to learn more about responses of the near-earth environment to processes occurring in outer space. The satellite's geostationary orbit would be in a region of earth's magnetosphere where many dynamic processes causing magnetic and ionospheric disturbances were believed to develop. Geos-2's exceptionally high real-time transmission rate (over 100kilobits per sec), combined with a continuous link with ESA's Odenwald ground station in Germany, would allow it to transmit about 100 times as much data as any previous European scientific satellite. Built as a qualification model at the same time as *Geos 1,* the spacecraft had been converted into a high-quality flight model. Based on *Geos 1* experience, only minor modifications were made in some Geos-2 subsystems. Geos-2's very sensitive experiment payload had put severe constraints on spacecraft design in the areas of electromagnetic and chemical cleanliness: for example, a complex system of 8 booms deploying in orbit had to be designed to put the sensitive experiment detectors measuring minute variations in the magnetosphere as far as possible from any electrical interference generated by the satellite. Prime contractor for Geos-2 was British Aerospace Dynamics Group, under the direction of industries in 10 European countries. (ESA Release, June 22/78; ESA Newsletter, June 78, 4)

June 23: NASA declared the launch from Cape Kennedy at 6:36pm Mar. 31 of *Intelsat IVA F-6,* last of the Intelsat IVA series, to be successful. NASA had launched the satellite on an Atlas Centaur into the desired transfer orbit, thus meeting all objectives. Orbital parameters had been 35 912km apogee, 548.8km perigee, 21.84° inclination. ComSatCorp had fired the apogee kick motor Apr. 1 to place the satellite in a near-geosynchronous orbit that would take it at mid-June to its final location at 63°E over the Indian Ocean. (MOR E-491-633-78-06 [postlaunch], June 23/78)

• NASA announced that the USAF Space and Missile Test Command at Vandenberg AFB would launch SeaSat-A on a modified Atlas-F booster no earlier than June 26. Anthony Calio, associate administrator for space and terrestrial applications, noted that NASA had established a launch date of May 17, 1978, early in the program, and had maintained it even when a resolicitation of bids became necessary because of adding a sensor to the original complement. The launch date had slipped from May 17 to June 10 because of a late start of thermal-vacuum testing, and subsequent replacement of a cracked alumina substrate in the power amplifier of the data link. The launch date had then slipped to June 24, then to June 26, because of Atlas-F booster problems [see June 12]. The USAF had grounded the entire Atlas-F fleet for analysis and testing, but this had not delayed NASA's schedule, as rework had continued while the vehicle remained on the stand. (MOR E-655-78-01 [prelauch] June 23/78; *Spaceport News,* June 23/78, 1; *Lewis News,* June 23/78, 1; *Marshall Star,* June 21/78, 1; *Langley Researcher,* June 30/78, 1; *DFRC X-Press,* June 16/78, 3; *JPL Universe,* June 23/78, 1; *Nature,* June 22/78, 586)

• JSC researchers confirmed that a meteorite found by Dr. William Cassidy last winter on an expedition sponsored by NSF to the Antarctic [see Feb. 17] had proved to be one of the rarest types ever seen. A team of JSC meteorite experts using a binocular microscope had examined the fragment inside a lunar-type glove box flushed with dry nitrogen gas, and had sent an 0.4-gram sample to Dr. Brian Mason at the Smithsonian Institution for petrographic analysis of thin sections.

The fragment, 4.5 to 4.6 billion yr old, had been a Type II carbonaceous chondrite (so called because of its high carbon content) of which only 15 other samples had been found, though not in so clean a condition or so well-preserved. These chondrites had been shown to contain amino acids of nonterrestrial origin, suggesting chemical formation of complex organic molecules in other regions of the solar system. The fragment would next be characterized, sectioned, and fully documented; subsequent analysis would then continue. (JSC Release 78-26; *JSC Roundup,* June 9/78, 1; June 23/78, 1)

• Jet Propulsion Laboratory announced its engineers had delivered the first hardware—a thermal-model tape recorder—for the earth-orbiting infrared astronomical satellite IRAS to their counterparts in the Netherlands for testing, before delivering the actual flight recorder in

1979 for installation in the IRAS satellite. A joint program by the U.S., the United Kingdom, and the Netherlands, IRAS had been scheduled for launch from Vandenberg AFB in Feb. 1981, carrying a single experiment: a liquid-helium-cooled telescope to map the sky in infrared frequencies from 1 to 100microns. Infrared astronomers would use the data to construct an infrared sky map and a source catalog with roughly 1 million sources. IRAS would make about 14 orbits of earth daily, but only two daily data-transmitting passes over the U.K. ground station. The tape recorder would therefore store all data for transmission every 12hr. Odetics Corp. of Anaheim, Calif., had built the flight version of the tape recorder for shipment to the Netherlands, where the spacecraft was being built. After completing construction and installation of the telescope, IRAS would return to the U.S. for tests at JPL. (*JPL Universe,* June 23/78, 1)

• LeRC reported that the center's flight operations group had modified a C-131 aircraft for data collection in monitoring earth resources (such as ice thickness on the Great Lakes and heat losses from residential and commercial buildings) by adding an auxiliary power unit, an electrical power-distribution system, special equipment racks, radar antennas, and an inertial-navigation system. Lockheed had constructed special windows and other modifications in the belly of the aircraft to accommodate aerial cameras and an 11-channel multispectral scanner system.

One of the C-131's remote sensing missions had conducted experiments over the Arctic ice ridge in the Beaufort and Bering Seas to see if airborne microwave instruments could obtain ice pressure information. LeRC had also used the aircraft to develop techniques and systems for reporting ice conditions in both fresh- and sea-ice areas for scientific and commercial applications and, cooperatively with EPA, to determine the effects of agricultural practices, strip mining, and land drainage. (*LeRC News,* June 23/78, 3)

• INTELSAT announced it had elected Canadian Marcel Perras chairman of its board of governors and Randolph Payne of Australia vice chairman. Perras, vice president of Teleglobe Canada, had been a member of the INTELSAT board since 1975. Payne, director of marketing for the Overseas Telecommunications Commission of Australia, had represented Australia on the INTELSAT board at various times since its formation in 1973. The board had also reappointed C. J. Steffen of Switzerland and N. Tuckwell of Australia as chairman and vice chairman, respectively, of the advisory committee on planning; and O. Schmeller of the FRG and K. Nosaka of Japan as chairman and vice chairman, respectively, of the advisory committee on technical matters. (INTELSAT Release 78-17-I)

June 26: NASA announced that following the June 16 launch of *Goes 3* on a Delta and successful placement in transfer orbit, it had fired the apogee boost motor at 11:22pm EDT on the same date to produce an orbit with 35 469.1km perigee, 36 679.2km apogee, and 1.7° inclination. The drift had been too rapid, and an orbit-adjust maneuver June 20 had reduced it to 2 deg/day West. *Goes 3* had been scheduled to arrive on station at 135°W July 16, 1978. All subsystems were on, except the visible-infrared spin-scan radiometer (VISSR); the main power bus was normal. The S-band and UHF equipment had functioned properly, but NASA was checking specifics of gain, margin, and sensitivity. Although the space-environment monitor (SEM) subsystem had appeared satisfactory, it needed further testing for possible interference from other subsystems. NASA expected to check out *Goes 3* and turn it over to NOAA for operation when it arrived at 135°W.

Goes 3 would be a key element of the Global Weather Experiment, a worldwide yearlong accumulation of meteorological and oceanographic data that had begun in Dec. 1977. Largest international scientific experiment in history, with 140 nations participating, GWE was collecting information from sources that included nine satellites and scores of ships and aircraft, plus thousands of daily surface and upper-air observations by several hundred buoys distributed in the southern hemisphere and by conventional observation methods elsewhere. The experiment, part of the Global Atmospheric Research Program (GARP) sponsored by the UN's World Meteorological Organization and the Intl. Council of Scientific Unions, would last through 1979 and provide scientists with millions of pieces of information from all over the world. *Goes 3* would not only provide the major western-hemisphere coverage in the experiment, but would also determine how much important meteorological information could be gathered from a data-sparse area of the world centering on the Indian Ocean.

Goes 3 was the last NOAA geostationary satellite NASA would launch on an expendable vehicle; three additional geostationary spacecraft planned for the next 8yr, beginning with GOES-D, would be launched on the Space Shuttle. (NASA Release 78-72; MOR E-612-78-01 [prelaunch summary] Apr 26/78, [postlaunch] June 26/78; NOAA Release 78-66)

• INTELSAT reported that telecasts of the 1978 World Cup soccer championship won by Argentina had been the world's biggest satellite television event. Although final figures had not been tallied, officials estimated the World Cup had surpassed the popularity of the 1976 Montreal Olympics, measured by numbers and hours of international satellite telecasts. INTELSAT said the number of transmissions and receptions of the games had totaled 1364; with an average 2-hr transmission, total time given to transmissions and receptions worldwide was 2728hr. INTELSAT's operations center in Washington, D.C. had estimated an additional 400 transmissions of World Cup material, other than the actual games, accounting for another 700hr. The World Cup thus ran to more than 3400hr compared to the 1976 Olympics, which had held the previous record for satellite television time, more than 2600 transmission/reception hr. (INTELSAT Release 78-18-I)

• General Electric Co. and Comsat General had presented to the convention of the Armed Forces Communications and Electronics Association divergent views on whether DOD should have its own satellite communications systems or should lease service from other organizations, *AvWk* reported. Lee Farnham, vice president of General Electric, said his company, as a hardware manufacturer, would evaluate a number of considerations before entering into any lease arrangement. Joseph O'Conner, vice president for finance and administration of Comsat General, said his firm, as a telecommunications common carrier, had been structured to establish system and lease communications services.

At the insistence of Congress, the Navy had issued an RFP for a Leasat system based on a leasing arrangement. Comsat General might submit two bids, one for a dedicated Navy system and one for a system to serve merchant shipping as well as the Navy. GE's Farnham said a manufacturer would have to consider (in addition to matters such as user requirements and hardware complexity) three factors in making a bid decision: financing, corporate impingement (corporate credit rating), and length of the lease. (*AvWk*, June 26/78, 23)

June 27: NASA declared the *Explorer 52* (Hawkeye) successful over its 4-yr lifetime of investigating particles and fields in the earth's polar magnetosphere out to large radial distances, measuring magnetic-field and plasma distributions in the solar winds, and detecting and measuring with direction-finders the Type III radio emissions from the outer solar corona. Performance of the spacecraft and instruments during the mission had been excellent; the only anomaly (in the optical-aspect system) had been compensated for with the data from the magnetometer system. The spacecraft had reentered the atmosphere Apr. 28, 1978.

Results of *Explorer 52* investigations had included direction-finding measurements showing intense kilometric radio emissions from the earth's magnetosphere, generated at about 1 to 3 earth-radii altitude over the evening auroral zones, with detailed angular distributions of the radiation. Its observations over the South Pole had provided the first information on polarization of these kilometer wave emissions. These data, of special interest to the theoretical space-plasma physics community, had not been explained satisfactorily. Other high-latitude observations had shown an unusual concentration of both electrostatic and electromagnetic plasma-wave turbulence in the 1 to 3 earth-radii altitude range along auroral geomagnetic-field lines, which might be related to generation of terrestrial kilometric radiation. (MOR S-863-74-04 [postlaunch] June 27/78)

• NASA announced it had named Dr. Adrienne Timothy assistant associate administrator for space science, effective June 18. Dr. Timothy would be chief scientist in the Office of Space Science, representing that office to the scientific community as the focal point for obtaining, evaluating, and using its expertise in planning NASA space science programs. She would succeed Dr. S. Ichtiaque Rasool, who had been named chief scientist for NASA's Office of Space and Terrestrial Applications.

Timothy had begun her NASA career in 1974 as staff scientist for solar physics, Physics and Astronomy Programs, in the Office of Space Science. In 1975 she had become chief of the Solar Physics Branch, where she planned and directed U.S. space science research in the solar physics discipline; in 1977 she had become program manager for advanced programs and technology in NASA's Solar Terrestrial Division of the Office of Space Science. Before joining NASA, Timothy had been leader for the Apollo telescope-mount reduction and analysis system at the American Science and Engineering Corp. (NASA Release 78-91)

• NASA announced it had selected Hughes Aircraft Co.'s Space and Communications Group for negotiations leading to a contract for development of a Jupiter atmospheric-entry probe, part of the Galileo mission scheduled for launch in Jan. 1982. The mission would consist of an orbiter to circle the planet for 20mo as well as a probe to plunge into Jupiter's atmosphere, surviving high-speed entry of 48km/sec (20 000mph) and a pressure of more than 10 earth atmospheres, to transmit information on Jupiter's composition, structure, and cloud physics. Estimated cost of the contract would be about $35 million, most of the work to be performed under subcontract by the General Electric Co. Hughes would design, develop, and fabricate the protoflight probe, integrate the science instruments, and test and provide systems integration and launch support. Project Galileo had been named in honor of the 16th-century Italian astronomer; it was a cooperative project of NASA's JPL and ARC, with the Federal Republic of Germany providing the orbiter's retropropulsion motor and some of the science instruments. (NASA Release 78-92)

• Mstislav Keldysh, Soviet scientist and mathematician who had been spokesman for the USSR's space program as head of the Soviet Academy of Sciences, died June 24 at the age of 67, the *Washington Post* reported. Soviet news agency Tass described the death as sudden. Dr. Michael DeBakey had traveled to the Soviet Union 5yr ago to operate on Keldysh's circulatory system.

Recognized in a wide range of scientific and mathematical disciplines bearing on aviation and rocketry, Keldysh had proved his administrative ability by rising to the top of his country's scientific establishment. President of the Soviet Academy of Sciences from 1961 to 1975, Keldysh had been not only USSR science spokesman but also a key figure in organizing and developing scientific research throughout the Soviet Union. He had won the Stalin Prize in 1942 for "The Theory of Calculating and Developing Methods of Reducing Various Types of Vibrations in Aircraft," and in 1946 for "The Front-Wheel Shimmy of the Tricycle (Aircraft) Landing Gear." In 1943, Keldysh had become head of a top secret aircraft-development institute with increasing administrative responsibilities. As head of the Academy of Sciences, he announced in Oct. 1969, shortly after the U.S. had put men on the moon, that the USSR had abandoned plans to do likewise and would concentrate on putting manned space stations into orbit around the earth. He acknowledged that the Soviet Union had been debating space expenditures, and said, "My personal view is that, when a man has taken his first step into space, you cannot stop further development." (*W Post,* June 27/78, B6; FBIS, Tass Intl Serv in Russian, June 26/78)

June 28: NASA declared the launch of two satellites had been successful, Japan's direct broadcast satellite *BSE* and ESA's orbital test satellite *Ots 2.* A Delta had launched *BSE* into a synchronous transfer orbit from ETR at 5:01pm EST Apr. 7, 1978, with 35 923km apogee, 166.1km perigee, 27.2° inclination. Satellite performance was satisfactory during the transfer orbit. The boost motor had fired at 7:34pm EST Apr. 8 to move the satellite over the South Pacific to a position at approximately 110°E above the equator south of Japan. All subsystem functional checks had indicated satisfactory satellite status.

NASA had launched Ots B (*Ots 2* in orbit) at 6:59pm EDT May 11 from ETR on a Delta into a synchronous transfer orbit with 35 946km apogee, 183.6km perigee, and 27.4° inclination. Satellite performance was satisfactory during the transfer orbit. The boost motor had fired at 8am EDT May 13, to move the satellite over the South Atlantic to a position 10°E above the equator. All subsystem functional checks had indicated satisfactory satellite status. This was the first flight of the Delta 3914 (Castor IV) configuration since failure of the OTS-A mission in Sept. 1977. (MOR M-492-212-78-01 [postlaunch] and M-492-210-78-02 [postlaunch], both June 28/78)

• MSFC reported that a new 0.8km (0.5mi) antenna range had operated satisfactorily in testing Space Shuttle antennas for range safety. The new range had a 27.4m (90ft) transmit tower and a 22.8m (70ft) model test tower. Antenna tests had begun in May on a 1/15-scale model of the

Space Shuttle orbiter, external tank, and two solid-fuel rocket boosters installed in the tower.

Engineers had measured radiation patterns of the six range-safety antennas (two on the ET and two on each of the 2 SRBs) to ensure that a signal from the range-safety officer would reach the vehicle at any altitude. The range-safety antenna system was interconnected to immediately relay to the ET and both SRBs a destruct signal received on any one of the six antennas. Future Shuttle tests possibly requiring antenna-pattern measurements included the teleoperator retrieval system, 25kw power module, large space structures, and satellite power systems. (*Marshall Star,* June 28/78, 1)

• ESA reported it had agreed to sign June 30 the first license it had granted for use of one of its patents: the Italian electronics company SELI would get industrial-production rights to an ESA-patented computer terminal that could display two texts written in different alphabets side by side on the same screen. Under the agreement, SELI would have exclusive production rights to the terminal for 2yr.

This invention was an example of technological spinoff from space activities put to practical use. One version of the system, the "Eurab terminal," could display on the same screen one text in Arabic and another in Latin characters; although originally developed for Arabic, the system could display any alphabet. The Eurab terminal could show the Arabic alphabet with vowels included, making reading easier and unambiguous, and leading to wider use of Arabic in airports, banks, schools, television programs, or wherever display screens might be used. (ESA Release June 28/78)

• The *Chicago Tribune* reported that the USSR had launched *Soyuz 30* June 27 carrying a Soviet commander, Pyotr Klimuk, and the world's first Polish cosmonaut, Miroslaw Hermaszewski. *Soyuz 30* would dock with the *Salyut 6/Soyuz 29* complex occupied since June 15 by Soviet cosmonauts Vladimir Kovalenok and Alexander Ivanchenkov. Hermaszewski had been trained under the Soviet Union's ongoing Intercosmos program to include eastern European countries in the Soviet space effort.

The international crew would study the manufacture of semiconductor materials under conditions of weightlessness, assess the effect of spaceflight on humans, observe and photograph land and oceans, and carry out technical experiments with individual onboard systems and the orbital complex as a whole. (*C Trib,* June 28/78, 1-3; FBIS, Tass in English, June 28/78)

• The *Washington Star* reported that, despite Soviet efforts to suppress news of the incident, western sources said that in Feb. a Soviet cosmonaut had narrowly avoided hurtling off to his death in space. The near-mishap had occurred during an unauthorized spacewalk by cosmonaut Yuri Romanenko on the record-breaking 96-day orbital flight aboard the *Salyut 6* space station. Only cosmonaut Georgy Grechko had been slated to make a spacewalk; however, both cosmonauts had been wearing a new type of spacesuit that included a radio and an hr's supply of oxygen, with only a simple tether to the spacecraft to keep the cosmonaut from drifting away. During the Grechko spacewalk, after the Salyut had passed over the western Pacific and was out of range of Soviet ground stations, Romanenko (who was untethered) had jumped out of the open hatch. Why he did this remained unknown, although a U.S. space official speculated he might have gotten "space rapture." Grechko had been able to grab the end of Romanenko's safety line just before it passed out of his reach. (*W Star*, June 28/78, A4)

June 30: NASA announced it had selected Pan American World Airways, Inc., Aerospace Services Division, for negotiations leading to award of a contract for support services to facility operations at the National Space Technology Laboratories in Miss. Estimated cost over the first 3yr of the cost-plus-award-fee contract would be $32 million, with two 1-yr unpriced options. The contractor would operate and maintain NSTL real and installed property; manage supplies and equipment; and provide institutional, construction, and installation services. (NASA Release 78-94)

• DFRC announced it had begun a 2 wk flight program with Calspan Corp., using the total inflight simulator (TIFTS) to investigate advanced flight-control systems like those used in the Space Shuttle. TIFTS, a small twin-engine commercial passenger aircraft with the nose modified to include an additional pilot's cockpit, was one of the largest airborne simulators equipped with variable stability and 6 deg of freedom. Test pilots had flown the simulated mission from the nose cockpit (with safety pilots in the normal cockpit) to improve understanding of the flying qualities of advanced craft when landing, especially in actual touchdown under differing pilot-task conditions. (*DFRC X-Press*, June 30/78, 2)

• On the day *Soyuz 30* cosmonauts had docked with *Salyut 6,* the USSR had launched from Plesetsk another satellite in what was believed to be its early warning satellite system, followed by launch of a new space-mission series, *Defense/Space Business Daily* reported. On June 8 Tass had announced launch of eight Cosmos satellites (*Cosmos 1013-1020*) on one booster rocket. The early-warning spacecraft, *Cosmos 1024,* had gone into a Molniya (communications satellite) type of orbit with parameters of 630/40 000km, 62.8° inclination, and 12hr 6min period. It was the first of this type of mission launched in 1978; three had been launched in 1977.

The USSR had initiated a new spacecraft-mission series with the launch of *Cosmos 1025,* put by observers in the general category of military monitors but following a regime used by special-research missions. This was the 45th Soviet space mission in 1978, three more than at this point in 1977. The 1978 missions had been 77% military (not including the four Soyuz missions that had direct military application in manned reconnaissance/surveillance). (*D/SBD,* June 30/78, 307; FBIS, Tass in English, June 8/78)

During June: NASA reported that the Society of Automotive Engineers (SAE) had awarded Dr. Jose Chirivella and Wesley Menard of JPL the 1977 Manly Memorial Award for the best annual paper dealing with engines. Dr. Chirivella, principal investigator of hydrogen enrichment in the aircraft piston-engine program, and Menard, supervisor of JPL's combustion research group, had achieved fuel savings of 10 to 20% in aircraft piston engines. The aeronautical-propulsion division of NASA's Office of Aeronautics and Space Technology had supported the research. (*NASA Activities,* June 78, 11)

• NASA's annual procurement report for FY77 said that agency procurements had totaled $3532 million, a 10.2% increase over FY76. Approximately 80% of the net dollar value had been placed directly with business firms, 5% with educational and other nonprofit institutions or organizations, 8% with the Calif. Inst. of Technology (for operations conducted by or through JPL), and 6% with or through other government agencies. Of procurements placed by NASA with other government agencies, 90% had resulted in contracts with industry; about 48% of funds under JPL contracts had been for subcontracts with or purchases from business firms. Thus, about 90% of NASA's procurement dollars had gone directly or indirectly to private industry.

Of total direct awards to business firms, 73% were competitive procurements. Small business firms had received $255 million (only 9%) of these direct awards, since most of them were for large continuing R&D contracts for major systems and items of hardware, generally beyond the prime contractor capability of a small business. However, small business had received $119 million (22%) of the $522 million in new contracts worth $10 000 and more with business firms. Small business had also received $496 million (18%) of subcontract awards from 87 of NASA's prime contractors, including $59.382 million awarded to minority small business enterprises.

During 1977, 48 states and Washington, D.C., had participated in NASA prime-contract awards of $10 000 and over, 83% of which were placed in labor-surplus areas located in 41 states. Of NASA's prime contractors, 86% reported their larger subcontract awards had gone to 1684 different subcontractors in 45 states and Washington, D.C. (NASA procur rpt FY77)

• The USAF announced that SAMSO had awarded three cost-plus-incentive contracts for the first three stages of the MX missile system. Systems-definition contracts had gone to Thiokol Corp.'s Wasatch Division, $4 635 775; Aerojet Solid Propulsion Co., $3 445 066 (stage two); and Hercules, Inc., $3 493 191 (stage three). A requirement to design explosive ordnance-initiation devices for stage separation and thrust termination accompanied the stage-one contract with Thiokol Corp.

The 3-stage system-definition contracts called for studies, analyses, systems engineering, and missile/stage-interface definition to be completed by the end of 1978. Options for full-scale engineering development would include design, development, fabrication, and testing of hardware for the three stages. The MX program, a follow-on to the Minuteman intercontinental ballistic missile, was aimed at ensuring the survivability and effectiveness of U.S. ICBMs. (AFSC *Newsreview*, June 78, 3)

• The USAF announced it had issued to industry an RFP to investigate the feasibility of using widebody aircraft to supplement the B-52 as cruise-missile carriers. RFPs to aid in selecting one or more contractors had gone to Boeing, Lockheed, and McDonnell Douglas, all producers of domestic widebody transport aircraft. Proposals called for concept-tradeoff and design studies to verify aircraft suitability for the carrier mission. The studies would include the prototype widebody aircraft built by Boeing and McDonnell Douglas for the advanced medium short-takeoff-and-landing (STOL) transport program. Subsequent developmental efforts were in the planning stage. (AFSC *Newsreview*, June 78, 16)

• ESA reported that its council and communications satellites program board had unanimously approved at March and April meetings a first series of five operational Ariane launchers [see Apr. 24]; two operational satellites for intra-European telephone, telegraph, and telex communications and television relay (ECS-1 and -2); and a second maritime-communications satellite (Marots-B). Three of the five Ariane launchers would be financed from ESA programs for the scientific satellite Exosat (to be launched in early 1981), Marots-B (mid-1981), and ESC-1 (end of 1981). One launcher had been earmarked for the French earth-observation satellite SPOT; the fifth launcher was a reserve.

The council also had agreed that a preliminary phase of the H-sat program should revise the proposed satellite design to require only minor modifications (reducing the time scale) to go from an experimental "heavy" platform and payload to an operating direct-broadcast satellite. Other design modifications required to move from a qualification to an operational Ariane had delayed the scheduled H-sat launch. (ESA newsletter, June 78, 2)

• The NAA newsletter reported that competition had become more intense for world-record flights in general aviation. Robert Mucklestone was attempting to regain his around the world title in Class C-1.c Group 1 (2204 to 3858lb) that he had first captured in 1975 from Dr. Alvin Marks, who in 1969 had beaten the record of British aviatrix Sheila Scott, who had traveled a 33 000mi route in 33 days with a total average speed of 36.15mph, including ground time. In 1977 Harold Benham and Jack Rood in a Beechcraft Bonanza Model 35 had bested Mucklestone's record. Whatever the outcome of Mucklestone's current flight, two Texans (William Wisner and Frank Haile) had planned a flight in July 1978 to break the record in that class. (NAA newsletter, June 78, 7)

• Researcher William Bainbridge, who had polled registered voters in the Seattle area about attitudes on the space program, reported widespread enthusiasm for the knowledge and practical benefits derived from the space program, *Astronautics and Aeronautics* magazine said. Bainbridge had offered his poll sample a set of 49 statements as "not good," "moderately good," or "extremely good" reasons for continuing the space program: communications satellite benefits had received most positive responses; scientific knowledge attained through space exploration ranked second. The unconventional idea that ranked highest — communication with extraterrestrial intelligence — was approved by 53% of those polled. Bainbridge felt his results showed public appreciation of the practical and scientific results of the space program, and an anticipation of more benefits from new projects. (*A&A,* June 78, 60)

July

July 2: The *Washington Post* reported that, as a result of belated discovery that Soviet radar in Europe would interfere with two NASA satellites scheduled for launch in 1980 to communicate with orbiting spacecraft, NASA would have to spend $100 million to redesign the satellites' electronic systems. NASA had awarded Western Union a $786 million contract to build six 5000lb tracking and data-relay satellites (giant orbiting transmitters and receivers, each carrying two umbrella-like antennas weighing 50lb apiece and unfurling in orbit to a diameter of 16.5ft) that would have replaced 60% of NASA ground-station antennas at an estimated saving of more than $100 million a yr. The 3-mo delay necessary for redesign was important because it would affect communications with the Space Shuttle; failure by the Pentagon and the CIA to warn NASA of the size and scope of Soviet communications interference in high-orbit regions had delayed discovery of the problem.

NASA testimony before two Senate committees revealed that Soviet radars from the Baltic to the Black Sea were emitting beams that converged high over the Atlantic and Pacific in the exact places NASA wanted to put the TDRS. Built to replace obsolete and expensive ground antennas at Ascension Island; Quito, Ecuador; Santiago, Chile; Guam; and Hawaii, the TDRS would improve communications by allowing ground controllers to talk to other satellites and Space Shuttle astronauts during more than 90% of each earth orbit. The electronic interference was said to have been unintentional. NASA "did not fully understand the environment and the effects it would have on the system," said C. Curtis Johnson, satellite project manager at Goddard Space Flight Center. "Otherwise, we would have been more careful in the specifications of the system." NASA had considered reducing the number of satellites on order from six to four to save $100 million, estimated cost of the redesign. (*W Post,* July 2/78, A1)

July 5: NASA announced it had awarded Computer Sciences Corp., Falls Church, Va., a $40 million contract for 3yr of institutional computer systems engineering, development, and production operations work at Johnson Space Center. Under JSC management, the corporation would perform systems engineering and definition, applied software development, and computer systems and related facility planning and development for the central computing facility, as well as computer operations, equipment operations, job scheduling and processing, dispatch services, job quality control, maintenance of tape libraries, production coordination, and other related tasks for the facility and mission control center. (NASA Release 78-95; JSC Release 78-29)

• NASA announced it had awarded Martin Marietta Corp. a $32 million contract to develop a teleoperator retrieval system (TRS), a reusable TV-equipped propulsive device for use by Space Shuttle crews to deliver, stabilize, and recover satellites in orbit. Managed by Marshall Space Flight Center, the contract called for delivery of flight hardware by September 1979.

The TRS, a low-thrust box-like spacecraft operated from the Space Shuttle by remote control, could be carried in the orbiter cargo bay or left in space after completion of a mission for retrieval by the next Space Shuttle. First mission for the TRS would be a docking with Skylab to boost it into higher orbit or to aid a controlled reentry; a camera on the forward end of the TRS would permit an astronaut to maneuver it for rendezvous and docking, or for other purposes. The TRS would return to the Space Shuttle using its own guidance and control system. (NASA Release 78-96; *Marshall Star,* July 5/78, 1; *DFRC X-Press,* July 14/78, 2; *JSC Roundup,* July 7/78, 1)

• Kennedy Space Center announced it had awarded Boeing Services International, Inc., a 1-yr $24 071 643 extension of a ground-systems operational contract, which brought the contract total to $67 143 012. Services by Boeing would include facility and utility operations and maintenance for propellants; cranes, doors, and platforms; elevators; services and shops; nondestructive evaluation; and life support and industrial operations. (KSC Release 63-78)

• MSFC reported that electronic simulators for Space Shuttle main engines, solid-fuel rocket boosters, and external tanks had passed tests designed to check the Shuttle's avionics during simulated missions. JSC would conduct Shuttle-flight simulations in a Shuttle avionics integration laboratory (SAIL) for which MSFC engineers had provided a mated elements system. All Shuttle avionics would work together in the SAIL for the first time in Shuttle-flight simulation that would include part of the countdown under normal, abnormal, and certain failure conditions. (*Marshall Star,* July 5/78, 2)

July 6: NASA reported that the 200kw experimental wind turbine generator managed by Lewis Research Center had completed more than 1000hr of operation for DOE in Clayton, N.M., and was ready for checks to verify design-life estimates. Twin rotor blades and other dynamic components that had rotated at 40rpm or more (making more than a million cycles) since the machine began operating in January had developed loose or missing rivets and three small 3-in cracks, and would undergo more detailed inspection. Although DOE had shipped the blades to LeRC for detailed inspection and analysis of their wear and life potential, the turbine had continued operation using rotor blades from the prototype wind turbine operated on an experimental basis near Sandusky, Ohio, since October 1975. The first U.S. federally sponsored wind system to produce electricity directly for a community power system, the turbine had supplied more than 100 000kw-hr of electrical power to the Clayton-owned utility system. (NASA Release 78-97)

• NASA announced that 35 new astronaut candidates would report to JSC July 10 to begin 2yr of training and evaluation, including life-support and ejection-seat training for the T-38 aircraft, aircraft physiological training, and T-38 aircraft systems and operations. In Aug., the candidates would take the standard USAF water-survival course at Homestead AFB in Fla., and would attend lectures on spaceflight history, technical-assignment methods and procedures in the astronaut office, manned-spacecraft engineering, Space Shuttle program, aerodynamics, flight operations, and the many disciplines associated with preparing and operating vehicles in space. Astronauts, engineers, and JSC management and support contractors would be instructors and lecturers. (NASA Release 78-98; *JSC Roundup,* July 7/78, 1; *Marshall Star,* July 19/78, 2)

• NASA announced that HiMAT (highly maneuverable aircraft technology), a flight-research vehicle to demonstrate advanced technologies for air superiority fighter concepts in the 1990s, would begin flight tests later in 1978. Engineering studies had indicated the HiMAT would have twice the turning capacity of the most maneuverable fighter, doing 8g turns at near-supersonic (6g turns at supersonic) speeds with no loss of speed or altitude.

NASA and the USAF had jointly designed the craft to hasten the transfer of advances in aeronautical design from research laboratories into flight testing and to expand design techniques leading to quantum jumps in performance from one generation of aircraft to the next. HiMAT would exploit the "undirectional" stiffness of advanced composite materials in its wings and canard (small forward-wing) surfaces, layering the composite material to capitalize on natural bending of the wing/canard during flight. The HiMAT design would combine for the first time the lift of the wing and canard to enhance maneuverability and controllability. NASA would use the remotely-piloted technique it had developed, to fly HiMAT in high-risk flight testing without risking a test pilot and to reduce the costs of man-rating the aircraft. Built by

Rockwell Intl. Corp.'s Los Angeles Aircraft Division, HiMAT was one of NASA's many aeronautics R&D efforts aimed at maintaining U.S. superiority in civil and military aviation. (NASA Release 78-99)

July 7: NASA announced that a USAF launch team, members of the 6595th Space Test Group, and a team from the Jet Propulsion Laboratory had launched NASA's SeaSat-A at 6:12pm June 26 from Vandenberg AFB on a USAF/General Dynamics Atlas F/Lockheed Agena. Launch sequence went smoothly, with orbit insertions at launch plus 57min over the east coast of Africa. The satellite, called *SeaSat 1* in orbit, had deployed its solar panels and communications and sensor antennas during the second and third orbits, and extended its synthetic aperture radar antenna about 10pm. Spacecraft and booster problems had delayed launch since May; program officials had feared additional postponements would jeopardize the use of *SeaSat 1* in ocean-monitoring programs to confirm the accuracy of the data it returned. Lockheed Missiles and Space Co. had built the spacecraft for NASA and integrated its sensors. (*JPL Universe,* July 7/78, 1)

• KSC reported that NASA's 4-engine Jetstar aircraft had completed the second of five flight tests of the microwave scanning beam landing system (MSBLS) installed at the orbiter landing facility. The first three flight tests were to verify the system's ability to land the unpowered Space Shuttle orbiter safely on the 15 000ft runway; the final two would be to commission (declare operational) the runway-landing system. Because orbiters could approach KSC from either the northwest or southeast, the center would have to commission the system twice. Each of the first tests had taken about a week, as the Jetstar had made daily 2hr passes over the runway at different altitudes and azimuths. Precision laser tracking system instrumentation that could locate an aircraft at distances of 14mi and altitudes of 100 to 20 000ft had locked on to special instrumentation in the Jetstar, continuously tape recording the Jetstar's position while the MSBLS tape recorded the same data. The tapes had been fed into a computer and compared; any positional errors in the MSBLS could then be corrected, aligning the landing system. (*Spaceport News,* July 7/78, 2)

• JSC had updated data from the two Voyager spacecraft, the *JSC Roundup* reported, to show them on course and (as of July 1) 425 875 313mi (*Voyager 1*) and 408 475 978mi (*Voyager 2*) from earth. *Voyager 1* was 152 714 714mi from Jupiter; *Voyager 2,* 172 112 418mi; *Voyager 1* was 651 024 027mi from Saturn; *Voyager 2,* 651 229 423mi. *Voyager 1* would encounter Jupiter on March 5, 1979; *Voyager 2* on July 9, 1979. JPL had been managing the Voyager program and providing deep-space tracking. Roundtrip communications had taken 76min for *Voyager 1* and 72min for *Voyager 2*. (*JSC Roundup*, July 7/78, 4)

July 10: The U.S. Navy had orbited two clusters of ocean-surveillance satellites to monitor radio communications and radar emissions from Soviet submarines and ships, and would add more spacecraft to expand coverage, *Av Wk* reported. The Navy had launched the first of the two clusters (each consisting of three small satellites in close proximity) in 1976, the second cluster in 1977, all from Vandenberg AFB into near-circular 63°-inclination orbits at 700mi altitudes. At that altitude, the spacecraft could receive signals from surface vessels more than 2000mi away.

The Naval Research Laboratory, which in Dec. 1971 had built and launched three spacecraft as a first demonstration of using multiple satellites to direction-find and eavesdrop on Soviet surface vessels and submarines, had designed and built the ocean-surveillance satellites (NOSS). Martin Marietta had begun production of additional satellites with E-Systems, Inc., as major subcontractor to provide the electronic intelligence (elint) receivers and antennas. For the future, the Navy had planned a more advanced spaceborne ocean-surveillance system using active radar to pinpoint the position of surface vessels, to be developed for launch by 1983 under a program called Clipper Bow. (*AvWk*, July 10/78, 22)

July 11: NASA announced it had selected General Electric Co., Valley Forge, Pa., to provide a Landsat-D mission system consisting of a flight segment (and backup) and a ground segment that would include a data management system, an operations control center, a Landsat assessment system, and a transportable ground station, all to be managed by GSFC. General Electric would fabricate the mission-unique equipment and integrate and test the flight segments, as well as build and install the ground segment in government facilities. Proposed cost of the cost-plus-award-fee contract was approximately $70.5 million, to cover the Landsat-D launch in late 1981 and early orbital operations.

The thematic mapper (TM) in Landsat-D's instrument module, a second-generation 7-band multispectral scanning radiometer with a ground resolution of about 30m (100ft), could extend the inventorying of earth resources. NASA had designed Landsat-D to improve activities such as crop forecasting, vegetation-health assessment, forest- and range-resources management, water-resources management, land-use mapping, and mineral exploration. (NASA Release 78-102)

• ESA announced that Europe's first comsat, *Ots 2,* had been operating correctly at the end of the 8-wk period covered by an insurance contract.

Ots 2, launched May 12 from Cape Canaveral, had reached its final station 35 900km above the equator at 10°E. Its six SHF antennas would cover western Europe, the Middle East, north Africa, the Azores, the Canary Islands, Madeira, and Iceland. *Ots 2* had successfully exchanged signals with four large earth stations and eight stations equipped with small antennas. Checks and preliminary tests would continue until the end of September, when ESA would turn over *Ots 2* to its users, national telecommunications administrations grouped in interim Eutelsat and European research units and centers. The experimental program would include telephone and television traffic-routing tests, propagation experiments, and a variety of experiments related to new comsat applications in Europe. (ESA Release July 11/78)

July 12: NASA announced it had awarded TRW Defense and Space Systems Group, Redondo Beach, Calif., a contract for the materials processing in space (Spacelab) program at an estimated cost of approximately $9.7 million. TRW would be prime contractor for various aspects of the Materials Processing in Space (MPS) program, the contract to cover the initial phase of the MPS Spacelab-payloads project expected to begin in September 1978 and continue through 1981. TRW would develop (and support the operation of) specialized and general purpose payload systems that would accommodate a variety of different materials processing experiments during recurring earth orbit missions of the Space Shuttle.

NASA had selected the first series of experiments from among proposals by scientists from the U.S. and numerous foreign countries. Objectives of the MPS program were to study new or improved processes and to identify candidate products of initial commercial interest: electronic materials, metals, glasses, and certain chemicals produced by processes such as crystal growth, solidification, and containerless processing. MSFC would manage operational aspects of the program. (NASA Release 78-103; *ARC Astrogram,* July 27/78, 3)

• Cosmonaut Valery Kubasov had spoken of the possibility of joint experimental flights by the USSR's Salyut-type orbital station and the U.S. Space Shuttle, according to a Tass report. Kubasov noted that such flights could use the joint approach and docking procedures worked out for the Apollo-Soyuz test project and tested during the joint Soviet-U.S. flight of Apollo and Soyuz in July 1975. A new agreement between the USSR and the U.S. on cooperation in space research for peaceful purposes had came into force the previous May, Kubasov pointed out. He added that millions of Americans, like the citizens of the USSR, had viewed the Soyuz-Apollo program as a display of goodwill by the two nations and the beginning of great cooperation between nations in space.

However, *Av Wk* reported that any U.S. Space Shuttle/Soviet Salyut space station joint mission was in deep trouble, if not already dead, because of increased U.S. concern over technology transfer to the Soviets and the current abrasive political climate between the two countries. A NSC interagency committee had reevaluated the entire joint mission idea, while U.S./Soviet discussions on the project were in limbo. The project probably would have been discussed during a planned technology-oriented trip to the USSR by presidential science adviser Frank Press, but the White House had canceled the trip in disapproval of the Soviet treatment of dissidents. (FBIS, Tass Intl Svc in Russian, July 12/78; *AvWk,* July 17/78, 13)

July 13: KSC announced that for the third consecutive yr it would host about 80 of the nation's leading atmospheric physicists and lightning researchers, representing 10 universities, 2 research laboratories, NOAA, and 3 other NASA centers, for a summer of combined study of the electrical properties of thunderstorms. The thunderstorm research international program (TRIP) would use the unique meteorological facilities built at KSC during the Apollo and Skylab programs, to assess thunderstorm hazards during launch operations and to monitor the high incidence of lightning-charged storms in the area during the summer. Participating investigators would bring their own instruments for many of the experiments; others would use KSC's instruments, like the field-mill system built for the Apollo program to detect buildup of electrical charges in overhead thunderclouds; KSC's instrumented aircraft, NASA-6, used in previous lightning studies; and the lightning detection and ranging (LDAR) system that detected electrification in distant clouds. NOAA's geosynchronous operational environmental satellite *Goes* would provide satellite photographs for the scientists.

Experiments would include an evaluation of prototype lightning sensors as part of the scientific package to be flown in 1982 on the Jupiter orbiter probe designated Project Galileo, first planetary spacecraft to be carried aboard the Space Shuttle; a study of evolution of lightning activity; examination of the structure of waveforms radiated from lightning, for characteristics indicating storm type; obtaining from ground-based systems range and azimuth data on lightning strokes; monitoring of electric-field and corona (surface discharge of electricity) current over water; and study of the relationship of precipitation-formation development in thunderclouds to lightning occurrences. (KSC Release 66-78; *Spaceport News,* July 21/78, 1)

• Langley Research Center announced it had selected Frank Godfrey, contract specialist, to participate in the Education for Public Management program of the John F. Kennedy School of Government at Harvard Univ. As part of a federal government effort to develop promising mid-careerists' talents and abilities for management in the public sector, the program would train individuals identified by their agencies as having potential to assume increased responsibility for agency programs and policies. Godfrey, who had been at LaRC since 1974, had served as a loaned executive to the 1977 Peninsula United Fund Campaign and was on the adjunct faculty of St. Leo College. (LaRC Release 78-35)

July 16: NOAA announced it had developed a technique that would permit national weather centers around the world to monitor rainfall over agricultural areas, detect potential flash-flood situations, and predict rainfall from hurricanes while they were still at sea. Once the technique had been perfected, automated, and put into operation, scientists could use earth-orbiting satellites to estimate rainfall from convective (vertically-developed) clouds and to monitor the movement and distribution of precipitation over the planet's entire surface.

Dr. William Woodley and Cecilia Griffith of the Commerce Department's National Hurricane and Experimental Meteorology Laboratory had adapted the experimental method from a similar system of estimating rainfall from convective weather systems in the tropics, applied successfully during cloud-seeding experiments in southern Fla. The imagery had revealed a difference between young rain-producing clouds and dying rainless systems, both of which had appeared in bright shades on satellite images. NOAA's National Weather Service had already made quasi-operational use of a similar method, applied by flood forecasters in predicting flood potential of approaching hurricanes and large thunderstorms. Woodley and Griffith had begun to use the technique in a real-time warning mode to detect potential flash flood situations. (NOAA Release 78-94)

July 17: Worldwide demand for international satellite communications would increase by more than 100% in the next 5yr, INTELSAT announced. The prediction came out of the fourth global telecommunications traffic meeting sponsored by INTELSAT, in which 172 international communications experts representing 93 telecommunications entities around the world participated. Purpose of the annual meeting was to forecast the demand for services provided by the INTELSAT system. INTELSAT would use these forecasts in configuring its system to meet user requirements with maximum efficiency, and in planning for future satellite systems.

Preliminary figures from the meeting indicated that global demand for satellite-telecommunications capacity would increase from 14 105 equivalent full-time simultaneous telephone circuits at year end 1978, to 29 282 circuits by the end of 1982, up 108%. The biggest demand was expected to be from countries surrrounding the Atlantic Ocean, where communications traffic by satellite was expected to rise by 114% over the next 5yr. Traffic in the Indian Ocean region was projected to increase by 98%, and in the Pacific Ocean region by 91%. (INTELSAT Releases 78-19-I, 78-20-I)

July 18: NASA reported that the Netherlands astronomical satellite *ANS* had been a success, based on comparison of mission results and prelaunch objectives. *ANS,* an Explorer-class x-ray and ultraviolet mission, was a cooperative program of the U.S. and the Netherlands, the U.S. providing a 2-part medium-energy x-ray instrument and the Scout launch vehicle, and the Netherlands providing the spacecraft, a 2-part soft x-ray instrument, and a small ultraviolet telescope with photometers. During its 20mo lifetime, *ANS* had measured positions, spectra, and time variations of galactic and extragalactic x-ray sources in the energy range 2 to 15keV; defined upper limits of silicon-line emissions around 2keV; and transmitted more than 18 000 observations of about 4000 objects in the UV range 1500 to 3300A, in addition to other unexpected discoveries.

Although an unplanned elliptical orbit had drastically limited the x-ray observations, particularly in the pointed mode, and had substantially increased the spacecraft's exposure to radiation, the data it had returned were of a consistently high quality, except for partial failure of half the soft x-ray instrument 10mo after launch. *ANS* had reentered the atmosphere June 14, 1977. (NASA MOR S-875-74-01 [postlaunch] July 18/76)

• NASA reported the U.S. Postal Service would commemorate the Viking missions to Mars by issuing a special postal stamp Thursday, July 20, second anniversary of the *Viking 1* landing on Mars. LaRC and the Postal Service would hold a First Day of Issue ceremony in LaRC's activity center, and Dr. Donald Hearth, center director, would buy the first stamp. Robert McCall, whose two large murals were on display in the National Air and Space Museum, had designed the stamp.

NASA had launched two Viking spacecraft toward Mars in Aug. and Sept. 1975; the *Viking 1* lander had touched down on Mars at 8:12am EDT July 20, 1976, and the *Viking 2* lander arrived at 6:59pm EDT Sept. 3, 1976. During the mission ending in Nov. 1976, the landers and their partner, the orbiter spacecraft, had conducted experiments in 13 scientific disciplines. NASA had extended a reduced Viking mission until May 1978 to gather more information on Mars's weather, seasonal variations, and soil characteristics; a further reduced mission controlled by JPL would continue through Feb. 1979. The Viking high-resolution photographs of the planet from the surface and from orbit, numbering more than 30 000, had mapped most of Mars and had provided scientists with valuable information. (NASA Release 78-106; LaRC Release 78-34; *Langley Researcher,* July 14/78, 1)

July 19: NASA announced that the Egyptian government had reserved four small self-contained payloads to fly on the Space Shuttle in the 1980s. At the NASA Hq ceremony July 13, Dr. Mohamed Shakar, minister of the embassy of Egypt, and Dr. Farouk El-Baz, research director for the Smithsonian Institution's Center for Earth and Planetary Studies, gave NASA officials a down payment to reserve Space Shuttle space.

The payloads called "getaway specials," each weighing no more than 90kg (200lb) and no larger than $0.5m^3$ ($5ft^3$), would fly on a space-available basis. NASA had received payments for about 240 small payloads so far; purchasers had included private individuals, commercial firms, and foreign countries. The Egyptian purchase marked the first foreign educational use of a Shuttle payload, as Egyptian students would compete in proposing experiments for the Shuttle missions. Dr. El-Baz would direct evaluation of the proposals. (NASA Release 78-107)

• NASA announced that Dr. Robert Frosch, NASA administrator, and Professor S. Dhawan, secretary of the department of space and chairman of the Space Commission, Government of India, had signed a memorandum of understanding (MOU) July 18 for launch in 1981 from the Space Shuttle of India's first national satellite, INSAT-1. INSAT-1, a multipurpose spacecraft in geostationary orbit, would provide India with domestic public telecommunications and direct TV broadcasting and meteorology services. This was the first MOU NASA had signed with a foreign government setting forth terms and conditions under which NASA would furnish Space Transportation System launch services on a reimbursable basis. (NASA Release 78-109)

• NASA announced that Puerto Rico's governor Carlos Romero Barcelo and officials from DOE, NASA Hq, and LeRC would dedicate on July 21 the federal government's second large wind-turbine generator to be field-tested by a U.S. utility, on the small Puerto Rican island of Culebra. NASA had designed the turbine to produce 200kw of electric power in winds of 29 to 60kph (18 to 34mph) measured 9m (30ft) above ground level, equal to 20% of Culebra's electric power, or about enough power for 150 island homes. The Puerto Rico Water Resources Authority would operate the wind turbine for 2yr, supplying test data to DOE and NASA on the performance and economics of large wind-energy systems used by utilities. Later in the yr NASA planned to begin operation of a much larger wind turbine rated at 2000kw (2mw) near Boone, N.C. (NASA Release 78-104; DOE Release R-78-266)

• MSFC reported that a failure in Skylab's power system July 9 had caused loss of vehicle attitude; MSFC engineers had reconfigured the system and fully charged the spacecraft's batteries, before performing reorientation maneuvers to return it to its minimum-drag attitude. Skylab was now about 240mi above earth, rolling at about 2 revolutions per hr. Engineers had planned maneuvers to stop the roll and place the vehicle first in a solar-inertial attitude, then in an end-on velocity vector that would encounter minimum atmospheric drag. However, Dr. Chris Kraft, JSC director, said in a copyrighted interview with the *Dallas Morning News* reported in the *Washington Post*: "I don't believe Skylab is going to live long enough for us to get to it with the Space Shuttle . . . We're going to have to live with the fact that Skylab is going to die a natural death." (*Marshall Star*, July 19/78, 1, July 12/78, 1; *JSC Roundup*, July 21/78, 1; *ARC Astrogram*, July 13/78; 1; *W Post*, July 21/78, A14)

- MSFC announced that ISEE-C, third and last of the spacecraft in a joint NASA/ESA study of near-earth environment, would be launched August 12. *Isee 1* and *2* built by ESA had, since they were launched into looping trajectories Oct. 22, 1977, been gathering detailed data on solar-wind particle control of the boundaries between earth space and interplanetary space. ISEE-C, at a point in a sun-centered orbit where gravitational forces of the earth and sun were exactly equal, should obtain nearly continuous data on the fluctuating solar wind and on special solar phenomena, about an hr before the solar particles would flow past the earth orbit of *Isee 1* and *2*. (*Marshall Star*, July 19/78, 4)

- GSFC reported that, after a 2mo search, its Copernicus satellite had located what might be a second invisible "black hole" in the universe. The satellite's x-ray experiment had found in the constellation Scorpio what was believed to be a super dense collapsed star with such strong gravity that even light could not escape it. Scientists had declared it much more active and convincing than the first black hole found, Cygnus X-1. The new black hole was orbiting Scorpio V-861, a supergiant star visible with the naked eye from earth, and gradually siphoning away the larger star's atmosphere.

Tracing the massive amounts of x-rays released from matter emanating from the giant star had led scientists to what they called the "best black hole yet"; discovery had given astronomers their first opportunity to study a black hole as it passed behind its companion star and to test some of their black-hole theories. Glen Pollard, coinvestigator on the project, said he hoped the new information would define the shape of the cloud of stellar material sucked into the black hole, reveal the dynamics of such systems, and provide a measure of the quantity of material being removed at speeds up to 2 million mph. Studying black holes would extend astronomers' knowledge of physics into areas not observable on earth—the physics of relativity. (*Goddard News*, July 19/78, 1)

• GSFC reported it had held in June the first regional remote-sensing applications training course for state-related users of remote-sensing systems. Participants represented agencies implementing or planning facilities to use Landsat image data. Developed by Dr. Nicholas Short, GSFC training officer, and Henry Robinson of CSC, the course had offered comprehensive explanations of image processing offered by expert lecturers drawn from NASA, outside agencies, and contractors. (*Goddard News*, July 19/78, 4)

• GSFC reported that its astronomers believed they were about to confirm for the first time the presence of gold on a star, perhaps as much as 100 billion tons of it. That amount of gold would indicate the star contained one part per 100 000, comparable to earth's one part per million. William Heacox and David Leckrone of GSFC's Stellar and Cosmic Astronomy Branch had planned additional observations to confirm the first signs of gold on Kappa Cancrii, a blue-white star visible to the naked eye in the constellation Cancer. The astronomers had found the gold traces on the star while using the IUE satellite to look at chemically peculiar stars rich in exotic elements. (*Goddard News*, July 19/78, 2)

• GSFC reported that U.S., Canadian, and other scientists had been studying the first data from the applications Explorer mission *Aem 1*, launched Apr. 26 into a polar orbit. Using information on day-night temperature changes, the scientists had examined black-and-white and false-color images for clues to the locations of mineral resources and to identify rock types, soil-moisture changes and plant stress, and the effects of urban heat islands. The first false-color image had covered a 700km swath of the eastern U.S. from Cape Hatteras to Lake Ontario. The scientists would correlate data from *Aem 1*, first of a series of satellites for NASA's heat-capacity mapping mission, with data from other satellites, especially Landsat. (*Goddard News*, July 19/78, 2)

• INTELSAT reported it had commissioned a $23 500 study in 1979 of ionosphere effects on communications-satellite transmissions. Cable and Wireless Ltd., U.K., in Hong Kong would conduct the study to discover the cause and extent of, and remedy for, fluctuations in satellite-signal strength noted by earth stations near the equator. The fluctuations in the 4 and 6GHz frequency band, believed to be caused by sunspot activity, had occurred only in the evenings, mainly during equinox periods, and had impaired international communications. As scientists had predicted

1979 would be a high sunspot-activity year, data gathered then would be vital to INTELSAT attempts to overcome such interference. (INTELSAT Release 78-21-I)

July 20: NASA announced that an Ames Research Center scientific team led by Univ. of Chicago astronomer Dr. Richard Miller and ARC astrophysicist Dr. Bruce Smith had been pioneering in computer simulations of cosmic events that might alter traditional ideas of how galaxies formed, what shapes they assumed, and what happened when they collided.

Using ARC's ILLIAC IV, world's most powerful computer, the researchers had recreated the birth and evolution of stars and galaxies, compressing 200 million yr into several hr of computer time and reducing the distance light traveled in 100 000yr to the length of a television screen, to witness galaxies colliding and gas clouds begetting embryonic stars. Computer simulations had shown elliptical galaxies, thought to be oblate (shaped like a frisbee), to have a 3-dimensional prolate shape (oblong); astronomers had found observational evidence confirming the computer discovery. The researchers had instructed the computer to create swirling systems of 100 000 computer points, each point representing the mass of about 1 million suns to give the simulated galaxy a realistic total mass. Each point had felt the gravitational pull of its neighbors and had absorbed energy from exploding supernova. Furnishing a computer with the power of ILLIAC IV with a detailed program on formation and dynamics of galaxies had proved computer simulation to be a useful approach to explaining basic astrophysical processes.

ARC scientists had filmed several computer simulations showing results unexpected in light of traditional astrophysical theories. For example, vast congregations of computer-generated stars had repeatedly evolved into stable prolate bars that rotated end over end to form what was perhaps the true shape of elliptical galaxies. When Miller and Smith had set two simulated 50 000-star galaxies on a collision course, the colliding galaxies had exhibited much greater interaction than was forecast. First they contracted, their gravitational fields reinforcing each other, then they bounced back in violent expansion flinging hundreds of stars out of the galaxies, resulting in merged galaxies containing considerably fewer stars. The two researchers' most recent efforts had centered on tracing simulated masses of gas as they coalesced and formed stars. (NASA Release 78-108)

• NASA announced that officials from DOE, the U.S. Forest Service, and LeRC had gathered atop a 2392m (7684ft) peak to dedicate a pioneer solar-cell-powered lookout tower that would demonstrate applications of solar-electric technology in the foreseeable future. The lookout tower, on Antelope Peak in Calif.'s Lassen National Forest, would use a 300w solar-cell system to power the tower's refrigerator, water pump, lights, radio-communications equipment, and small appliances used by a fire lookout who would live round the clock in the tower. Electricity generated by the solar cells would charge eighteen 12-volt lead/acid batteries, storing up to 36kw-hr of electric energy and supplying normal amounts of power for about a mo before needing recharge.

The tower's solar electric power system had been designed to operate continuously for a yr without fuel or periodic maintenance. LeRC had built the solar electric-power system funded by DOE through the Test and Applications Project managed by the center as part of DOE's photovoltaic conversion program. LeRC officials noted that, at current prices, electricity generated by solar-cell power systems over a 20yr system life would cost about $1.75 per kw-hr, which compared favorably with the price of electricity generated in remote areas by alternate methods such as diesel generators. (NASA Release 78-110; DOE Release R-78-270)

• The Naval Research Laboratory reported that a team of its space scientists had evaluated two new types of far-ultraviolet electrographic cameras. Designed for possible use in the Space Shuttle program to investigate radiation spectra of distant stars and other celestial objects, both cameras would use alkaline halide photocathodes having more sensitivity in the vacuum ultraviolet below 2000A, and higher angular resolution, than cameras previously used.

One camera, the internal-Schmidt-optics type, had an opaque photocathode like those NRL had used in previous sounding-rocket flights, in the lunar camera on the *Apollo 16* mission, and on *Skylab 4*. The new camera, flown on two sounding-rocket missions, had on the first flight obtained far-ultraviolet images of the Andromeda galaxy and on the second had recorded the North America nebula in Cygnus and far-ultraviolet spectra of stars in Cygnus. The other camera developed by NRL, a more conventional end-window type, had a semitransparent photocathode intended for possible use at the focus of a large telescope or for making spectrographs on the Spacelab. This large-format electrographic camera, instead of using a phosphor screen, would record directly on film the electrons emitted by the photocathode, allowing better resolution and photometric accuracy. NASA had supported development of the electrographic cameras. (NRL Release July 20/78)

July 21: After 2 decades as key booster in NASA's space program, the Atlas Centaur would remain "old reliable" for a variety of planetary, commercial, and military payloads launched in the 1980s, *Lewis News* reported. The Centaur project had begun in November 1958 when DOD's Advanced Research Project Agency (ARPA) awarded the first contract to General Dynamics after a study recommending a high-energy upper stage for the Atlas launch vehicle. Before the study, Pratt and Whitney had received a contract for a hydrogen/oxygen rocket engine. Since Centaur would be the first space-launch vehicle to use a liquid-hydrogen/liquid-oxygen combination, a whole new technology was needed for handling materials, behavior, fabrication, and testing of liquid hydrogen in a space environment, not only to make Centaur a success but also because this fuel was to be used for the upcoming Apollo program.

In 1958 Centaur received its first mission assignment, definition of geosynchronous orbit, and in 1960 its first deep-space mission. The creation of NASA resulted in transfer of the Centaur program from ARPA to NASA in July 1959. LeRC, which had pioneered research in hydrogen technology and had fired an experimental liquid-hydrogen/liquid-oxygen engine with 5000lb thrust as early as 1953, received technical management of Centaur in 1962. The first successful flight of an Atlas Centaur in Nov. 1963 included the world's first inflight ignition of a hydrogen/oxygen rocket engine. Centaur had proved its operational capability in 1965 by successfully injecting a dynamic model of the Surveyor spacecraft into a prescribed transfer trajectory, resulting in a simulated lunar-target impact.

After successfully launching *Surveyor* (for the first soft landing on the moon) and *Pioneer 11* (for a flyby of Jupiter), NASA in the mid-1960s had directed its efforts to integrating the Centaur with the USAF Titan booster. General Dynamics had built an improved Centaur adaptable to both the Atlas and the Titan boosters; Martin Marietta had conducted studies for NASA on integrating the Titan booster with an improved Centaur, and on modifications needed to the Titan launch facility at ETR.

Atlas Centaur missions had included earth-orbiting spacecraft such as the applications technology satellite, orbiting astronomical observatories, Comstar, high-energy astronomical observatories, and FltSatCom. Atlas Centaur, used for future missions with HEAO and FltSatCom spacecraft, would remain one of NASA's prime launch vehicles until the Space Shuttle became fully operational. (*Lewis News*, July 21/78, 1)

• In 1977 JPL scientists Donn Lynn and Jean Lorre had agreed to apply image-processing techniques to photographs of the Shroud of Turin, but had not expected their results to be published in newspapers and magazines around the world, *JPL Universe* reported. Their "mild curiosity" about the relic had introduced them to other American scientists who proposed using a variety of sophisticated tests to determine how the shroud had been imprinted. The result was a trip planned for Oct. that would take Lynn and Lorre to Turin, where church authorities would allow the American science team an unprecedented 24hr examination of the shroud itself.

Coinciding with a public exhibition and international conference about the shroud, the tests would entail infrared and ultraviolet photography as well as conventional black-and-white pictures. Other team members would make radiographic and x-ray fluorescence examinations, all to investigate the mechanical formation of the "negative" image on the cloth. The shroud had been stitched to a protective cloth backing, but the team would use a flexible optical instrument to examine the reverse side of the fragile fabric. Major business corporations, including photographic and electronic-equipment companies such as Polaroid, Kodak, and the Brooks Institute of Photography had agreed to lend or donate facilities and equipment for the tests. Lynn and Lorre expected to provide a quick-look report of their findings soon after the Turin exhibition. (*JPL Universe*, July 21/78, 1)

• The *JSC Roundup* printed a report by Dr. Thornton Page on his experiences during a friendship tour to China. Page, who had used the extreme-ultraviolet camera during Apollo missions, had asked for and received tours of Chinese astronomical facilities; in discussions with Chinese, Page learned most of their data were of western origin because of the limited quality and size of Chinese astronomical instruments. "They were delighted to have the books, reports, pamphlets, and photos I presented to them, particularly the NASA photos of Shuttle, Skylab in orbit, and ALSEP instruments deployed on the moon," Page reported. "They spoke about the large telescope under construction in Nanking, and their own telescope planned for a mountain site some 30mi northeast of Peking. Drs. Pan How-Ren and Wu Jian-Cheng are calibrating x-ray and far-UV detectors for use on a satellite, but they could not tell me when it will fly."

The Chinese told Page that a major change in national policy on science had taken place during the past yr, since the "Gang of Four" had been removed from power. Chairman Hua had given high priority to science and technology and had set a goal of doubling by 1982 the number of scientists and technicians in China. "None of the scientists I met," Page added, "was involved in teaching, and none of our other visits to universities and high schools in Manchuria or Shanghai revealed classroom instruction in these subjects." (*JSC Roundup*, July 21/78, 1)

July 24: NASA announced that mission controllers at ARC had completed final course adjustments for the *Pioneer 11* encounter with Saturn

Sept. 1, 1978. Controllers reported *Pioneer 11* had locked on a trajectory that would bring it within 30 000km (18 000mi) of the edge of Saturn's outer ring; any approach closer to the ring edge would risk impact with fragments in the ring. *Pioneer 11* would then swing under the plane of the rings to a point 25 000km (15 000mi) from the planet's surface. Without the course correction, *Pioneer 11* would have flown by Saturn at a much greater distance (100 000km or 60 000mi). Pioneer data transmitted to NASA tracking stations indicated the spacecraft had responded perfectly to a day-long series of commands calling for timed rocket thrusts to alter its trajectory. (NASA Release 78-113)

• NASA announced the ground breaking on a mountaintop near Boone, N.C., for the largest U.S. wind-turbine generator, scheduled for completion later in 1978. Using two 30m (100ft) steel rotor blades, the wind turbine would produce 2000kw of electric power in winds of about 39 to 69km (24 to 40mi) per hr, enough to meet the needs of about 500 homes in and around Boone, and would be the largest system field-tested by a local utility under DOE's wind-systems development program managed by LeRC. General Electric Co. had been prime contractor for the Boone machine, which would be field-tested by the Blue Ridge Electrical Membership Corp. (NASA Release 78-105; DOE Release R-78-270)

• LaRC announced that Dr. Peter Calder, project director for the Olympus 593 engine at Rolls Royce Ltd., Bristol, England, would be guest speaker at a center colloquium July 31 on "Engine Options for Supersonic Cruise Aircraft." Calder would discuss the Olympus 593 twin-spool turbojet engine produced by Rolls Royce in Britain and SNECMA in France, which had powered the Concorde supersonic jet transport. He would describe development of the engine and its inlet, and discuss possible modifications to improve propulsion system operating efficiency and environmental acceptability. Concorde supersonic passenger service from Paris and London to South America and the Middle East had been under way for 2.5yr and to North America for 2yr, initially to Washington, D.C., and then to New York. (LaRC Release 78-38; *Langley Researcher,* July 28/78, 2)

• INTELSAT announced that it, together with the Federal Republic of Germany and Messerschmitt-Bolkow-Blohm (MBB), had funded development of a new high-efficiency solar-power collector. MBB had built the lightweight deployable solar array for use in satellites of the 1980s, with AEG Telefunken as subcontractor for solar cells on the array surface. ComSatCorp laboratories in Clarksburg, Md., had tested the array. Its weight, considerably less than that of current counterparts, would mean lower launch costs, and its higher power output would handle more onboard satellite electronics, enlarging its communications capacity. At the beginning of a 7yr mission, the new solar array would provide up to 38watts of power per kg of weight, a 300% increase over current models. Incorporating thin solar cells in the design might increase output to 50watts of power per kg of weight. (INTELSAT Release 78-22-I)

July 26: NASA announced that 16 astronaut candidates from JSC, including six female candidates, would undergo three days of training beginning July 31 at the Homestead Air Force water survival school in Fla. [See July 6.] More than half the 35 current candidates had had water survival training before entering the NASA program. Each day's activities would include classroom lectures on water-survival techniques and actual training in the water. The candidates, wearing a tethered parachute harness, would slide down a wire from a tower and be towed through the water in the harness to practice parachute release. In other exercises, candidates towed aloft under a parasail canopy would land in the water to be picked up by boat. On the final plunge from the parasail into the water, candidates wearing full survival gear would be picked up from their liferaft by helicopter. The Air Force's 3613th Combat Crew training squadron operated the water survival school. (NASA Release 78-114)

• NASA announced that effective July 30, the Legislative Affairs Division would become a separate organizational entity reporting to the Office of the Administrator. Experience over the past 8mo indicated that this change would better balance the workload of the senior staff and assure close and continuous liaison with the Legislative Branch. (NASA special anno July 26/78)

• MSFC announced it would issue a request for proposals (RFP) from industry to develop automatic welding for commercial use. Under NASA's requirement to transfer new technology to industry, it had decided to seek a new line of equipment for economical and more productive automatic welding of single and double contour welds. Automotive, shipbuilding, small watercraft, military hardware, aerospace, petroleum, and solar heating and cooling industries might use such an invention.

MSFC had built a prototype model with real-time computer-controlled weld speed and torch-angle control for making a variety of contour welds. The unit, equipped with digital closed-circuit television welding-torch guidance, could automatically maintain accurate alignment of the torch and weld joint during welding. Values for up to 5 constantly changing variables, measured and fed into the control computer to maintain constant speed and angle, could easily make contour corrections at every millimeter of travel or more frequently. The contract would call for development on a cost-sharing basis of commercial versions of the equipment for purchase and use by industry. MSFC's computer-weld skate with automatic torch guidance had three patents, two on the computer control and one on the closed-circuit television guidance; NASA had applied for an additional patent for the entire computer-weld skate. These patents could be licensed by the successful proposers. (*Marshall Star*, July 26/78, 4)

• The Dept. of Transportation announced it had issued an updated plan for development of an all-weather landing system that would eventually replace the current instrument-landing system worldwide. An FAA update of the original 1971 national plan had set forth requirements for completing the prototype phase of the program, on terms set by the International Civil Aviation Organization (ICAO) earlier in 1978. ICAO had selected the U.S./Australian time-reference scanning-beam microwave-landing system (MLS) to be the standard approach-and-landing guidance system of the future. The updated "National Plan for Development of the Microwave Landing System" called for completion of the prototype development and testing phase by the end of 1982. The plan made allowance for a transition phase between system development and implementation. Although it made no firm recommendations on implementation of the system, it described alternate strategies. Work had been completed on two of the six MLS configurations: a small-community airport version and a basic system for most commercial fields. Still to be developed were the "expanded" version for large hub airports and various tactical systems for military use.

DOT also had drafted a document for adoption by ICAO on Standards and Recommended Practices (SARPS) that would prescribe technical and operational characteristics of MLS to ensure quality of system performance and compatibility of air and ground components throughout the world. (FAA Release 63-78)

July 27: NASA announced it would launch on or about Aug. 7 from Cape Canaveral the second of two Pioneer spacecraft designed for direct measurement of Venus's dense searing atmosphere, making the 354 million km (220 million mi) flight to Venus on an Atlas Centaur rocket. Pioneer/Venus 2, 20da out from Venus, would separate into five atmospheric-entry craft. Four would enter Venus's atmosphere at points over the earth-facing hemisphere, two on the day side and two on the night side, and the fifth (the transporter bus) would enter on the day side. The Pioneer/Venus 2 entry probe would arrive five days after the Pioneer/Venus 1 orbiter.

The 30 experiments aboard the two Pioneers, a coordinated observational system, would employ the most spacecraft (6) ever devoted to one planet. NASA has designed the flights to study atmosphere and weather on another planet on a global scale as an aid to understanding the forces driving earth weather, and had issued a fact sheet (NASA Release 78-100) giving details on the surface, atmosphere, and stratosphere of Venus. (NASA Release 78-101)

• NASA announced that JPL engineers had commanded the shutdown of the *Viking 2* orbiter on July 25, when it ran out of attitude-control gas after circling Mars 706 times over the past 2 yr. Several mo previously, orbiter 2 had sprung a leak in one of its attitude-control jets; engineers had slowed the leak but could not stop it completely. *Viking 2*, launched Sept. 9, 1975, had traveled more than 643 735 000km (400 million mi) to reach Mars orbit Aug. 7, 1976. The *Viking 2* lander, riding piggyback on the orbiter, had landed Sept. 3, 1976, in the Utopia Planitia region of Mars's northern hemisphere. Plans called for the *Viking 1* orbiter and both Viking landers, functioning well almost 3yr after launch, to operate through Feb. 1979. The Viking project would continue through Sept. 1979 to allow scientists and engineers to process a backlog of data from the spacecraft and experiments. NASA had designed the orbiters to operate for 150da after planetary encounter and to orbit for at least 50yr, in hopes of avoiding planetary contamination before they disintegrated. (NASA Release 78-115)

• The *ARC Astrogram* reported that the quiet short-haul research aircraft (QSRA) built by Boeing under a contract from NASA had actually been a remanufactured twin-engine C-8A "Buffalo" airplane originally produced by de Havilland Aircraft of Canada, Ltd., that had been modified as an experimental aircraft to demonstrate technology for quiet short-haul commercial airliners of the future with short takeoff and landing capabilities.

The remodeled aircraft had made its first flight July 6 at Boeing Field, Seattle, Wash., strictly to test airworthiness and not for high-performance takeoffs or landings. A failure in the lateral stability-augmentation system (an automatic system to keep the plane stable during flight) had caused initial unsteadiness. The pilots had switched off the system and had flown the plane manually, correcting the motion problem. Pilot Tom Twiggs commented: "I didn't see any surprises. The QSRA simulation at Ames was one of the most accurate simulations I've flown. The failure modes we studied in the simulator saved the day in the first few moments of the flight." The plane had reached its maximum altitude of 6500ft and a speed of 120knots, maximum allowed by engineers for the first flight.

As of July 20 the plane had completed 10 flights with a total flight time of 13hr, and had accomplished all the objectives of the Boeing pre-ferry flight program except the noise tests. NASA would expect the QSRA at

ARC in early August. (*ARC Astrogram,* July 27/78, 1; *AvWk,* July 17/78, 21; *Aerospace,* Summer 78, 14)

July 28: NASA announced that a water pump and grain mill powered by solar cells, which had been installed in a remote Upper Volta village, would soon relieve the daily burdens of the villagers. Under a cooperative agreement between the U.S. Agency for International Development (AID) and NASA, AID would purchase all hardware including the solar-cell array, pump, and grinder; LeRC would provide manpower and technical management to design, fabricate, install, and monitor the system. The project would last about 1.5yr—several mo to set up the system and a yr for experimental operation, after which the villagers would run the system.

Dr. Louis Rosenblum, chief of the Solar and Electrochemistry Division at LeRC, said the solar-cell power system, after its completion in Dec. 1978, would be able to supply a peak 1.8kw of electricity; the water pump would be able to supply 4542 liters (1200 gal) of water per day. "The grinder will cut the time required to produce flour from 2hr to 10 minutes per family," he added. The cooperative project, part of an AID study of energy needs in food systems of less developed countries, would also address the socioeconomic effects of reducing the time needed by the women to grind grain and draw water. "There are more than three million villages without electric power in the world today," Dr. Rosenblum continued. "Success of the experiment in Upper Volta is bound to influence the direction taken by other developing countries to satisfy the pressing energy needs of their large rural populations." (NASA Release 78-111)

• LaRC reported the White House had issued a statement July 20, 1978, commemorating the ninth anniversary of man's first walk on the moon. The statement read in part: "Nine years ago today, the world paused to watch two brave men tread the surface of the moon. It was a moment without precedent in human experience, a moment when terrestrial life reached out to touch another world. It is a source of pride for us that those men were Americans. Today, the lunar surface is criss-crossed in a half dozen places with the footprints of American astronauts and implanted with a variety of American scientific instruments.

"The Space Shuttle, our next major manned space project, will begin regular, routine economical operation in the early 1980s. . . . In the deeper reaches of space, we will continue to seek to expand our knowledge of the solar system and the universe of which we are a part.

"As time and technology take us ever more deeply into the space age, it will continue to be our policy to conduct operations in space as required for our national well being and to support the right of all nations to do likewise. In so doing, we remain committed to the underlying principle of the exploration and use of space for peaceful purposes and for the benefit of all mankind." (*Langley Researcher*, July 28/78, 2; *Spaceport News*, July 21/78, 1; *JSC Roundup*, July 21/78, 4; *DFRC X-Press*, July 28/78, 4)

July 30: The *Washington Post* reported that two Soviet cosmonauts had set a new Soviet space walk record on July 29, spending 125min outside the *Salyut 6* space station testing a new type of suit and working with equipment mounted on the station, the accompanying *Progress 2* supply ship, and their own *Soyuz 29*. Cosmonauts Vladimir Kovalenok and Alexander Ivanchenkov were the fourth team of cosmonauts to successfully board the *Salyut 6* after it reached orbit the previous Sept. Kovalenok had filmed his colleague during part of the walk and had relayed the color footage to earth for broadcast on Moscow television.

Tass, the Soviet news agency, said the two cosmonauts had dismantled instruments and devices on the outer shell of *Salyut 6*, loading them into *Soyuz 29* for shipment to earth, and had mounted new instruments, including devices to measure radiation, on the Salyut. Instruments being returned had measured micrometeor activities and had aided research on "polymers, optical and other structural material used in building advance spacecraft," Tass said. The report did not give details on the new space suit, but the cosmonauts said that their "semistiff" clothes were well designed and that life-support systems had been improved to facilitate movement and "better fixation outside the station."

U.S. experts had interpreted the length of the Salyut mission and repeated trips to it by Soviet cosmonauts as indicating a focus by the Soviet space program on building permanent orbital platforms. The report had not indicated when the two cosmonauts would return to earth, but observers suggested that the next mission (*Soyuz 30*) would probably come soon and would be a marathon stay aboard the orbiting station. (*W Post*, July 30/78, D10; FBIS, Tass in English, July 29/78)

July 31: NASA announced that its scientists had been studying radar images of the North American coast from *SeaSat 1*, first ocean-monitoring satellite, launched in June 1978 and now in a 805km (500mi) high polar orbit. One of the satellite's five microwave sensors, a powerful radar system called a synthetic aperture radar (SAR), could take pictures of earth's surface day or night under any weather conditions. A typical SAR operation had produced a continuous swath of radar images 97km (60mi) wide by 4023km (2500mi) long, extending from the Mexican west coast to Alaska. JPL, manager of the SeaSat project, had processed the images. Scientists from NASA, NOAA, the Natl. Environmental Satellite Service, the U.S. Navy, and USGS had received images of the

Arctic pack, the Gulf Stream off the east coast of Fla., the Caribbean off the northern coast of South America, and a hurricane zone near the Baja Calif. peninsula. (NASA Release 78-116)

• ESA announced that its *Geos 2* scientific satellite, launched from Cape Canaveral July 14 on a Delta and placed in geosynchronous orbit, had reached its scheduled position of 6°E on July 26. It would move during its 2-yr mission from its first operational station in geosynchronous orbit 35 900km above the equator through a zone extending from 0° to 37°E longitude. On July 26 the European Space Operations Centre at Darmstadt, West Germany, had switched on the S-band telemetry system and the magnetometer experiment, and would switch on six other experiments progressively until the satellite had become fully operational in early Aug. All data confirmed that *Geos 2* was functioning correctly. The satellite, positioned on magnetic field lines linked to the earth's auroral zones, could correlate those data with similar data acquired from associated auroral stations by ground-based balloon-borne or rocket-borne instruments. NASA had expected that the *Geos 2* data, correlated with data acquired by the already launched *Isee A* and *B* and by the ISEE-C scheduled for launch in Aug. 1978, would constitute a substantial body of information on the magnetosphere to aid in understanding the effect of interplanetary-space processes on the near-earth environment. (ESA Release July 31/78)

During July: The USAF announced it had launched a second prototype satellite in DOD's NavStar global-positioning system from SAMTEC facilities aboard an Atlas F booster. The 955-lb navigational satellite working with a previously orbited NavStar satellite would be part of the program's concept-validation phase using six satellites to test navigation capability at the Army's Yuma, Ariz., proving grounds. When fully operational in the 1980s, the 24-satellite NavStar system would provide precise worldwide navigation coverage 24hr a day. The Air Force's global-satellite control network, managed by SAMSO's Air Force Satellite Control Facility at Sunnyvale, AFS, Calif., would track and operate the satellite. (AFSC *Newsreview*, July 78, 12)

The USAF announced it had launched a third defense meteorological satellite from SAMTEC facilities at Vandenberg AFB on a Thor LV-2F into a near-polar sun-synchronous orbit 450mi above the earth. The 1131-lb satellite carrying visual and infrared instruments had joined two previously orbited spacecraft in providing high-priority weather forecasts to U.S. military commanders worldwide. The sensors could detect and observe developing weather patterns and could track existing weather systems over remote areas, including oceans.

After 30 days of on-orbit tests, the satellite was orbiting the earth every

101min, scanning a 1841-mi area on each pass and photographing the entire earth over a 12-hr period. It would transmit weather imagery in real time to ground and shipborne terminals around the world for military purposes and, through NOAA, to the civilian community. (AFSC *Newsreview*, July 78, 1)

• NASA announced that President Jimmy Carter had presented 1977 presidential management-improvement awards to Curtis Helms and Thomas Winstead of MSFC's structures and propulsion laboratory. The two were among 7 awardees, selected from 50 nominees, honored at a May 23 ceremony in the White House rose garden. Their redesign of the Space Shuttle external fuel tank, eliminating one of the two vent-relief valves, had reduced vehicle weight and program costs while still meeting venting requirements and reliability standards, resulting in total savings of $5.6 million. (*NASA Actv.*, July 78, 27)

• NASA announced it had made several personnel changes. Richard Smith, deputy director at MSFC, had been appointed deputy associate administrator for Space Transportation Systems at NASA Hq for a one-yr tour of duty beginning August 15. Smith had transferred to NASA in 1960 from the rocket research and development team at the Army's Redstone Arsenal, when the development operations division of the Army Ballistic Missile Agency had become the nucleus of MSFC. He had managed the Saturn program, and in January 1974 had become director of science and engineering, and later deputy director, of the center. Smith had received NASA's exceptional service medal for contributions to the Apollo and Skylab programs as well as the medal for distinguished service. (NASA Release 78-112; NASA anno July 19/78)

NASA had named Harry Sonnemann deputy chief engineer at Hq, responsible to the chief engineer for managing program assurance, safety and environmental health, and the systems engineering divisions. Sonnemann had come to NASA in October 1977 from the office of the assistant secretary of the Navy, where he was special assistant for electronics and special assistant for antisubmarine warfare and ocean control from 1968 to 1977. He had also worked on the development of large seismic arrays as assistant director for field engineering in DOD's Advanced Research Projects Agency's nuclear test detection office from 1964 to 1968. (NASA anno July 20/78)

Scientist-astronaut Dr. Joseph Allen had returned to active flight status at JSC's astronaut office. From August 1, 1975, until his return to JSC, Allen had been director of the office of legislative affairs at NASA Hq; and he would continue those duties in addition to his astronaut duties until a replacement was named. He had returned to JSC as senior scientist-astronaut and was eligible for selection as a Space Shuttle crewman. NASA had selected Allen as a scientist-astronaut in 1967; he had completed the initial academic training and a 53-wk course in flight training at Vance AFB, Okla. Allen had been mission scientist as a member of the astronaut-support crew for *Apollo 15*, and was staff consultant on science and technology to the President's Council on International Economic Policy. (NASA Release 78-33)

NASA announced it had appointed Dr. Edward Ifft chief of international programs policy at Hq. Dr. Ifft had been a U.S. Foreign Service reserve officer, whose last assignment was director of the State Department's office of disarmament and arms control. While a graduate student, Ifft spent a yr at Moscow State Univ. under a U.S.-USSR cultural exchange program doing research on low-temperature physics. He had joined the U.S. Arms Control and Disarmament Agency in 1967; then moved to the State Department where he had been deputy chairman of the U.S. delegation that negotiated the threshold test-ban treaty in Moscow in 1974; and had participated in negotiations with the U.S. and USSR for a comprehensive test ban. (*NASA Actv,* July 78)

• The USAF announced that its ESD Spacetrack System Office had awarded Bunker Ramo Corp. of Westlake Village, Calif., a $927 000 contract for equipment to record and store data for military and authorized scientific use on all man-made objects orbiting the earth. The system would record information received at the Space Computational Center in NORAD's Cheyenne Mountain Complex, Colo., on positions of the thousands of satellites and other man-made objects in space, and would store the data for 3yr in the computer's active file. It would also microfilm historical data for permanent storage. The new system had become necessary because the automatic data-processing equipment formerly used by the Aerospace Defense Command lacked permanent-storage capability and was very expensive to maintain. (AFSC *Newsreview*, July 78, 6)

August

August 1: NASA announced that Chimex Systems, Inc., of Houston, Tex., had received a subcontract from Computer Sciences Corp. of Falls Church, Va, for institutional computer systems engineering, development, and production operations under CSC's 5-yr contract with Johnson Space Center. A major factor of CSC's winning bid had been its plan to use minority subcontractors. This was the fourth NASA contract won by Chimex, which had provided keypunch services at Kennedy Space Center and engineering support and keypunch services at JSC. Under the new subcontract, Chimex would employ 40 people primarily in keypunch and data-distribution at an estimated $800 000 per yr for 5yr. NASA's minority business awards for the first half of FY1978 had amounted to more than $37 million. (NASA Release 78-117)

August 2: NASA announced it had turned over *Goes 3*, third in a series of geostationary operational environmental satellites, to NOAA for operation on July 5, 1978, at 1600Z. Since its launch on June 16, NASA engineers, wishing to ensure proper operation, had completed an orbital checkout in 14da, rather than the 30da required for previous GOES missions. Goddard Space Flight Center's newly installed GOES test laboratory had accelerated the schedule by allowing full-time communications with *Goes 3*; NOAA's Wallops Island facilities had been available to previous missions only part-time. Engineers had checked out *Goes 3* during its westward drift at 2 deg/day toward its operational station at 135°W, and had corrected the orbit inclination from 1.7° to the specified 1°.

First pictures from the satellite's visible and infrared spin scan radiometer (VISSR), received June 29, were of excellent quality. More than 52lb of the initial 80lb load of hydrazine was available, enough for more than 5yr of normal operations. A minor problem with the spacecraft data-handling equipment that relayed signals from earth-based data-collection platforms had resulted from interference by VHF equipment that could be turned off during most of the mission. The GOES laboratory would monitor operating data to detect any variables in performance with time and temperatures.

NASA had advised NOAA that, based on preliminary analyses, the spacecraft was operating under no constraints. According to NOAA's plan for final positioning, *Goes 3* would replace *Goes 1* at 135°W longitude; *Goes 1*, relocated over the Indian Ocean, would support the first GARP global experiment. (NASA MOR E-612-78-01 [postlaunch] Aug 2/78)

• Marshall Space Flight Center announced that its test engineers had successfully completed the first series of live firing tests of the Space Shuttle's main propulsion system, with all major test objectives achieved. Rockwell Intl.'s Space Systems Group had conducted the test program for MSFC. The test version of the propulsion system had consisted of three Shuttle main engines mounted on an orbiter aft fuselage with a flight-type external propellant tank and associated systems. The first series of tests (Apr. 21, May 19, June 15, and July 7, 1978) had recorded systems operation at thrust levels from 70 to 90% of rated power for up to 100 sec; in the final test, engines were throttled down from 90% thrust to 70% and back up to 90%. The Shuttle would be the first spacecraft with throttleable engines.

After removal from the test stand at the National Space Technology Laboratories, the engines had undergone modifications for more firings later in 1978. The next test series would run the engine at rated power levels (357 000lb, or 1.67 million newtons at sea level) for more than the 500sec required to lift the Shuttle into orbit. Results of the tests would verify main propulsion system operation before the first manned orbital flight, scheduled for 1979. (*Marshall Star*, Aug 2/78, 1)

• Work on the dynamics test stand had begun in preparation for the next series of Space Shuttle mated vertical ground-vibration tests, the *Marshall Star* reported. After changing the configuration inside the stand from a "hanging" to a "sitting" arrangement, MSFC engineers would adjust the platform to fit the vehicle assembly. For the first series of tests, the orbiter and external tank had been suspended from airbags on a truss structure high in the stand, to simulate the Shuttle after the solid-fuel rocket booster had separated and before external-tank jettison. The next series would test the ET, orbiter, and SRBs mated in the liftoff configuration; the SRBs would sit on a hydrodynamic support system to which the ET and orbiter had been attached. (*Marshall Star*, Aug 2/78, 1)

• GSFC reported that faint radio signals from 10 million billion miles away had given astronomers the first proof of the long suspected presence of methane (natural gas) outside the solar system. GSFC scientists had detected methane in three locations in deep space: in the Orion A gas cloud, and in gas clouds surrounding variable star RX Boo and carbon star IRC-plus-10216. "The discovery of methane outside our solar system could have an important role in determining the abundance of carbon in the universe," noted Dr. Kenneth Fox. Methane previously had been detected only on earth, Jupiter, Saturn, and its moon Titan, Neptune, and Uranus.

The discovery had confirmed that interstellar space was not devoid of complicated molecules; at least 40 types of molecules had been discovered suspended in space, and the findings were important to scientists attempting to construct models of the origin of the universe. Dr. Fox and Dr. Donald Jennings had searched for the emissions for 3yr using radiotelescopes at the Natl. Radio Astronomy Observatory at Greenbank, W. Va., Kitt Peak near Tucson, Ariz., and the Haystack observatory in Westford, Mass. "Methane now appears to be one of the most abundant molecules in Orion A, perhaps second only to hydrogen [the most abundant gas in the cosmos]," Fox said. Fox and Jennings planned to map the source of methane in Orion A for its precise location and any associations with stars or stellar objects. (*Goddard News*, Aug 2/78, 2)

- *Aerospace Daily* reported that the Senate Appropriations Committee had approved a FY79 $4.359 billion funding bill for NASA, restoring $20.5 million cut by the House-passed companion bill for development of the teleoperator retrieval system (TRS) that NASA wanted to use to reboost or deorbit Skylab. Although the full committee had not changed the NASA funding approved earlier by Sen. William Proxmire's subcommittee, Proxmire indicated that when the omnibus bill reached the Senate floor, he would move to cut all its appropriations, including NASA's, by 2%. The Appropriations Committee also approved a second FY78 supplemental funding bill that included $58.7 million to acquire the Navy's fourth and fifth Fleet Satellite Communications (FltSatCom) spacecraft, as recommended earlier by the defense subcommittee. (*A/D*, Aug 2/78, 138)

August 3: NASA announced it had selected astronauts Dr. Owen Garriott and Dr. Robert Parker to be mission specialists on Spacelab 1, scheduled for launch in the early 1980s. Mission specialists would be responsible for coordinating Shuttle operations with the commander and pilot in crew-activity planning, consumables usage, and Space Transportation Systems/payload interaction. The 7-day flight, a verification test of Spacelab systems and Spacelab/orbiter interfaces, would carry about 40 experiments. Garriott had flown on the second manned Skylab mission of 56da duration; Parker had been a mission scientist and spacecraft communicator during Apollo and Skylab.

Ten European nations, under agreements with the European Space Agency, had developed and financed Spacelab to be carried in the cargo bay of the Space Shuttle orbiter as a fully furnished laboratory adapted to the weightless environment of space and pressurized for working without space suits. Spacelab 1 would also carry two payload specialists (one European and one U.S. citizen) to operate the scientific instruments. (NASA Release 78-121; JSC Release 78-34; *JSC Roundup*, Aug 4/78, 1; *Marshall Star*. Aug 16/78, 3; *DSFC X-Press,* Aug 11/78, 3; *Langley Researcher*, Aug 11/78, 1)

• NASA announced it had selected four American scientists as payload specialists for the second Spacelab mission scheduled for 1981. Those named were Dr. Loren Acton of Palo Alto, Calif., research scientist at Lockheed's Palo Alto research laboratory; Dr. John-David Bartoe of Reston, Va., and Dr. Dianne Prinz of Alexandria, Va., both research physicists at the U.S. Naval Research Laboratory; and Dr. George Simon of Alamagordo, N.M., chief of the solar-research branch at the Air Force Geophysics Laboratory. Two of these scientists would be selected to fly on the orbiting space laboratory to operate the scientific experiments; the others would operate ground-based experiment equipment and assist the two in orbit. The Spacelab investigators' working group (IWG), composed of Spacelab 2 principal investigators having experiments on the mission, had selected the payload specialists, each of whom was a coinvestigator on one of the Spacelab 2 experiments.

Scientific investigations on Spacelab 2 would be mainly in astronomy, high-energy astrophysics, and solar-physics research, with others in plasma physics, botany, medicine, and space technology. Scientific instruments on the mission would be "pallet only," exposed to space in the orbiter's cargo bay. Payload specialists would operate their experiment equipment from the orbiter's crew cabin, working in shifts to run the experiments 24hr a day. The mission, scheduled for launch from KSC in 1981, would orbit at an altitude of about 450km (250mi) for 9da. MSFC had been responsible for payload specialist training as part of its overall management responsibility for the Spacelab mission. (NASA Release 78-120; *Marshall Star*, Aug 9/78, 1)

August 4: MSFC announced it had awarded Lockheed Missiles and Space Co., Sunnyvale, Calif., a $2.7 million contract to develop and deliver a flight experiment solar-array wing by May 1980 for a Shuttle orbital-flight test in November 1980. A significant step toward producing large amounts of power in space, the experiment would verify structural and dynamic characteristics of the solar-array wing, its electrical performance, and the readiness of solar-array technology for planetary- and earth-orbit Space Shuttle payload applications.

The solar-array wing, 32m (105ft) long and 4m (13.5ft) wide, folded and stored in the Space Shuttle's cargo bay during launch, would be extended to its full length and retracted several times during the test. When its solar-cell surface was fully extended, the array's 82 panels would convert solar energy to produce 12.5kw of power. Only three of the panels would be active in the experiment flight. The experimental solar-array wing would demonstrate that use of this technology to augment power for Shuttle and Spacelab could extend mission duration and lead to a solar-electric propulsion stage. Solar-array technology could have Shuttle-payload applications such as space-base construction, satellite power systems, and power modules. NASA's Office of Aeronautics and Space Technology had directed the solar-array program as part of a larger effort on solar-electric propulsion for long-term missions in the mid-1980s. (NASA Release 78-123; *Marshall Star*, Aug 9/78, 1; *DSFC X-Press*, Aug 11/78, 4; *A/D*, Aug 4/78, 151; *AvWk*, Aug 14/78, 15)

• Lewis Research Center announced that NASA Hq had joined with the FAA in sponsoring a 3-day aircraft-icing workshop at LeRC. Attended by more than 100 icing specialists, sensor/instrument specialists, meteorologists, and weather forecasters, the workshop had assessed fixed-wing and rotorcraft operational icing, evaluated facilities requirements for R&D and certification, examined ways to improve icing forecasts, and identified gaps in knowledge of aeronautical icing. LeRC had been chosen for the conference because its research on icing in the 1940s and 1950s was still important in aircraft certification; it also had the largest icing-study facility currently actively operating in North America. As a result of needs defined at the workshop, LeRC would begin a new program of research on icing. (*Lewis News*, Aug 4/78, 2)

• NASA reported that the Natl. Business Aircraft Association at its annual convention in St. Louis Sept. 13 would present to Dr. Richard Whitcomb, aeronautical scientist at Langley Research Center, its Meritorious Service to Aviation Award for 1978. Whitcomb would be honored for "significant contributions to the field of aviation." Among those contributions were the area-rule design (a "Coke bottle" airplane shape that reduced drag and increased aircraft speed without the need for additional power, incorporated in every U.S. supersonic airplane now flying); and the "supercritical" wing developed during the 1960s to increase the speed and range of subsonic aircraft without an increase in required power or fuel consumption. Whitcomb had begun his career in 1943 with the National Advisory Committee for Aeronautics (NACA), NASA's predecessor, and was head of LaRC's Transonic Aerodynamics Branch. (NASA Release 78-119)

August 7: *Av Wk* reported that the Defense Dept.'s high-energy laser program had been encouraged by a demonstration earlier in 1978 of the use of laser radiation to destroy a high-speed antitank missile in flight. The test, using a chemical laser built by TRW Systems under sponsorship of the Navy and the Defense Advanced Research Projects Agency (DARPA), occurred at a TRW facility in southern Calif. near San Juan

Capistrano; Hughes Aircraft Co. had supplied the laser-beam pointing/target-tracking system.

DOD officials cautioned that, despite encouraging progress, high-energy lasers would not begin to replace conventional weapons, at least in the near future, because the Pentagon could not consider working on a high-energy laser-weapon system before FY82, and that date could slip. The most probable near-term application would be close-in air defense for large ships such as aircraft carriers, whose value would justify the cost of a laser-radiation weapon. Another potential near-term application would be mobile short-range air defense of high-value Army targets (known as the high-energy laser-tactical defense system). If the Pentagon decided to develop a laser-radiation weapon system, it would be only one in a mix of air-defense weapons, because of what appeared to be intrinsic range and weather limitations of lasers. (*AvWk*, Aug 7/78, 14)

August 8: NASA announced that data returned by the American-British-European international ultraviolet explorer (IUE) satellite had suggested the existence of a massive black hole at the center of globular clusters in earth's galaxy. A group of scientists headed by Dr. Herbert Gursky and Dr. Andrea Dupree, both of the Harvard-Smithsonian Center for Astrophysics, Cambridge, Mass., had examined six clusters, three of them x-ray sources.

NASA, in cooperation with ESA and the British Science Research Council, had launched IUE in Jan. 1978 into a modified near-equatorial synchronous orbit to study a wide range of celestial objects in the ultraviolet region of the electromagnetic spectrum. Dr. Gursky said the onboard UV instruments had delivered unexpected information when they penetrated the background denseness of the clusters, which were like miniature galaxies 15 000 light-yr away, to observe the cluster core. What they saw, according to Gursky, was probably radiation from a group of 10 to 20 bright-blue stars orbiting the core: "These stars may well be orbiting a massive black hole the size or mass of one thousand solar systems," he said. However, existence of a black hole was not certain, as rotation of the stars in relation to the center of the million-star cluster had to be studied first. If the stars were indeed orbiting a massive black hole, Gursky believed they were right on the edge or, if not, might be providing their own gravitational equilibrium.

What surprised the observation team, Gursky said, was the ability of the short-wavelength instrumentation to cut clearly through the million-star cluster. "For the first time we are seeing in a clean way the center or core of those globular clusters, and we were surprised. You can see the clusters in visible light, there are lots of red giant stars there, which mask what is going on in the center. Now we have a tool, the IUE's shortwave ultraviolet, capable of going for the first time right to the core through the whole cluster," said Gursky. Although the team would continue to observe the six globular clusters, Gursky doubted their findings could be definite until NASA put the space telescope in orbit from the Space Shuttle in 1983. That telescope, using much more powerful instrumentation including short-wavelength ultraviolet, could study the blue stars in more detail. (NASA Release 78-122)

• In a report on Soviet space activities, *Defense/Space Business Daily* said *Progress 2* had crashed into the Pacific Aug. 4, 2da after it had undocked from the Salyut station where it had been for 25da. This Progress resupply spacecraft (a modified Soyuz) had not been designed as a recoverable reentry and descent vehicle.

The Soviets had orbited *Cosmos 1028*, believed to be another in a series of long-duration high-resolution reconnaissance/surveillance satellites; orbit was 182/272km, 67.1° inclination, 88.7min period. Another in the series, *Cosmos 905*, launched Apr. 26, 1977, had remained in orbit 30da and was recovered. Others in the series had been *Cosmos 758*, in orbit for 20da after launch Sept. 5, 1975; *Cosmos 805*, in orbit for 20da after launch Feb. 20, 1976; and *Cosmos 844*, in orbit 39da after launch July 22, 1976. (*D/SD*, Aug 8/78, 171)

August 9: NASA announced that *Landsat 3*, an earth-monitoring satellite equipped to measure earth's emitted and reflected radiance, had suffered a failure in its fifth (or thermal infrared) band. Launched Mar. 5 into a near-circular 917km (570mi) high polar orbit, *Landsat 3* had returned excellent imagery from its other four bands. NASA engineers believed the thermal-sensor problem had resulted from moisture buildup on a glass screen through which the instrument viewed earth. Efforts to correct the situation had been unsuccessful, and NASA had convened a special review board to examine the problem.

Scientists had hoped to use *Landsat 3*'s thermal data to detect crop stress and to recognize "heat islands" associated with urban and industrial development. However, NASA's heat-capacity mapping mission (HCMM), launched in Apr. into an orbit more favorable to thermal-infrared data collection, had been returning excellent day-night data that could serve that purpose. (NASA Release 78-124; *D/SD*, Aug 14/78, 203; *A/D*, Aug 23/78, 236)

• JSC and the U.S. Geological Survey jointly announced that the Survey had published a surface-fault map of the southeastern greater Houston area. JSC geologist Uel Clanton and Earl Verbeek of the USGS had created the map covering about 200 sq mi to evaluate the magnitude of faulting in metropolitan Houston. The map had displayed 91 faults with a total length of 110 mi, cutting through residential and commercial areas in many cities. Faulting was extensive east and southwest of Hobby Airport and in the Mykawa oil field in both northwesterly and northeasterly directions. The mapmakers had used a combination of aerial photography and ground-level confirmation of faults. (JSC Release 78-35; *JSC Roundup*, Aug 18/78, 1)

• Payload specialist candidates had begun training at MSFC for the late-1980 first Spacelab mission, the *Marshall Star* reported. The first 5-day session had covered training approach and plans. Trainees were Michael Lampton of the Univ. of Calif., Berkeley; Byron Lichtenberg of the Mass. Inst. of Technology; Ulf Merbold of W. Germany's Max-Planck Inst.; Claude Nicollier, a Swiss, from the European Space Technology Center; and Wubbo Ockels of Groenigen Univ. in the Netherlands. One American and one European would fly on the mission; the others would assist on the ground. Later in the training cycle, NASA would select those to fly. (*Marshall Star*, Aug 9/78, 2)

• The Senate had cut $5 million from NASA's $3.87 billion R&D budget for FY79 before passing a package appropriations bill that included the agency, *Aerospace Daily* reported. The Senate had first rejected by 44-43, then accepted by 45-42, an amendment offered by Sen. William Proxmire (D-Wis.) to cut NASA's budget and reduce budgets of five other departments and agencies by $805 million. Proxmire's move might have spared NASA a far deeper cut, because Sen. William Roth (R-Del.) had proposed a 2% across-the-board cut for all agencies and departments covered by the bill (which, for NASA, would have been $80 million). Proxmire had offered his more selective amendment as a substitute, to which Roth had agreed. The amendment did not specify where NASA should cut its R&D. (*A/D*, Aug 9/78, 169)

August 10: NASA announced that the first major symposium on crop monitoring based on space-age technology, scheduled for JSC Oct. 23-26, would discuss results of the Large-Area Crop-Inventory Experiment (LACIE). The symposium would give participants from government, industry-agriculture, and university communities around the world more information about LACIE's pioneering effort and how it could best be used to improve the world food situation. LACIE, a 3yr program of NASA and the U.S. Department of Agriculture working with university and industrial research groups, had been designed to estimate wheat production in major growing areas throughout the world by using satellite data and the global weather-observation network. LACIE had begun in the fall of 1974 when DOA had recognized a need for information on wheat-growing areas outside the U.S. Landsat satellites continuously scanning agricultural regions had provided the electronic imagery for area estimates. Combined daily with ground-acquired data and information received from 8000 worldwide weather stations, the imagery had made it possible to predict crop area, yield, and production of domestic and foreign wheat-growing regions and to give an early warning of problems.

Earth resources scientists had learned to identify the "signature" (appearance) of wheat in the satellite data; others had been able to estimate the growth stage and potential yield of wheat. Computer programs combining records of weather conditions in past growing seasons with figures on crop yields in the past had enabled experimenters to estimate yields for the current growing season. Results from LACIE had shown the effectiveness of new technology in improving knowledge of global wheat production generally applicable to other crops. USDA was considering use of the data source to give early warning of significant changes in prospects for global commodity production. (NASA Release 78-125)

- LaRC announced it had awarded two computer contracts: one to the Charles Stark Draper Laboratory, Inc., Cambridge, Mass., for an engineering model of a fault-tolerant multiprocessor (FTMP) computer. Draper Laboratory, which had designed the FTMP computer under a previous NASA contract, had intended it for safety-critical avionic and flight-control systems on future civil-transport aircraft. The FTMP could start and operate several aircraft functions (flight control, autopilot, navigation and guidance, and display and master alarms) without manual intervention and could also assess flight procedures including takeoff, climb, cruise, descent, and landing. The 2.5-yr contract would cost about $1.8 million.

The second contract award was to SRI International, Menlo Park, Calif., for an engineering model of a software-implemented fault-tolerant (SIFT) computer. Conceived by SRI under a previous NASA contract, SIFT would perform the same functions as FTMP. SRI would work under the $1.6-million 2-yr contract in Menlo Park and at a subcontractor facility in Teterboro, N.J. LaRC would manage both contracts. (LaRC Release 78-42)

- The USAF had budgeted $664 million for space procurement in FY80, up from $379.6 million in FY79, *Defense/Space Business Daily* reported. Major increases were in the defense support program (early-warning satellite by TRW and Aeroject), up $48 million; the satellite data system (Hughes), up $80 million; the defense satellite-communications systems (DSCS), up $65 million; and the Space Shuttle (Rockwell Intl.), up $73 million. The USAF had planned to buy five satellite systems in FY80: an RCA defense meteorological satellite program (DMSP) spacecraft; a defense support program spacecraft; a satellite data system spacecraft; and two DSCS spacecraft. One spacecraft to be purchased in FY79 would be a DMSP. The Air Force had last purchased boosters (three

Martin Marietta Titan IIIs) in FY78 and planned none for FY79 or 80, "assuming Shuttle stays on schedule," a USAF official said. (*D/SBD*, Aug 10/78, 188)

August 14: NASA announced it had flown its quiet short-haul research aircraft (QSRA) to Ames Research Center to begin a flight-research program on design and operation of future quiet short-haul transport. The QSRA [see July 27] had been the first to achieve truly quiet operation of jet aircraft, through high-performance special treatment of the engines and nacelles to restrict landing and takeoff noise to relatively small areas. Operating less noisily than most small private aircraft, the 4-jet QSRA had a 90-decibel noise-impact area, smaller than $1.3km^2$ ($0.5mi^2$). Scaled up to the equivalent of a 150-passenger aircraft, the 90-decibel noise-impact area would be less than $2.6km^2$ ($1mi^2$), compared to the approximately $78km^2$ ($30mi^2$) noise impact of comparable-size aircraft.

High performance of the QSRA resulted from a design concept in which four jet engines mounted on top of the wing directed fan air from the engines across the upper surface of the wing and flaps, significantly increasing lift, particularly at lower speeds. Compressed air from the engines, fed through an ejector system, provided boundary-layer control-blowing at the wing leading edges and ailerons, further enhancing lift and control. Future QSRA-derivative aircraft the size of the Boeing 727 transport could carry the same payload at the same speeds but could operate from small airports with short runways (about 1/2 to 1km [1500 to 3000ft], compared to current requirements for lengths of 1.6km [1mi] or more) and could operate without disturbing surrounding communities. (NASA Release 78-126; *DSFC X-Press*, Aug 11/78, 2; *AvWk*, Aug 14/78, 20)

• NASA announced that Dr. Lo I. Yin, a scientist at GSFC, had demonstrated his invention of a hand-held x-ray device at the annual convention of the National Medical Association in Washington, D.C. The Lixiscope (low-intensity x-ray imaging scope), a gun-shaped instrument weighing about 0.67kg (1.25lb), was powered by a single penlight 2.7v battery and a small radioactive source. Although Dr. Yin had not claimed his invention could replace x-ray equipment, he saw its value in mobility—in military situations or sports, for example. Dr. Yin had done this work under NASA's technology utilization program, aimed at identifying and transferring aerospace technology benefits to the general public. NASA had patented the Lixiscope, and the license would be available at no cost. (NASA Release 78-127)

GSFC reported that 100 representatives from U.S. industry and research institutions had met there to discuss possible applications of the Lixiscope, which had originated out of a need for single-photon imaging in x-ray astronomy. The sensitivity of the Lixiscope was so great that it could significantly reduce the dose of radiation received by a patient. Pulling a trigger on the device would unshield the radioactive source and send into the object being examined a low dosage of x-rays to be absorbed by a phosphor screen that would convert them to visible light. Six companies had applied for a license to manufacture or distribute the Lixiscope, not yet commercially available. NASA had estimated the cost of a mass-produced instrument composed entirely of existing technical components at about $5000. NASA was considering use of the device on the Space Shuttle. (*Goddard News*, Aug 16/78, 1)

August 15: Dr. Christopher Kraft, Jr., JSC director, had invited 31 former astronauts to take part in technical briefings and updates on the status of NASA programs at JSC Aug. 21-22, NASA announced. NASA Administrator Dr. Robert Frosch would address the group; John Yardley, associate administrator for Space Transportation Systems, would review the Shuttle program. Others giving briefings would be Robert Thompson, manager of the Space Shuttle program; Aaron Cohen, manager of the Orbiter project; Glynn Lunney, manager of the Shuttle payload integration and development program; Lt. Gen. Thomas Stafford, former astronaut now deputy chief of staff for research, development, and acquisition, on the USAF Shuttle program; John Young, chief of the astronaut office; and flight crews of the orbiter approach and landing tests (ALT). Those attending would take conducted tours of the crew systems area, remote manipulator system, full-scale orbiter mockup, and the orbiter aeroflight simulator/Shuttle mission simulator.

Astronauts invited were Edwin Aldrin, Jr.; William Anders; Neil Armstrong; Frank Borman; M. Scott Carpenter; Gerald Carr; Eugene Cernan; Michael Collins; Charles Conrad, Jr.; L. Gordon Cooper; Walter Cunningham; Charles Duke, Jr.; Donn Eisele; Anthony England; Ronald Evans; John Glenn; Richard Gordon, Jr.; James Irwin; James Lovell, Jr.; James McDivitt; Edgar Mitchell; William Pogue; Stuart Roosa; Walter Schirra, Jr.; Harrison Schmitt; Russell Schweickart; David Scott; Alan Shepard, Jr.; Thomas Stafford; John Swigert, Jr.; and Alfred Worden. (NASA Release 78-128; JSC Release 78-37)

August 16: GSFC reported that two Voyager spacecraft heading for Jupiter had discovered low-frequency radio waves naturally emitted from the planet behaving in a manner directly opposite to that of its high-frequency emissions. The Voyagers, now slightly more than halfway to their 1979 Jupiter encounter, had looked back at the earth for new clues to the origin of natural low-frequency radio waves flooding into space from earth. "The polarization plane of the low- and of the high-frequency radio waves from Jupiter seems to be exactly opposite," said Joseph Alexander, one of a team of GSFC scientists working on Voyager experiments. Observers could distinguish the source of the waves from the polarization plane (within which radio waves vibrated as they traveled through space) according to whether it varied randomly, systematically rotated clockwise or counterclockwise, or stayed constant. The Voyager spacecraft had permitted first-time measurement of the polarization plane of low-frequency radio waves, because the earth's ionosphere had heretofore shielded the waves from ground telescopes. Scientists for years had observed the high-frequency waves, not similarly blocked.

The GSFC team called the new data from Jupiter more puzzling than enlightening, with no explanation of why the waves differed so radically. One theory was that the two kinds of waves originated from entirely different kinds of processes in Jupiter's atmosphere; another theory suggested the same emission mechanism for both, with waves generated from Jupiter's southern hemisphere rotating clockwise at low frequency and those from the northern hemisphere turning counterclockwise at high frequency. As the Voyagers moved closer to Jupiter, the team hoped to find new clues by studying the planet's response to the solar wind. (*Goddard News,* Aug 16/78, 1)

• GSFC reported that several Goddard-controlled satellites had observed the largest solar flare in recorded history July 11. The observations had furnished a new data base for scientists investigating the causes of solar flares and how they affected the earth. Explosions of solar flares on the sun's surface had generated huge shock waves in the solar wind (ionized particles flowing at supersonic speeds over the earth and other planets) and had blasted into it new free-flying highly energetic particles that disorganized interplanetary magnetic fields. The flare had disrupted earth's ionosphere and had blacked out shortwave radio communications for 2hr throughout the entire sunlit hemisphere.

Two interplanetary monitoring platforms (IMPs) orbiting the earth had recorded the interplanetary magnetic field disturbances and the intensification of particle flow from the sun. *Helios 1,* a sun-orbiting satellite observing the sun from the opposite side of the earth July 11, had recorded changes in and composition of the radio noise, plasma, magnetic fields, and energetic particles streaming from the sun. The flare had saturated the high-sensitivity detectors on orbiting solar observatory *OSO 8,* which had previously obtained data on smaller flares associated with the large-flare region. *Atmosphere Explorer C* had noted atmospheric heating from solar-wind particles deposited on the earth at high latitudes. The international sun-earth explorers A and B *(Isee 1* and *2)* had obtained data on the accompanying magnetic fields and energetic particles, and *Applications Technology Satellite 6* had recorded data on high- and medium-energy plasma particles. (*Goddard News*, Aug 16/78, 2)

• GSFC reported that scientists using astronomy satellite *Sas 3* had pinpointed just beyond earth's galaxy the nearest of some 600 quasars found in the universe so far. The object, first discovered a yr ago, had never been identified as a quasar; besides being the nearest, the new quasar (named 0241) had been the third in recent mo found to radiate x-rays. The quasar had been 800 million light-yr away, obscured by thick dust in the galaxy and overlooked previously. Scientists hoped the nearby quasar might explain the mysteries of quasars' origin, either as eruptions in the centers of galaxies or as long-dead objects associated with the beginning of the universe, whose light was only now reaching the earth. (*Goddard News*, Aug 16/78, 4)

• MSFC announced it had awarded to Mass. Inst. of Technology a $65 000 contract to study the economic and environmental advantages of large space-system material delivery and construction from extraterrestrial sources. The 7-mo study, assuming previous mining and preprocessing of the materials on the lunar surface, would consider final space processing to produce commercial-grade materials for manufacturing, and would explore manufacturing processes required to produce large space-system components.

Past studies had postulated economic and environmental advantages to the use of lunar materials for large space systems because such structures would require tremendous quantities of materials. One satellite power-system structure, for example, could occupy a 3mi^2 area in space, and a large number of systems would be required to provide power to U.S. cities and towns. Materials mined and preprocessed on the moon and refined and manufactured in space would reduce the need to ferry large quantities of material from earth into space, avoiding the transportation costs and and potential environmental impact of heavy-lift launch vehicles passing through earth's atmosphere. (*Marshall Star*, Aug 16/78, 2)

• *Aerospace Daily* reported that *Progress 3*, launched by the USSR to dock with *Salyut 6*, had carried 617lb of food, 992lb of oxygen-regeneration equipment, and 41gal of water to cosmonauts Kovalenok and Ivanchenkov, according to cosmonaut Konstantin Feoktistov, one of the designers of the Progress resupply vehicles. UPI quoted a *Novosti* report that consumption of food, water, and air by a 2-man space crew would range from 44 to 66lb per day; a 2-man space station would need about 10tons of these supplies to stay in full operation for a year. (*A/D*, Aug 16/78, 202)

August 17: ESA's *Geos 2* had apparently suffered a collision in space that did not affect its payload performance but had badly disturbed experiments studying the low-energy plasma, *Nature* magazine reported. *Geos 2* had become fully operational Aug. 2, with all onboard instruments functioning well; prospects for a full 2-yr. mission had been excellent until 07.13 GMT Aug. 5, when UHF telemetry temporarily shut off. The fault had seriously affected the low-energy plasma experiments: the electrical-reference potential of the spacecraft appeared to have jumped 12v for half of each 6-sec spin cycle. This seemed to be the result of slight damage to a solar-array panel that apparently shorted the output of a string of cells to the structure whenever they became illuminated. The design of *Geos 2* had attempted to eliminate differential charging; the conductive indium oxide coating of the solar cells was supposed to offer a shorting path in the case of mechanical damage.

The Space Shuttle might some day serve as a vacuum cleaner to gather the many manmade objects floating in space. In the meantime, the geostationary corridor had rapidly become crowded, and spacecraft designers would have to consider the probabilities and consequences of space collisions. (*Nature*, Aug 17/78, 631)

August 18: *Pioneer Venus 2*, launched from Cape Canaveral Aug. 8 on an Atlas Centaur, would take about 5mo to arrive at Venus, encounter being predicted for Dec. 9. The second craft of this mission was referred to as the multiprobe bus. *Pioneer Venus 1*, launched previously, was an orbiter that would study the Venus atmosphere and other planet characteristics for 243da.

In the multiprobe phase of the mission, a large entry probe would make detailed soundings of the lower Venus atmosphere and clouds, while three small probes descending through the planet's atmosphere measured atmospheric conditions at widely separated points during a 60min effective life before impact. NASA had designed the 30 experiments aboard *Pioneer Venus 1* and *2* as a coordinated observation system; the mission would use six spacecraft—largest number ever devoted to one planet—and would make the most measurements at the greatest number of locations.

The multiprobe spacecraft consisted of a spin-stabilized 2.4m (8ft)-diameter cylindrical bus weighing 904kg (1990lb) and carrying 51kg (116lb) of scientific instruments. A thermal-controlled compartment in the cylinder would house instruments, communications, and data-handling systems, as well as navigation, orientation, thruster, and power systems. The large probe and three small probes would be launched from the bus, which was covered with solar cells.

The high density, high temperatures, and corrosive constituents of the Venus atmosphere had offered designers of the entry craft a difficult problem, complicated by high entry speeds of about 41 600km/hr

(26 000mi/hr). Also, all instruments in the probe pressure-measuring vessels required either observation of or direct-sampling access to the hostile atmosphere. (*JPL Universe*, Aug 18/78, 1; *Spaceport News*, Aug 4/78, 1; *Lewis News*, Aug 4/78, 1; *AvWk*, Aug 17/78, 23)

• JSC announced it would issue an RFP for management-consulting firms to answer questions such as: Would controlling 40 to 60 Space Shuttle flights per yr limit JSC's ability to perform its basic R&D job? How could JSC keep costs low to encourage maximum use of the space transportation system? How would increased Shuttle activity affect JSC relationships with other NASA centers and contractors? If Shuttle launches reached the expected rate of one or two per wk by 1985, JSC officials wanted to be sure these operations did not deplete already limited personnel and other resources JSC would need to remain a major R&D arm of the space agency.

JSC planners had drafted an operations plan for the full-use era of the Shuttle, whose onboard capabilities had exceeded those of previous spacecraft. The plan would reduce the number of flight controllers now assigned to real-time ground operations, and would increase the effort put into flight planning. Because JSC flight control work would affect NASA-wide operations and JSC would have large responsibilities in future spaceflight R&D, center management wanted an outside examination of the plan and alternatives before making a commitment. (*JSC Roundup*, Aug 18/78, 1)

• KSC reported that Joseph Malaga, center director of administration and management operations, had accepted appointment as vice chairman of the 16th Space Congress to be held Apr. 25–27, 1979. Sponsored annually by the Canaveral Council of Technical Societies, the congress had originated as a weekend seminar for engineers and technicians and had evolved into an event drawing speakers and delegates from technical communities throughout the world. Malaga would also be NASA's senior representative to the Congress. (*Spaceport News*, Aug 18/78, 1)

• The remote location of the White Sands Test Facility (WSTF) just east of Las Cruces, N.M., had made it ideal for testing space-propulsion and power systems and for investigating the behavior of propellants and other hazardous chemicals, said the *JSC Roundup* in a feature article. "We have no environmental limits for testing Shuttle programs," said WSTF Technical Manager Louis Gomez. "We recently added seven square miles of buffer along the western boundary of the facility as we saw various types of land use creeping our way from Las Cruces." Valued at $75 million in 1973 dollars, WSTF had been designed to operate with 1000 people; current facility population was 670, of which 74 were NASA employees, most of the others being Lockheed Electronics Co. employees on a facility-maintenance and test-operations contract. Rockwell and McDonnell Douglas also had employees at WSTF for orbital-maneuvering and attitude-thruster testing.

Currently on the test stands were a version of the orbital-maneuvering subsystem pod and the orbiter's forward and aft reaction-control subsystems. OMS engines had been rated at 6000-lb thrust. Steam ejectors that could produce a vacuum of 120 000ft-equivalent altitude in two of the propulsion test chambers would simulate engine firing in a near-space environment. Liquid oxygen and alcohol fed to three modified X-15-type rocket engines would produce a plume quenched with water, making high-velocity steam to aspirate the altitude chambers. Another lab would burn, stretch, pound, heat, and immerse complex samples of materials and substances that were candidates for use in space, to record how the materials responded to such punishment. Some tests were on flame propagation in pure oxygen at varying pressures; others might be for flashpoints and impact ignition.

When the tracking and data relay satellite (TDRS) designed to work with the space transportation system (STS) would arrive on station in geosynchronous orbit, a new ground station at WSTF would collect data and voice transmissions to and from orbiters and other spacecraft. Three 60ft-diameter Ku-band TDRS antennas and a control center/support building were under construction there. WSTF would relay Shuttle data and voice to other geosynchronous satellites; the transmission would bounce back from an adjacent RCA ground station to the two 30ft-diameter RCA dishes at JSC, then to mission-control at Houston. Western Union would employ about 100 people at the ground station after it had become operational in 1980. (*JSC Roundup*, Aug 18/78, 4)

• NASA's Office of Legislative Affairs had reported on a letter from the National Taxpayers Union to Sen. James Abourezk regarding S.2860, a bill that would commit the U.S. to the development and demonstration of solar power satellites. The letter read in part:

" . . . Passage of this bill would commit the nation's taxpayers to $25 million . . . in fiscal year 1979. Conservation estimates indicate capital investment reaching a . . . price tag of $2.5 trillion! The proponents of this legislation indicate that passage would commit the nation to a program plan to study the feasibility of utilizing solar energy to generate electricity for domestic purposes. This is not the first phase of this study. We respectfully point out that such a program has already been underway for sometime by the Department of Energy and NASA. . . .

"Boeing estimates that the cost of a single launch vehicle would be $10 billion alone while the cost per satellite is in the $20.5 billion neighborhood. These are just a sampling of the costs in terms of dollars;

they do not reflect the potential dangers to the environment that are inherent in this program. While the use of solar energy is always viewed as benign, this bill would spawn a . . . malignancy on the nation's already fiscally ill taxpayers . . . " (NASA Legis Actv Rept. Aug 22/78)

August 22: NASA announced it had officially designated Space Shuttle orbital flight tests (OFT) as Space Shuttle 1 (SS-1), Space Shuttle 2 (SS-2), etc. Following orbital flight tests, the same designation system would continue for Space Transportation System operational flights. For example, if the OFT phase had consisted of six flights, the next flight (first flight of the STS operational phase) would be Space Shuttle 7 (SS-7). John Yardley, associate administrator of STS, had requested that emphasis be placed on a gradual conversion to the new official designation system. (NASA anno Aug 22/78)

August 23: NASA announced that two radio astronomers at the Jet Propulsion Laboratory had discovered major changes deep in the atmosphere of Uranus. Dr. M.J. Klein of JPL and Dr. J.A. Turegano, a visiting research associate from the Univ. of Zaragoza, Spain, found that radio emissions from Uranus had become 30% stronger over the past 10yr. Klein and Turegano, who had used NASA's 64m (210ft) radio antenna at Goldstone, Calif., explained that radio waves emanating from dense clouds on Uranus had originated deep in the atmosphere, where pressures might be more than 10 times greater than at earth's surface.

Observations in the spring of 1978 had confirmed the two astronomers' discovery a yr earlier; the atmosphere of Uranus could be either warming or becoming clearer to the passage of radio waves, although the investigators doubted that temperature so deep in a planet's atmosphere could warm more than 30% in only 10yr. They thought the change was more probably caused by the planet's unique orientation: Uranus, unlike any other planet, spun on its side as it orbited the sun, and every 84yr (the length of one Uranus yr) the sun would shine directly on the North Pole. Forty-two yr later, the northern hemisphere would be dark with the South Pole pointed sunward.

Ammonia gas in the planet's atmosphere probably blocked radio emissions from Uranus's deep interior. The scientists suggested that ammonia might have been preferentially depleted in the planet's polar regions, either by convection currents or chemical reactions. If so, then the radio telescopes would be registering hotter temperatures deeper in the atmosphere than those recorded 10yr previously when the planet had a different orientation to the sun. Theoretical studies to discover the relationship of the radio measurements to global changes in Uranus's climate, weather patterns, and atmospheric chemistry would help in drawing general conclusions about planetary atmospheres applicable in turn to specific studies of earth's atmosphere. (NASA Release 78-130)

• ESA and INTELSAT had prepared a joint offer of Marecs (formerly Marots) satellites, plus maritime packages on Intelsat V comsats, for an international joint-venture attempt to set up a pre-Inmarsat maritime-communications system for the early 1980s, *Aerospace Daily* reported. The joint venture of Europe's 17-nation Eutelsat group and other countries, including Japan and the Soviet Union, had accounted for about 74% of Inmarsat shares. The U.S., with 17%, had not participated.

At its July meeting, the joint group indicated preference for a space system consisting of three Marecs satellites dedicated to maritime communications, plus maritime packages to be added to the last three of the seven planned Intelsat Vs. If costs or other factors made the option unsatisfactory, the joint venture had two others: a four-Marecs system, or a combination of two dedicated Intelsat spacecraft and three Intelsat V maritime packages. (An option using four dedicated Intelsat satellites was no longer under consideration.) INTELSAT's board of governors at its mid-Sept. meeting would decide on pricing and other details, and the joint venture would receive the ESA-INTELSAT proposals at its late-September meeting. (*A/D,* Aug 23/78, 234)

August 24: NASA reported it had launched *Isee 3* from ETR on a Delta rocket at 11:12am EDT Aug. 12, 1978. Eleven days into the mission, the spacecraft had traveled approximately 900 000km in a transfer trajectory out from earth to a halo orbit about the sun-earth libration point. All scientific instruments had been activated and were operational; all spacecraft appendages had been deployed except the $+/-$ Z-axis antennas, which had partially deployed.

The international sun-earth explorers (ISEE) project, a NASA/ESA joint contribution to the International Magnetospheric Study (IMS), consisted of two missions using three spacecraft to study solar-terrestrial relationships, NASA providing ISEE-A and -C and ESA providing ISEE-B. NASA had launched ISEE-A and -B (*Isee 1* and *2*) on Oct. 22, 1977, on a single Delta vehicle into the same highly elliptical earth orbit, where they were both operating. *Isee 3,* just launched toward a heliocentric orbit near the sun-earth libration point, had carried a scientific payload of 16 detector systems and associated electronics provided by 12 principal U.S. and European investigators to measure the solar wind and its fluctuations in detail.

Whereas ISEE-A, -B, and -C were capable of contributing individually to scientific knowledge, the greatest return would result from acquiring data from all three spacecraft: ISEE-C in its heliocentric orbit would measure the sun's input function unperturbed by the earth's influence, while ISEE-A and -B were measuring the effect of this input on the region surrounding the earth. At the time the ISEE-C entered its halo orbit in November 1978, it would become the world's first libration-point satellite. (MOR S-862-78-03 [postlaunch] Aug 24/78, [prelaunch] July 31/78; *Marshall Star*, Aug 30/78, 2; *Spaceport News*, Aug 18/78, 1; ESA release Aug 10/78; *Nature*, Aug 17/78, 630; *D/SD*, Aug 17/78, 630; Aug 14/78, 199)

• NASA had decided to launch its Venus-orbiting imaging radar (VOIR) late in 1984 rather than mid-1983 to reduce schedule risk, *Aerospace Daily* reported. Officials said the change, not meant to enhance VOIR's prospects as a new-start candidate for FY80, would reduce the program's first-yr and peak-yr funding needs. A 1984 launch funded in FY80 would permit Phase-B definition studies of the mission's primary instrument—a synthetic-aperture radar (SAR)—to precede those of the spacecraft by about 6mo and to produce data for use in the spacecraft work. Work on both the SAR and the spacecraft development would begin simultaneously in the second quarter of 1981. Although the Dec. 1984 launch opportunity was less favorable than that of May-June 1983, officials considered it acceptable and weight margins, based on completed Phase-A studies, adequate. (*A/D*, Aug 24/78, 241)

August 25: The gondola of Double Eagle II, first balloon to successfully cross the Atlantic, remaining parts of its balloon, and some of its equipment would soon become an exhibit at the Smithsonian's Air and Space Museum, the *Washington Star* reported. The U.S. embassy in Paris had notified the museum that the three balloonists wanted the Air and Space Museum to be the repository of their craft. They had planned also to present documents and onboard equipment that had not been jettisoned, as well as the balloon envelope (the shell containing the gas).

While the balloon was still aloft, Melvin Zisfein, acting director of the museum, had cabled the U.S. embassies in London and Paris to approach balloonists Maxie Anderson, Larry Newman, and Ben Abruzzo on the Smithsonian's behalf. When Lindbergh was still in the air on his historic flight, Paul Garber had sent a cable to the airfield at Le Bourget asking for the plane, the Spirit of St. Louis, which had become one of the museum's premier exhibits. "I thought it would be nice to continue in that tradition," Zisfein said. (*W Star*, Aug 25/78, A-6)

• NASA announced it had appointed Terence Finn director of legislative affairs. Finn had served since Dec. 1974 as senior analyst for energy, science, and space on the staff of the Senate budget committee, where he had been responsible for the budgets of NASA, National Space Foundation, and DOE. He was teaching a graduate course, Congress and

Budget, at Catholic Univ., and also had taught at American Univ. and the Dept. of Agriculture Graduate School. Finn was a member of the American Political Science Association and a life member of the American Aviation Historical Society. (NASA Release 78-131)

• The *Langley Researcher* reported that, under a program sponsored jointly by NASA and the American Society of Engineering Education (ASEE), a team of 19 university professors and two legal interns had spent the summer at LaRC on a study, An Integrated Air-Cargo Transport System. The team had reviewed forecasts of the air-cargo industry and had recommended changes in aircraft, air-freight terminals, air-transportation networks, and government regulation, to best serve national needs between 1990–2000.

Recommendations had included designing a new medium-range aircraft for cargo service and a long-range aircraft that constituted an improved or derivative version of today's craft; arrangements for all aircraft and major freight terminals to handle 20ft intermodal containers; continuation of air-cargo industry deregulation for at least the immediate future; and permission for airlines to own interstate trucking companies. Dr. Griffith McRee of Old Dominion Univ. had directed the study. (*Langley Researcher,* Aug 25/78, 4)

• Dr. Donald Hearth, LaRC director, had accepted an invitation to serve on the board of governors of the Natl. Space Club, the *Langley Researcher* reported. The club, a nontechnical nonprofit organization, had been founded in 1957 "to stimulate the advancement and application of space flight and related aerospace technologies for the benefit of all mankind." It had sponsored luncheons with speakers from the aerospace community and had granted scholarships for postgraduate and doctoral studies in the aerospace sciences. The board of governors would offer advice and counsel and annually nominate and elect the recipient of the Goddard trophy, premier award of the aerospace community. (*Langley Researcher,* Aug 25/78, 3)

August 28: Soyuz 31, carrying East Germany's first cosmonaut and a Soviet commander, had docked with the *Salyut 6* space station, the *Washington Post* reported. The USSR had launched *Soyuz 31* Aug. 26 at 17:51 Moscow time. Sigmund Jähn of East Germany and Valery Bykovsky had joined the *Soyuz 29* cosmonauts who had been circling earth in the space station since June 16. Tass said the crew was well and had begun implementation of tasks assigned to them. It added that the launch of *Soyuz 31* had been part of the preparations for an historic event, the 30th anniversary of the creation of the first German socialist state. The flight had been intended as an example of the use of space for peaceful purposes, based on "disinterested international cooperation and comradely mutual assistance." (*W Post*, Aug 27/78, A 21; FBIS, Tass in English, Aug 26/78)

August 29: Wallops Flight Center announced that an attempt to recover in midair an experiment launched on a single-stage meteorological rocket had been successful. The experiment (to obtain vertical profiles of nitric-oxide concentrations in the middle and upper atmosphere between 50 and 30km up) would later provide in situ measurements for use by Nimbus-G satellites from launch sites in the northern hemisphere. The payload launched Aug. 29 on a Super Arcas rocket at 3:38pm EDT had reached an altitude of 66km (216 000ft); the 8.7kg payload ejected at apogee deployed the parachute immediately as programmed. Radars at WFC had vectored the fixed-wing Sky Van aircraft to an intercept position to snare the package in midair about 4km (13 000ft) up and about 13km (7 nautical mi) offshore from Wallops Island. (WFC Release 78-13).

August 30: Preliminary results of a yr-long energy-conservation study at LaRC had indicated that a family could save money and energy using new technical systems developed in the space program, NASA announced. A family of four that volunteered to live in a conventional house equipped with unconventional technical systems designed to save energy and water, had achieved the following results: total energy used for all purposes, including heating and cooling, was a little less than half the amount used by a conventional all-electric home under the same conditions. Total dollar savings from use of less energy and water had amounted to more than $1200 for the yr, an average of more than $100 a mo.

The Technology Utilization House (called Tech House) was a contemporary-style home built at LaRC to test equipment available or to be shortly available. The family of Dr. Charles Swain, professor at Florida State Univ. in Tallahassee, had volunteered to live in the house for a year to test the technical systems under normal living conditions. Swain, his wife, daughter of 18, and son of 13 had moved into the house in August 1977 to live as normally as possible, while many of their actions were monitored, sensed, measured, and turned into computer printouts of data numbers.

Early research results had indicated that all the systems incorporated in the Tech House had been put to use. Although some systems worked better than predicted, and others did not work as well, the technical information collected would be useful in redesigning or modifying some of the systems to make them more efficient. Analysis of the experiment data would tell engineers much more about how to customize systems to fit particular kinds of houses, to suit areas of the country with different geographical and climate conditions, or to meet special needs of individuals. One possible "sleeper" system incorporated in the Tech House—the water recycling equipment—might greatly benefit future homeowners if the U.S. continued to experience periodic water shortages.

Preliminary performance data from some of the major systems indicated that thermal-design features such as improved insulation and double-door entries had accounted for about 60% of energy savings. Solar collectors on the roof and a solar-supplemented heat-pump system had supplied more than 50% of the energy required to heat the house, amounting to about $300 in energy-cost savings. The Tech House, funded by NASA's Office of Technology Utilization, had been completed in June 1976 and opened to the public for a yr before the Swains moved in. (NASA Release 78-135)

• NASA announced that the United Nations would use the world's most powerful comsat, the *Cts* communications technology satellite operated jointly by NASA and the Canadian Dept. of Communications, in a demonstration of remote simultaneous interpretation of a conference and of transmission by facsimile of conference documents for remote translation. Cooperating in the demonstration would be ComSatCorp and ENTEL, an Argentine state-owned telephone company.

The UN Conference on Technical Cooperation Among Developing Countries currently meeting in Buenos Aires, Argentina, through Sept. 12 would be the demonstration subject. Pictures and voices of the conferees sent to a portable NASA terminal at UN Headquarters in New York would be interpreted, translated into five official UN languages, and returned through *Cts* to a portable ComSatCorp terminal in Buenos Aires, where conference attendees could select the language of their choice. The New York terminal, a NASA-developed portable earth terminal bus, was a mobile broadcast-receive studio with a 2.4m (8ft) dish antenna designed for use with *Cts*. Another phase of the experiment would be high-speed simultaneous transmission on other *Cts* channels of texts of speeches and other documents from Buenos Aires to UN. Hq for translation and return the next day. *Cts* earth terminals could transmit at a frequency of 14GHz and receive at 12GHz. (NASA Release 78-134)

• FAA announced it had awarded a $1 989 193 contract to Bendix Corporation's Communications Division, Baltimore, Md., for prototypes of an advanced microwave landing system (MLS) offering precision all-weather approach-and-landing guidance at large airports. Called the "basic wide" MLS because of its antenna system, it would provide more precise guidance than simpler lower-cost versions already demonstrated at locations around the world—the small community airport and "basic narrow" versions adequate for most commercial fields. The new equipment, installed by FAA at NASA's Wallops center, would undergo tests for approximately 6mo as a joint FAA-NASA-USAF effort. The International Civil Aviation Organization had adopted the U.S./Australian-developed MLS earlier in 1978 as the standard approach-and-landing guidance system of the future [see April 24]. Actual installation and use of the equipment at airports worldwide would follow a transition phase demonstrating MLS-unit operation in the field. (FAA Release 78-71)

August 31: DOD's top command, control, and communications official, Dr. Gerald Dinneen, had characterized the use of satellite-borne lasers to communicate with submarines as costly and inefficient, *Aerospace Daily* reported. In testimony before the House Committee on Armed Services, Dinneen had said such a scheme would require "a number of satellites." Spreading out a laser beam would lose energy, so the beam would have to be relatively narrow to maintain contact with an antenna on the surface of the water. This would require either a large number of satellites, or scanning the beam around the surface of the water to communicate with other submarines. Dinneen had downplayed the idea of piggybacking blue-green laser kits on satellites (suggested as a possibility by 1981 or 1982) because very few satellites, communications or navigation, had excess capacity: "If you are going to put something on," he added, "you have to take off something else." (*A/D*, Aug 31/78, 276)

During August: The USAF reported that *NavStar 2*, within seconds of reaching orbit after its May 13 launch, had been tracked by one of SAMSO's stations. "About one hour after launch we made initial contact with the satellite," said 2nd Lt. Joe Fury, satellite operations controller. "Controllers then tested the satellite's systems to assure their operation before beginning other on-orbit checks to make the spacecraft function." In its initial 6-hr orbit, the NavStar had "talked" with ground controllers around the world, each of whom fine-tuned the satellite as it passed overhead.

Two days after launch *NavStar 2*'s internal rocket motor, activated by a command from a satellite controller, had shifted it from an egg-shaped orbit into a circular 12-hr orbit; as it arrived there, the NavStar was spun up to 89rpm to stabilize it. For 1.5da, controllers sent computer-coded signals to gradually slow the spin rate to zero. After stabilization, controllers had deployed the two folded solar-array panels and began to warm up the navigation payload to receive computer data transmitted to the satellite, giving it the intelligence it needed to carry out its military mission. Three wk after launch, controllers had positioned *NavStar 2* in a 10 900-nautical-mi drift orbit, slowly moving it toward the final position most suitable for the NavStar program. By the mid-1980s the NavStar system of 24 satellites would give users highly accurate data on position, velocity, and time in all weather, anywhere in the world. (AFSC *Newsreview*, Aug 78, 3)

• The USAF reported that the E-3A airborne warning and control system (AWACS), "leader" of the USAF command and control structure, had arrived at Wright-Patterson AFB for low-level flight tests conducted by the 6750th Aerospace Medical Research Laboratory to evaluate effects of aircraft noise. Tests with the E-3A would complete the 1978 measurement phase of NOISEMAP, a computer program to assess environmental impact of aircraft noise that had included the F-15, A-10, C-141, and KC-135. The assessments would serve in making alternative aircraft assignments and flight schedules, deciding on appropriate noise suppressors, and discovering potential noise problems for aircraft under design. Civil engineers had used the program in fixing sites for new buildings, determining requirements for adding noise-reduction materials to existing structures, and in orienting trim pads to minimize aircraft-noise impact on the community.

AWACS was a modified Boeing 707 used as an airborne command post to coordinate fighters, close air-support aircraft, and strategic bombers in a combat arena. Its friend-or-foe antenna could identify enemy craft in a battlefield setting, allowing the 17-member AWACS crew to relay information to friendly aircraft. (AFSC *Newsreview*, Aug/78, 7)

• The AFSC *Newsreview* reported that the USAF had awarded Harris Corp.'s Electrical Systems Division, Melbourne, Fla., a $4.7 million contract to design and produce transportable weather-satellite receiving terminals. The company would build and deliver one operational prototype and three production units to support SAMSO's Defense Meteorological Satellite Program (DMSP).

The terminals (called Mark IV, a follow-on to the Mark III weather terminals currently in operation) would give military commanders precise weather information accumulated from DMSP spacecraft orbiting the earth every 101min. The terminals, half the size of present terminals, could fit on C-130 aircraft or M-55 trucks and be operational within 6hr, as opposed to the present 48hr, automatically tracking a satellite appearing over the horizon and receiving visual and infrared imagery in real time.

An operator using magnification and other techniques could choose and enhance a specific area of coverage, or could process the entire 4500-nautical-mi-long 1600-nm-wide satellite image. The Mark IV project officer, Capt. James Fowler, said that the operator, after choosing the area most suitable for the mission, could produce a hard-copy print in 2min and transmit the photo to four remote laser-image processors 10mi from the terminal. "This capability, unique to the Mark IV system, will provide battlefield unit commanders with weather data within 20 minutes after the satellite's pass," he said. The prototype Mark IV terminal had been scheduled for delivery in Feb. 1979; production units were to be in the field by Oct. 1980. (AFSC *Newsreview*, Aug/78, 1)

• The USAF reported that a wind-tunnel test at Arnold Engineering Development Center using a new type of model and new testing techniques had demonstrated selection of an optimum aircraft-wing shape. Previous efforts to determine the best wing shape for a particular aircraft had compared features of several (sometimes many) models representing educated guesses; even after selecting one shape, testers could not be certain there was not another more efficient.

The latest test had used simulated flight conditions with a new computer-controlled wing model able to change shape automatically. Three-dimensional images made by a stereophotographic technique defined the actual wing shape during each flight change, making it possible to consider numerous shapes. The model, one of the most complex tested in AEDC's 16-ft transonic wind tunnel, was a product of General Dynamic's Convair Division, San Diego, that had been redesigned and reassembled by AEDC's operating contractor ARO, Inc., after its arrival at AEDC. (AFSC *Newsreview*, Aug/78, 12)

• *Air Force Magazine*, in an article on Goddard Space Flight Center, called it the jack-of-all-trades among NASA's space centers. It was NASA's communications hub, with a system of landlines, undersea cables, and satellites reaching around the world and connecting the 21 sites of the Spaceflight Tracking and Data Network (STDN) that had served all NASA's spaceflight activities. GSFC also had one of the world's largest assemblies of scientists, who theorized either scientific or applications satellite missions and made investigations into disciplines such as astronomy, planetary atmosphere, solar activity, and near-earth physics. About a third of scientific experiments on U.S. spacecraft were conceived, designed, and/or built by Goddard scientists.

The Goddard-developed earth-resources survey programs used orbiting Landsats capable of scanning the planet (except for the polar regions) every 9da. The photographs they provided had allowed detailed study of the earth faster, cheaper, and more accurately than aerial photography. Its impact on agriculture worldwide could be enormous; Landsat would make possible, for example, an inventory in Calif.'s Imperial Valley of more than 25 separate crops in nearly 9000 fields scattered over 458 000 acres. Landsat data would identify specific crops and indicate their condition, permitting very accurate agricultural production forecasts.

NASA planned to launch in the early 1980s a pair of Goddard-developed satellites containing space-tracking equipment to replace all but six of the present ground-based tracking stations, supplying data

85% of the time with greater efficiency and manpower savings. GSFC scientists were also making experimental and practical applications of laser technology. One program would eventually establish laser-satellite systems on each continent, to measure the most minute movements of the plates forming earth's crust. (*AF Mag*, Aug/78, 68)

• *Transportation USA*, in a story on 75 years of powered flight, quoted the Dec. 17, 1903, telegram from Orville and Wilbur Wright to their father in Dayton, O.: "Kitty Hawk, North Carolina. Success four flights Thursday morning. All against 21 mile wind. Started from level with engine power alone. Average speed through air 31 miles. Longest 59 seconds. Inform press. Home for Christmas." Newspapers had not been impressed; not even the Dayton *Journal* had mentioned the achievement. Ridiculed by the press and snubbed by their government, the two brothers had taken their craft to Paris and London; when stories of their successes and honors reached the U.S., the two were finally prevailed upon to come home.

Aviation had then become big business. The value of airline property and equipment, not counting general aviation, had reached $125 million at the end of World War II; by the end of 1955, the investment had multiplied tenfold to $1.4 billion. By 1965 it had reached $5 billion, and in 1977 $17 billion. In 1977 U.S. airlines carried 242 million passengers, and general aviation another 100 million. The FAA predicted that in 1987 U.S. airlines would carry upwards of 500 million passengers annually, and that privately owned aircraft would carry 200 million. The future might see aircraft using energy other than petroleum; of particular interest was hydrogen, having pound for pound 3 times the energy output. If storage were not a problem, hydrogen might make an excellent aviation fuel.

The article concluded: "We've come an unbelievably long way in the past 75 years—from a fragile craft airborne for 59 seconds to walks on the moon. No telling where our lust for flight may take us in the next 75 years. The sky's no longer the limit." (*Trsp USA*, Summer 78, 3)

• ESA reported it had scheduled 3mo of propellant-mockup tests of the European launcher Ariane to begin at the French Guiana Space Center launch site [see June 16]. Tests would check general conditions for launcher-assembly compatibility of ground facilities (platform, tower, etc.) with the vehicle, and for technical functioning of the fueling and draining facilities and systems both on the ground and on the vehicle. The latter three of four qualification flights scheduled from June 1979 to Oct. 1980 would launch Amsat (telecommunications for international radio amateurs), Firewheel (science/Germany), Meteosat-2 (meteorology/ESA), APPLE (communications/India), and Marots-A (maritime communications/ESA). (ESA newsletter Aug/78, 3)

September

September 1: Johnson Space Center reported that 11 of 35 current astronaut candidates had undergone parasail training Aug. 28 at Vance AFB, Enid, Okla., learning how to land by parachute in the event of an emergency ejection from the T-38 aircraft over land. Other astronaut candidates had taken the course before entering the NASA program. The six female astronaut candidates along with five male counterparts had taken the 1-day training exercise. (*JSC Roundup*, Sept 1/78, 1)

• The latest in a family of small-diameter 8-bladed propellers had completed wind tunnel testing at Lewis Research Center, the *Lewis News* reported. Third in a series tested and designed for new high-speed fuel-conservative aircraft, these propellers had unique extreme curve to the blades designed to improve performance and minimize cabin noise during flight. At simulated cruise conditions of 855km/hr (530mph) and 9145m altitude (30 000ft), the propellers proved less noisy and more efficient than previous models tested. NASA estimated that, at those cruise conditions, an advanced turboprop engine using the propellers could save 20 to 40% in fuel over current turbofan engines and 10 to 20% in fuel over advanced turbofan engines.

LeRC work on the propellers was part of a NASA-wide effort toward 50% fuel savings in future U.S. aircraft, with testing expected to be completed in later 1980. Hamilton Standard Division of United Technologies Corp. had been contractor for design and fabrication of the advanced propellers. (*Lewis News*, Sept 1/78, 1)

September 5: NASA announced plans to launch TIROS-N from WTR Sept. 15 as part of a third generation of new meteorological satellites. First of a series of eight polar-orbiting environmental spacecraft, TIROS-N would carry new environmental monitors representing significant technological advances over those on the current NOAA-series spacecraft. NOAA's National Environmental Satellite Services (NESS) would operate TIROS-N, intended to improve weather analysis for more accurate weather forecasts; to locate ocean currents and areas of upwelling, important to fishing and shipping interests; and to acquire more precise data on snow cover, snowmelt, and rainfall, essential to water-resource management and flood forecasting. Also possible would be more accurate alerts of high-energy solar-radiation levels above the atmosphere, important to space missions, high-altitude commercial aircraft flights, long-range communications, and electrical power-distribution networks.

Its advanced data-collection and platform-location system meant that TIROS-N would have for the first time an operational capability to collect and transmit environmental data from platforms on land, at sea, and in the air; and to determine the geographic locations of those platforms in motion on the surface of the sea, land, or aloft. TIROS-N's scanning radiometer, most versatile to be carried aboard an environmental satellite, could gather and store for later playback visible and infrared measurements and images in 4 channels for more precise evaluation of land, ice, surface-water, and cloud conditions, and sea-surface temperatures, transmitting them in real time to both automatic picture-transmission (APT) and high-resolution picture-transmission (HRPT) users in more than 100 countries around the world. The vertical-sounder subsystem, containing three instruments, could register temperatures to within 1°C and moisture data from earth's surface up through the stratosphere. Even under obscuring cloud cover, TIROS-N could obtain some data because one of the three instruments would be able to record microwave radiation regardless of clouds.

TIROS-N would be a primary data source for the Global Atmospheric Research Program's first global experiment, an international cooperative project of 140 countries scheduled to begin Dec. 1, 1978. Its instruments would meet FGE requirements for collecting quantitative data on earth's atmosphere and sea surface to construct numerical atmospheric models for improved long-range weather forecasts. Sensors were multinational: the stratospheric-sounding units were from Great Britain, the data-collection platform-location system from France, and other sensors from the U.S. Ground-support systems were also multinational: the U.S. would acquire and process most data through NOAA's command and data-acquisition (CDA) stations near Fairbanks, Alaska; Wallops Island, Va.; and a central processing system at Suitland, Md. France's Centre National d'Études Spatiales (CNES) would process and distribute data associated with collection and the platform-location system.

TIROS-N instruments had drawn heavily on technology from NASA's Nimbus-spacecraft program: TIROS-N's bus was a modified USAF Block 5D spacecraft bus with a 25% growth allowance in payload-carrying capability. NASA had designed and funded TIROS-N; Goddard Space Flight Center would manage it during checkout (about 2mo) before turning it over to NOAA. Launch vehicle would be a standard reconditioned Atlas E/F. (NASA Release 78-132; NASA press conf. Sept 8/78; *Spaceport News*, Sept 15/78, 2)

September 6: NASA announced that *Voyager 1* on its way to Jupiter and Saturn had completed transit of the asteroid belt, leaving the region Sept. 8. The belt, a band of rock and dust 360 million km (233 million mi) wide between the orbits of Mars and Jupiter with its inner edge about 105 million km (65 million mi) from earth's orbit, was formerly believed an impassable hazard to spacecraft bound for outer planets and beyond. *Voyager 1*, launched after *Voyager 2*, had entered the asteroid belt Dec. 10 a few hours behind *Voyager 2* on a master trajectory that would carry it steadily beyond *Voyager 2*, which it would overtake Dec. 15. *Voyager 1* would make its closest approach to Jupiter Mar. 5, 1979, at 286 000km (178 000mi); *Voyager 2* would pass 643 000km (399 000mi) from Jupiter in July 1979. Both spacecraft would continue on to Saturn where *Voyager 1* would arrive in Nov. 1980, *Voyager 2* in Aug. 1981. (NASA Release 78-137)

• Sen. William Proxmire (D-Wis.) said he would try his best to do away with $5.7 million in "federal pork" for lunar-rock analysis and "to force scientists to compete for lunar sample support with other basic research proposals in the geological sciences," *Defense/Space Business Daily* reported. Proxmire said that NASA had spent a total of $41.2 million on lunar-sample analysis and had requested another $5.7 million for FY79. Although the House had approved the request, the Senate had adopted Proxmire's recommendation to delete the funds, leaving the issue to be resolved in conference during Sept. Proxmire had maintained that lunar-sample research should not enjoy privileged status over other basic research in geology, geochemistry, and geophysics, but should "compete with all other research in these areas, preferably at the National Science Foundation—our principal basic research agency. If the lunar samples have the value claimed for them when the curatorial facility was built, this should be no problem."

Proxmire said the Senate Appropriations Committee had directed NSF to seek proposals for lunar analysis next year as NASA phased out its research program. His correspondence had charged that the primary purpose of lunar analysis appeared to be training graduate students and keeping university laboratories well equipped. "I do not in any way mean to show disrespect for these motives or the participants in the lunar sample program," Proxmire noted. "However, we have programs for training graduate students. We have programs for providing research equipment to universities. NASA's moonrock program should not be used for these purposes." (*D/SBD*, Sept 6/78, 4)

September 7: NASA announced it had selected Computer Sciences Corp., Falls Church, Va., to negotiate a 2yr cost-plus-award-fee contract for engineering support and related services at Wallops Flight Center. The contract would include support of instrumentation, computer operations, programming, data support, aerodynamics, mechanical design, and systems integration, both for flight articles and ground-instrumentation facilities and for related specializations. The company's bid was approximately $3.8 million; the contract, effective Oct. 1, would contain options for three additional 1-yr periods. WFC would administer the contract. (WFC Release 78-14)

• Tokyo Univ.'s Institute of Space and Aeronautical Science had successfully launched a satellite to study the phenomenon of northern lights, the *Washington Post* reported. The 200-lb *Exos B*, one of three artificial moons to be launched in Japan's international magnetic-sphere observation program, would orbit the earth in an elliptical trajectory every 8hr 45min. The northern lights (aurora borealis) was a luminous phenomenon best viewed in the Arctic region. (*W Post*, Sept 7/78, A-16)

September 8: NASA announced plans to launch Nimbus-G, first satellite designed to monitor manmade and natural pollutants in earth's atmosphere and the last of seven Nimbus environmental-research spacecraft launched since Aug. 1964, from WTR no earlier than Sept. 18, 1978. A 2-stage Delta launch vehicle would put the 907-kg observatory toward a near-polar orbit.

Nimbus-G would for the first time provide scientists throughout the world with continuous environmental data on physical conditions in the global atmosphere, the oceans, the dynamic atmosphere-ocean interface, and the earth's heat balance—information vital to understanding climate, oceanography, atmospheric pollution, and regional and global weather patterns. NOAA and several U.S. university scientists would work with NASA scientists on the Nimbus-G project, along with scientists from the United Kingdom, Denmark, Switzerland, Canada, South Africa, West Germany, France, and Belgium. For the first time ESA would bring on line a Nimbus-G direct reception and data-processing capability at Lannion in France that would eventually receive, process, and distribute on a regular basis Nimbus-G data to approved European investigators.

The 8 highly complex sensors on Nimbus-G (seven from the U.S. and one from the U.K.) were improved versions of instruments carried on six previous Nimbus flights to answer three questions: Was ozone in the earth's upper atmosphere changing? Was the earth warming or cooling? What was the extent of pollution in the world's oceans? Nimbus-G sensors would include four to measure atmospheric gases and pollutants and their interrelationships in the earth's stratosphere and mesophere (the upper atmosphere from 10 to 90km [6.2 to 56mi] up). The coastal zone color scanner, sensing colors of the oceans, would enable oceanographers to map chlorophyll concentrations, sediment distribution, and salinity in large areas of coastal or ocean water, and to detect and track water pollution such as oil spills, sewage and industrial waste dumpings, and river sediment.

GSFC would manage the spacecraft, Delta launch vehicle, and global network for tracking and 2-way communications. Kennedy Space Center's unmanned-launch operations team would supervise the Delta launch. General Electric Co.'s Space Division, Nimbus prime contractor with more than 50 major subsystem contractors, would be responsible for sensors and components in the spacecraft, launch vehicle, and ground-reception equipment. More than 1000 subcontractors and vendors had participated in the program. Cost of Nimbus-G spacecraft and instruments had totaled $79 million, $2.6 million less than budgeted. (NASA Release 78-136 [press kit]; MOR E-604-78-08 [prelaunch] Sept 1/78, [prelaunch summary] Sept 1/78; *Langley Researcher*, Sept 22/78, 4)

• NASA announced it had appointed Frank Simokaitis (Maj. Gen., USAF, Ret.) director of its DOD affairs division, Hq Office of External Relations. During World War II, Gen. Simokaitis had been an instructor pilot and bomber pilot in the European theater. Later he had held a series of special investigative, legal, planning, administrative, and staff posts, including executive assistant to the Secretary of the Air Force from 1969 to 1973. Until his retirement, Gen. Simokaitis had been commandant of the USAF Inst. of Technology at Wright Patterson AFB and commandant of the Defense Institute of Security Assistance Management. He had received LL.B. and J.D. degrees from St. Louis Univ. and was a member of the bar of the State of Missouri, admitted to practice before the U.S. Supreme Court and the U.S. Court of Claims. (NASA Release 78-139)

• Langley Research Center announced that Dr. Leonard Hayflick, senior cell research biologist at Children's Hospital Medical Center of Northern Calif., would speak on "The Biology of Human Aging" Sept. 18 as part of LaRC's colloquium series. According to Dr. Hayflick, the ability of the human body to maintain itself had been demonstrated from its large-scale immune and endocrine systems down to the cellular level; its highly adaptable self-maintenance ability had appeared in biological responses of astronauts to sustained zero-g environments. Why did this self-maintenance ability give way to the aging process and eventually death? Could the biological clock be altered, and would it be desirable to do so? Dr. Hayflick would discuss aging both from a demographic and a biological point of view. After explaining cell-culture technology and its implications, Dr. Hayflick would summarize some modern theories on the aging process. (LaRC Release 78-45)

• The U.S. House Committee on Science and Technology reported that Rep. Olin Teague (D-Tex.) had called for hearings on the current status of NASA programs. Convened by the subcommittee on space science and applications and chaired by Rep. Don Fuqua (D-Fla.), the hearings would review NASA projects, with particular emphasis on the Space Shuttle and the tracking and data relay satellite system. Fuqua said it had been subcommittee practice to hold hearings in the fall before considering NASA requests for new and continuing work in the next year's budget.

John Yardley, associate administrator for space transportation systems, and William Schneider, associate administrator for space tracking and data systems, would testify Sept. 25. Dr. Anthony Calio, associate administrator for space and terrestrial applications, and Dr. Noel Hinners, associate administrator for space science, would testify Sept. 26. Dr. James Kramer, associate administrator for aeronautics and space technology, and Raymond Kline, associate administrator for management operations, would both testify Sept. 27. (H Comm on Sci & Tech Release 95-114)

September 9-14: FBIS reported that the USSR had launched *Venera 11*, an automatic interplanetary station, on Sept. 9 to continue exploration of Venus. Purposes of the mission were to study solar winds, cosmic rays, ultraviolet and roentgen rays, and (with apparatus designed by the USSR and France) gamma rays. *Venera 11* should reach the vicinity of Venus in Dec. 1978. (FBIS, Moscow Dom Svc in Russian, Sept 9/78)

The USSR had launched *Venera 12* Sept. 14 with a design and purpose similar to those of *Venera 11*, Tass announced. Onboard systems were functioning normally, and trajectories of both stations were close to original calculations. *Venera 12* would also reach the vicinity of Venus in Dec. 1978. (FBIS, Tass in English, Sept 14/78)

September 11: NASA announced it had tentatively selected five scientific experiments for its gamma ray observatory, a planned 1984 mission to explore a variety of the most energetic forms of radiation known: very high-energy gamma rays from pulsars, nuclear gamma rays, and gamma-ray bursts. Observatory data should provide a much better understanding of the nature of supernovas, pulsars, quasars, radio galaxies, the character of the early universe, and the possible existence of antimatter in the universe. Scientists also hoped to learn more about incredibly dense matter composing objects such as neutron stars, matter so compressed that a teaspoonful would weigh millions of tons. Data from the GRO might also tell more about gamma-ray bursts, erratic gamma-ray pulses of unknown origin that flashed across the solar system every month or so. Although Congress had not yet approved GRO, early solicitation of scientific participants and investigations would allow for mission definition and a prompt start after approval.

GSFC would manage the observatory to be carried by the Shuttle into a planned orbit of 400 to 500km (250 to 300mi). NASA had selected principal investigators and experiments: Dr. James Kurfess of NRL, broad line gamma-ray spectrometer; Dr. Laurence Peterson of the Univ. of Calif., San Diego, narrow line gamma-ray spectrometer; Dr. Carl Fichtel of GSFC, high-energy gamma-ray detector; Dr. V. Schoenfelder of the Max-Planck Inst., Garching, W. Germany, medium-energy double Compton telescope; and Dr. Gerald Fishman of Marshall Space Flight Center, transient gamma-ray monitor. (NASA Release 78-140)

• NASA announced that tests by LeRC and General Electric Co., Peebles, O., on two experimental aircraft engines might result in major reductions in noise and air pollution. Technology of both engines, known as quiet clean short-haul experimental engines (QCSEE), had been directed to future short-haul commuter-type aircraft in the 450 to 800km (300 to 500mi) range customarily used by the smaller airports. NASA intended to apply QCSEE technology also to engines in the 80,700-newton (40 000lb) thrust size used in the largest planes of the U.S. commercial passenger fleet.

The two 40 300n (20 000lb)-thrust power plants had run 8 to 12db quieter than the quietest engine (the CF6) on widebodied DC-10 and 747 civil transports, about 16db below current FAA noise-level standards and some 9db below tougher FAA requirements set for the 1980s. Reduction in noise had resulted from slowing engine-exhaust velocity by increasing the ratio of air passing through the engine fan to air passing through the core of the engine, and by other design features including use of acoustically absorbent materials.

Aeronautical engineers also reported substantial air pollution reductions for the research engines, compared with engines then flying; carbon monoxide emissions, for example, had decreased 84% in one engine and 81% in another. Unburned hydrocarbons had decreased 98% for one engine, 97% for the other. LeRC engineers attributed air pollution improvements to new designs incorporated in the combustion system and said these lower emission levels would more than meet stringent EPA standards set for 1979.

An additional bonus would be fuel savings of about 10% compared with current engines of similar size. Fuel savings had resulted from replacing normally heavy metal components (the engine cowling, frame, and fan blades) with lightweight nonmetallic composite materials of equal or greater strength. The experimental engines had internal differences, but the most visible was their location on the aircraft wing: one had been under the wing as in conventional installations, the other over the wing. Wing flaps in both positions deflected jet exhaust downward to add lift for short-run operations.

QCSEE was a $30 million program to develop and demonstrate

powered-lift engines (under and over the wing) that could reduce congestion, noise, and air pollution around airports. General Electric Co. had been prime contractor for both engines. (NASA Release 78-138)

• Hughes Communications Services, Inc., Culver City, Calif., had received a $335 million Navy Systems Command contract to provide 5yr of Leasat (leased satellite) communications services at each of four locations in space starting in 1982, *Aerospace Daily* reported. Hughes, which had proposed to use its Syncom-4 Shuttle-optimized spacecraft, had won out over Comsat General Corp. and TRW Inc. Leasat would replace the Gapfiller service currently leased from Comsat General's Marisat system, and would complement the TRW-built Navy-owned FltSatCom system by serving both Navy ships and Army, Marine Corps, and Air Force ground-mobile forces. Choice of the Hughes system, dedicated to Pentagon use, should enhance prospects of a ESA and INTELSAT joint offer of a civilian system to be used by the pre-Inmarsat joint venture. (*A/D*, Sept 11/78, 38)

• NASA had dropped supersonic transport-oriented research aircraft, previously considered as a new start for FY83, from its latest 5-yr plan covering FY 1980-1984, *AvWk* reported. Inclusion of the 75 000lb-class aircraft in an earlier 5yr plan had angered Congressional foes of supersonic transport. A senior NASA official acknowledged it had been a mistake to include the aircraft in the earlier plan. (*AvWk*, Sept 11/78, 13)

• ESA announced it had awarded to the European consortium Cosmos headed by Messerschmitt-Bolkow-Blohm, W. Germany, a contract for phases C and D of the scientific satellite Exosat to include fabrication, integration, and testing of the observatory intended to measure position, structural aspects, and spectral and temporal characteristics of x-ray sources. It would carry four experiments: two imaging telescopes for low-energy x-rays, a large-area proportional-counter array, and a gas-scintillation spectrometer for medium-energy x-rays.

The observatory, mounted on a 3-axis-stabilized platform, would travel in a highly eccentric orbit whose apogee (roughly above the North Pole) would be at 200 000km altitude. ESA planned to launch Exosat in the second quarter of 1981 on an Ariane equipped with a fourth stage to put the spacecraft into the required orbit. (ESA Release Sept 11/78)

September 12: LaRC announced that James Patton, Jr., head of the pilots' office, would receive the Iven C. Kincheloe award at the annual symposium of the Society of Experimental Test Pilots in Los Angeles Sept. 30. The award would honor Patton's professional accomplishment in initiation, formulation, and conduct of the NASA comprehensive general aviation stall/spin flight research program. John Reeder, chief of LaRC's TCV program office and a Society fellow, had nominated Patton for the award.

In his nomination Reeder noted, "During the (stall/spin) program formulation, Mr. Patton provided invaluable leadership and technical guidance to establish flight safety procedures and ensure a rational and productive flight research program. As a result of his efforts, Langley Research Center has established a stall/spin flight test facility, procedures, and equipment which provide the most sophisticated system to date for general aviation stall/spin research." Patton had been in his present position since 1968; he had started as a research pilot with NASA in Apr. 1966. (LaRC Release 78-46)

- INTELSAT announced that earth stations handling satellite communications in countries bordering the Indian Ocean would begin switching to a new larger-capacity comsat, *Intelsat IV-A*, which would carry about 6000 2-way telephone circuits (25% more capacity than its *Intelsat IV* predecessor) and would eventually provide increased capacity for international satellite television.

The Indian Ocean Intelsat had provided communications for and between countries in an area bounded approximately by Britain and Europe in the west and Japan and Australia in the east. The transition of 47 earth stations through 150 steps was scheduled to start Sept. 18 and to be substantially completed by the end of 1978. No interruption or disruption of telecommunications services was anticipated, except that television transmission facilities would be unavailable during the latter half of Nov.

The switchover operation would position *Intelsat IV* and *Intelsat IV-A* in orbit at the same location and synchronize their harmonic motion (a sway induced by natural forces acting on the spacecraft). The proximity of the satellites would allow earth stations to see both simultaneously, permitting changes in the operational configuration without a break in service. (INTELSAT Release 78-25-I)

- A delegation from NASA Hq meeting in Paris with ESA's Spacelab program board would attempt to overcome W. Germany's objections to a proposed barter agreement, under which NASA would provide four free Space Shuttle launches instead of paying cash for Spacelab hardware it had agreed to buy, *Aerospace Daily* reported.

The barter deal, proposed by ESA in 1977, had met with opposition from ESA-member governments. Main objection, stated most strongly by W. Germany, was that a European commitment to four Shuttle flights would be premature because European plans to use the Space

Shuttle and Spacelab ought to be definite before a barter would make sense. NASA had agreed to the barter instead of paying cash for a full set of Spacelab hardware it had agreed to buy years before. ESA felt that, under the barter, the industrial consortium headed by VFW-Fokker Erno could continue to work with European money at a far more satisfactory pace than could be expected if they relied on NASA money. In exchange for the roughly $100 million the Europeans would pay, NASA would provide four free flights for Spacelab, two for ESA and two for Germany.

Observers said that W. Germany's reservations centered on uncertainty about ESA's Spacelab program, and that its own commitment to two flights was premature. If (as was likely) W. Germany's Spacelab missions would not use the Shuttle-Spacelab system's full capability although it had agreed to the dedicated launches, Germany would have to fill up the Space Shuttle or absorb the resulting losses. Other concerns were the rising costs of Space Shuttle flights, as launch costs in deutschmarks increased while the value of the dollar decreased. (*A/D,* Sept 12/78, 43)

September 12-17: NASA announced it had received word from NORAD that the *Pegasus 1* spacecraft assembly launched in 1965 would reenter earth's atmosphere on or about Sept. 17. Predicted dispersion (footprint) of surviving pieces was 200km (125mi) cross range, 3500km (2175mi) down range. The Pegasus orbital band had extended 31.7°N and 31.7°S of the equator over an area 75% water. NASA estimated that reentry heat would destroy approximately 9705kg (21 400lb) of orbital hardware, and that about 726kg (1600lb) might survive reentry in several pieces. Normal breakup of reentering spacecraft had not so far resulted in personal or property damage on earth; the possibility of a Pegasus fragment's striking an individual or structure was considered to be less than that resulting from meteorites impacting the earth.

Pegasus 1, launched Feb. 16, 1965, had gathered micrometeoroid data for use in designing spacecraft; data collection had ceased Jan. 13, 1968. The *Pegasus 1* assembly weighed about 10 430kg (23 000lb) and was 21m (70ft) long. The spacecraft, weighing about 1900kg (3200lb), had remained attached to the empty S-IV stage and the instrument unit of the Saturn I launch vehicle. Neither the satellite nor the attached S-IV contained radioactive nuclear power sources or materials. *Pegasus 2* was still in orbit; *Pegasus 3* had reentered Aug. 4, 1969, over the Pacific Ocean.

Aerospace Daily reported that *Pegasus 1* and associated hardware had reentered earth's atmosphere Sept. 17 over Angola. A NASA spokesman said there were no sightings or reports of debris. (NASA Release 78-141; *Marshall Star,* Sept 13/78, 2, 20/78, 2; *DFRC X-Press,* Sept 22/78, 2; *A/D,* Sept 20/78, 91)

September 13: A space processing applications rocket, SPAR V, was launched at 11am CDT September 11, from White Sands, N.M., the *Marshall Star* reported. The SPAR V science payload contained 4 experiments to be processed during about 4min of near-zero gravity achieved by the suborbital flight. The flight appeared, from limited data, to have gone extremely well, lasting about 15min and reaching a height of around 102.5 statute mi. The payload was recovered about 55 to 60mi down range in a low mountain area. The payload had no external damage and had been returned to the missile range.

Two of the experiments studied resolidification of molten materials in low gravity, from photographs made at 1-sec intervals during the process. Another experiment verified results of a SPAR II experiment during which an immiscible liquid processed in an electric furnace exhibited unexpected massive separation. The fourth experiment both melted and crystallized several samples of materials during the low-gravity processing period, while a camera recorded particle behavior. (*Marshall Star*, Sept 13/78, 1; 6/78, 1)

September 15: Jet Propulsion Laboratory announced it had installed a new RCA domestic satellite (domsat) earth station with 10m antenna, now fully operational as part of the NASA Communications Network (NASCOM). RCA had received a contract to provide three 56-kilobit-per-sec (kbps) wideband data channels between JPL and GSFC, primarily to serve the Voyager project. The three channels, used with new NASCOM-installed "tri-channel" end terminals, would provide a 168kbps end-to-end data transfer capability.

The antenna was the completion of a high-data-rate link to JPL from the Deep Space Network's 64m tracking station at Madrid; Voyager high-rate telemetry received at Madrid would be beamed up to the Intelsat satellite, down to an East Coast receiving station, then to GSFC for transmission to JPL over the new domsat link. Although the 56kbps rate had formerly sufficed for telemetry received from overseas stations, the Voyager encounters with Jupiter scheduled to begin in the spring would need a higher (115.2 kbps) rate. (*JPL Universe*, Sept 15/78, 1)

• JPL reported that solar cell roofing shingles made for it under contract by General Electric Co. could deal with the weather in two ways: they could both generate electricity and constitute a roof impermeable to wind and rain. The 50 shingles, developed and tested at JPL under a 1yr $200 000 contract as part of Task IV of DOE's large solar array project, consisted of a rectangular synthetic-rubber base with a tempered-glass hexagon coating containing 19 photovoltaic cells. Solar-cell arrays would butt against each other when the modules were laid in the conventional overlapping manner. The shingles produced an average of 98w/m³ according to GE tests, and an array of about 1900 shingles could supply 90% of the power used by an all-electric home in the southwest U.S. However, the cost of the shingle-generated electricity was about $30 per watt, far higher than DOE's $8-a-watt target. Environmental and field tests at JPL would verify GE's results and study weathering effects, to help GE develop cost-cutting design refinements. (*JPL Universe*, Sept 15/78, 1)

• *Aerospace Daily* reported that House-Senate appropriations conferees would allow NASA to reprogram FY79 funds from Space Shuttle production into development as it had done with $100 million in 1978, if the agency first asked Congress for a budget supplement to restore the production money. Reporting on a Sept. 14 compromise $4.35 billion NASA money bill, the conferees promised that NASA would receive "prompt consideration" of a supplemental request.

The reprogramming suggestion seemed to address the major concern: whether or not NASA could increase the rate of development spending ahead of supplement approval. Under a reprogramming, production would be slowed pending restoration of funds, but development could proceed at a rate consistent with the expected extra money. This arrangement had followed up on an earlier House provision dropped by the conferees, allowing NASA to set up a reserve fund for Shuttle development by taking $30 million from three science programs.

The conferees had approved NASA's $20.5 million request for development of a teleoperator retrieval system, with the proviso that NASA get approval from both appropriations committees before obligating more than $10 million of the money, and that the balance of $10.5 million be available "only for Space Shuttle funding requirements in the absence of committee approval." Both committees had been skeptical of the TRS proposal because its use in a Skylab mission might be impossible in view of Shuttle delays and the deterioration of Skylab's orbit.

Conferees also had agreed to add $500 000 to the money bill to start a Stereosat program; the House had added $4 million, the Senate nothing. The conference had compromised on an add-on of $2 million for a solar-power satellite system, ratifying the House cut of $5 million from expendable launch vehicles. It had also eliminated funds for the SETI (search for extraterrestrial intelligence) program. (*A/D*, Sept 15/78, 65)

September 18: Onboard equipment problems had delayed launch of Nimbus-G and TIROS-N from Vandenberg AFB, *Av Wk* reported. Tape recorder problems had caused cancellation of the Nimbus-G launch scheduled for Sept. 18; the new launch would not be before Oct. 2. Replacement of a malfunctioning processing unit had delayed the TIROS-N launch for 2wk. The problem appeared to be isolated and should not affect the unit's design or the AF meteorological satellites then in orbit.

Aerospace Daily reported that NASA had returned the three tape recorders in Nimbus-G to the builder, RCA Corp., for cleaning and removal of loose solder that could cause short circuits; five essentially identical tape recorders on TIROS-N also needed cleaning. Nimbus-G and TIROS-N represented first uses of the NASA tape recorder, developed over several years by RCA as standard equipment for a variety of spacecraft. The problem had apparently arisen in soldering connectors to the mother board (a printed circuit board into which other PCBs were plugged); upon inspecting a tape recorder that had failed during tests, NASA found that loose solder had shorted pins. While drawing up a cleaning requirement for units not yet delivered, NASA had dismantled three tape recorders and had found loose solder capable of causing shorts in all three. (*AvWk*, Sept 18/78, 33; *A/D*, Sept 18/78, 73)

September 19: NASA reported that about 7da after a Delta rocket was scheduled to launch Nimbus-G into polar orbit from WTR, the second stage of the Delta would separately release two chemicals. Project Cameo (chemically active materials ejected in orbit) would consist of a single 90sec release of lithium over northern Scandinavia and a separate release of barium over northern Alaska, to study solar energy and plasma flows and electric fields in Arctic regions. Release of the lithium would create a huge bright-red cloud visible from Greenland, Scandinavia, and other parts of Europe for a maximum of 5min. Release of the barium would create a bluish white cloud visible from the Soviet Union, Canada, and northwestern parts of the U.S. for a maximum of a half hr. NASA had timed the releases during Arctic twilight or early morning, to ensure best lighting for ground-based and aircraft photography to pinpoint the position and configuration of each cloud as a function of time.

Primary objective of the barium experiment was to trace the flow of plasma (ionized particles) in and above the earth's ionosphere, with particular emphasis on the transition in earth's electric field between the polar ice cap and the auroral zone over Alaska and Canada. Primary objective of the lithium experiment was to detect electric fields oriented parallel to the earth's magnetic field. (NASA Release 78-142)

• LaRC announced that Dr. Donald "Deke" Slayton would be guest speaker Oct. 5 at a NASA-LaRC colloquium and a public lecture to commemorate NASA's 20th anniversary. Slayton, one of the original seven astronauts selected in 1959 for Project Mercury and the only one still active in the space program, was orbital flight test manager for JSC's Space Shuttle Project Office. He had been chief astronaut and director of flight crew operations before going into space in July 1975 with the Apollo-

Soyuz Test Project. Title of Slayton's colloquium lecture would be "The Space Shuttle"; his public lecture, "Manned Space Flight: Past Triumphs, Future Challenges." (LaRC Release 78-47; *Langley Researcher*, Sept 22/78, 2)

• JSC announced it had signed a contract with Itek Optical Systems Division of Lexington, Mass., for the Space Shuttle orbiter camera payload system. The $4 890 000 cost-plus-incentive-fee contract would cover the first flight system with an option for a second. Planned for spaceflight on pallets and test racks built by NASA or ESA in the orbiter's 15 by 65ft cargo bay, the system would be a flexible photographic array for earth-imaging surveys, remote sensing, and supplementing nonimaging systems. The second option (to produce a large-format camera for the USGS) would, if exercised, increase the contract value by $1.5 million. (JSC Release 78-42)

• MSFC reported that the final airdrop test of a system to retrieve Shuttle solid-fuel rocket boosters by parachute had demonstrated that the system could save the boosters for reuse. The sixth and final airdrop Sept. 12 had successfully tested a main parachute under forces exceeding those expected during flight. A test vehicle released at about 6100m (20 000ft) altitude from beneath the right wing of a B-52 aircraft had deployed a test parachute when vehicle speed reached about 660km per hr (410mph).

Although actual booster recoveries would use three main parachutes each, the test had used only one because the test vehicle weighed only about a third as much as a booster would during recovery, and the B-52 aircraft could lift only 22 700kg (50 000lb). Use of only one parachute had sufficed for the main parachute structural test, however, as it had sustained the necessary load. Actual weight of the boosters to be recovered after Shuttle launches would be about three times the present world-record weight of 24 948kg (55 000lb) for parachute recovery; each empty booster would weigh about 78 000kg (172 000lb) during descent. MSFC engineers had conducted all six drop tests at the National Parachute Test Range, El Centro, Calif. (*Marshall Star*, Sept 19/78, 1)

September 20: NASA reported that a flight-configured Space Shuttle engine fired in tests at NASA's Bay St. Louis, Miss., facility had exceeded 5000sec, the level to be met by a production engine to certify the main propulsion system for manned flight. Testing of the Space Shuttle main engine had increased dramatically, both in number of tests and time accumulated, since Aug. 12, 1978. Two engines had each accumulated approximately 7100sec of testing time, putting total engine-testing time over 25 000sec in 342 tests. Among these were five consecutive runs of

520sec, the length of time needed to put the Shuttle into orbit, on one engine at rated power level. Preliminary flight certification, which required a flight engine to have run 5000sec, was expected in the spring of 1979. Full-duration testing of the complete main propulsion system (a cluster of three engines) would resume in early 1979 when the first manned orbital flight configuration engines would become available. (NASA Release 78-144; JSC Release 78-40; *Marshall Star*, Sept 13/78, 1)

• NASA Administrator Dr. Robert Frosch and Dr. John Keys, assistant deputy minister of Canada's Dept. of Energy, Mines, and Resources (EMR), had signed an agreement Sept. 19 in Ottawa to establish a Canadian ground station at Shoe Cave, Newfoundland, to receive SeaSat data and to study data use, NASA announced. Under the agreement, Canada would build and operate a ground station to collect from SeaSat's five sensors the data needed to support their own surveillance satellite project (SURSAT) and to furnish SeaSat data and related surface-truth information to NASA at no cost. NASA would be responsible for SeaSat data transmission to the station and for necessary technical information. Data from SeaSat would support a number of Canadian experiments to assess the usefulness of its synthetic aperture radar and other sensors for oceanographic research and coastal management. (NASA Release 78-143)

September 21: INTELSAT announced it would hold open its offer of a new satellite system for world maritime communications, since its board of governors had authorized continued development of a maritime communications package to be carried on a number of its Intelsat V international comsats as well as further discussions on provision of maritime services by a global-satellite system of three modified Intelsat V satellites and three European MARECS satellites, beginning in 1981. Observers expected that the International Maritime Satellite Organization (INMARSAT), after its formal establishment, would eventually take over maritime communications.

The board also had authorized Ford Aerospace and Communications Corp. to continue studies on maritime-communications equipment sets for its fifth, sixth, and seventh Intelsat V satellites. Although the packages could not increase international communications capacity, they would offer the equivalent of 30 voice circuits for maritime shore-to-ship and ship-to-shore communications and would improve the quality of on-demand voice, telex, telegraph, and data communications.

Aerospace Daily reported that the choice between NASA's Space Shuttle and ESA's Ariane to launch Intelsat V had been put off until Dec. The board had apparently voted on at least two options, one of which was a compromise using both launch vehicles. The board had deadlocked, because of "a lot of abstentions," said one source. (INTELSAT Release 78-27-I; *A/D*, Sept 25/78, 110)

• *Soyuz 29* cosmonauts Vladimir Kovalenok and Alexander Ivanchenkov had set a space endurance record in *Salyut 6* Sept. 20 after more than 96da aloft, the *Washington Post* reported; the USSR had not indicated how much longer their flight would last. At 9:17am Moscow time (2:17am EDT), the two had surpassed the previous record of 96da 10hr, set by *Soyuz 26* cosmonauts Georgy Grechko and Yuri Romanenko Feb. 11 aboard the same spacecraft. The two missions had been similar, consisting of scientific experiments and maintenance activities. Grechko and Romanenko had broken the 84day 1hr 16min endurance record set in 1978 by *Skylab 4* astronauts Gerald Carr, Edward Gibson, and William Pogue. (*W Post*, Sept 21/78, A-16; *W Star*, Sept 20/78, A-13; FBIS, Tass in English, Sept 20/78; *AvWk*, Sept 25/78, 13)

September 22: NASA reported that President Jimmy Carter had named six astronauts to receive the Congressional Space Medal of Honor, first such medal ever awarded by the U.S. The six were Neil Armstrong; Frank Borman; Charles Conrad, Jr.; John Glenn; Virgil Grissom (posthumous); and Alan Shepard, Jr. The President would present the medals Oct. 1 during a visit to KSC on the 20th anniversary of NASA's establishment. Congress had authorized in 1969 the award "to any astronaut who in the performance of his duties has distinguished himself by exceptionally meritorious efforts and contributions to the welfare of the Nation and of mankind." Citations were as follows:

—Neil Armstrong, for actions during *Gemini 8* in Mar. 1966 to overcome problems and land his spacecraft safely, and for "steady cool professionalism, repeatedly overcoming hazards" on the *Apollo 11* mission in July 1969, when he became the first person to walk on the moon.

—Frank Borman, who had commanded *Gemini 7* in Dec. 1965 and *Apollo 8* in Dec. 1968, both of which "significantly hastened and facilitated achievement of the manned lunar landing objective." On *Apollo 8* he had commanded the first manned spacecraft to escape the earth's gravity.

—Charles Conrad, Jr., who from Aug. 1965 to June 1973 had participated in four spaceflights of increasing duration, complexity, and achievement. His contribution had culminated in the first manned Skylab mission in May and June 1973, when he commanded the crew that performed "lengthy, dangerous, and strenuous activities that were necessary to repair damage inflicted on the orbital workshop during launch and thereby save the two-billion-dollar program."

—John Glenn, Jr., first American to orbit the earth in the third manned mission of Project Mercury in Feb. 1962, when his professional handling of extreme difficulties with the spacecraft "demonstrated the value of the human pilot in space. . . . He returned to a nation and a world that seized on him as a major hero. This difficult role he handled with the same polite dignity that he brought to all his assignments."

—Virgil Grissom (posthumous), second American in space who from July 1961 to Jan. 1967 had participated in Mercury and Gemini spaceflights and had lost his life during preparation for the first Apollo flight. Experience gained from the first manned Gemini flight in March 1965, which he commanded, led to "procedures necessary for the support of subsequent long-duration and rendezvous missions."

—Alan Shepard, Jr., first American in space aboard a Mercury spacecraft in May 1961, which "demonstrated that his country lacked neither the courage nor the technology to compete in the new arena of space." He also was cited for showing in Feb. 1971 "the highest qualities of leadership" as commander of *Apollo 14*, the third lunar landing mission. (NASA Release 78-146)

• JSC announced it had selected General Electric Co., Space Division, Valley Forge, Pa., for negotiations leading to a contract for the Space Shuttle orbiter food system galley. The estimated $1.2 million cost-plus-fixed-fee contract, beginning December 1978 and ending January 1981, would be for a facility including food-preparation equipment (hot and cold water dispensers, oven, water heater) and serving equipment (meat trays). (JSC Release 78-43)

• Dryden Flight Research Center reported that the KC-135, testbed aircraft to be modified with winglets in early 1979, had made its first flight at DFRC as a fully instrumental research aircraft. The data package would be ferried by air to Tinker AFB where it would undergo fatigue inspection. The transport-sized aircraft would return to DFRC in mid-Nov. for reinstallation of the data package. Baseline-data flights with the conventional wing would begin in 1979, followed by installation of the winglets for full-data flights. (*DFRC X-Press*, Sept 22/78, 2)

• DFRC announced it had returned its unmanned spin research vehicle (SRV) to flight status, following several mo of downtime for aligning the aircraft's nose and adding tufts to study airflow. A B-52 had dropped the SRV at 45 000ft; it had gone into a spin down to 20 000ft and a lakebed landing. DFRC had scheduled a second flight test to be followed, if successful, by additional nose modifications. Changing nose shape would alter aircraft-spin characteristics. (*DFRC X-Press*, Sept 22/78, 4)

September 25: NASA reported that John Yardley, associate administrator for Space Transportation Systems, had testified before the

House Subcommittee on Space Science and Applications on Space Shuttle status. Yardley said a detailed program review of cost, schedule, and performance had indicated substantial progress that year. He cited successful completion of approach and landing tests of Orbiter 101 (Enterprise), which had been shipped to MSFC for mated ground-vibration tests; mission-duration test firings of the main engines at the rated power level; and completion of the first phase of the 3-engine main-propulsion configuration testing.

A review of the Shuttle schedule had shown that all program elements could be ready for a Sept. 28, 1978, first manned orbital flight (FMOF), if all planned tests were successful and if adjustments to the orbiter's maneuvering-system pod and the solid-fuel rocket motors were complete. Unforeseen problems or unsuccessful tests could delay the schedule; however, NASA believed the probability of flying the FMOF during CY1979 was high.

Significant Shuttle problems disclosed by the review had been in the main engine and in vehicle weight. Engine development, although slower than desired, had produced a design that, if tests went well, could be flight-certified by Sept. 1979. The weight problem had not affected early flight tests, but might hamper both the Galileo mission to Jupiter and certain USAF missions. Measures to reduce weight in the orbiter and external tank could satisfy all mission requirements until mid-1984, and performance augmentation options now under study could meet Shuttle mission requirements after that time.

NASA would need additional funding to support the revised FMOF schedule, and to complete the Shuttle design and development test and evaluation program. Funding requirements had arisen not from any single program element but from several items (including the main engines, solid-fuel rocket boosters, external tank, and thermal-protection system), on which more work was needed than had been estimated. Shuttle development costs in FY79 and FY80 had exceeded previous plans, and current estimates of total Shuttle development costs were 8 to 9% higher than the earlier estimated $5.2 billion (1971 dollars). NASA had discussed Shuttle funding with OMB as it affected the FY80 budget, and had reviewed with DOD the potential impact if additional Shuttle development funds were unavailable in FY79. Preliminary assessments had shown the first FMOF would be delayed an additional 6 to 9mo beyond the Sept. date, and delivery of production orbiters would be delayed up to 1yr. (NASA Release 78-145; NASA testimony, US Subcomm on Sp Sci & Appl, Sept 28/78; *JSC Roundup*, Sept 29/78, 1; *Marshall Star*, Sept 27/78, 1)

• The Dept. of Transportation announced that the FAA would install under contract early in 1979 a warning system at 17 additional U.S. airports to help pilots avoid hazardous low-level wind shear. Work under the $1.7 million contract would raise to 24 the number of airports with such equipment; the FAA, which had tested wind shear alert systems at seven other airports for a yr, had begun full-time operation of that equipment in early Sept. Wind shear (a change in wind direction and/or speed caused by thunderstorms and fast-moving weather fronts) could be particularly hazardous during takeoff and landing. An aircraft at those times might not have sufficient speed or altitude to recover from sudden and dramatic wind shifts.

The low-level wind shear alert system (LLWSAS) developed for FAA by EMR Telemetry, Sarosota, Fla., used a minicomputer to measure wind speed and direction from as many as 5 wind sensors located on the airport periphery, and to compare these readings with wind data from the airport center. Detection of a significant difference in wind velocity (approximately 15 knots) would trigger aural and visual alarms in the control tower. Alerted by air traffic controllers, pilots could request an alternate runway, adjust approach or takeoff procedures, or execute a missed approach. (FAA Release 77-78)

• The People's Republic of China had launched eight spacecraft, some of them military reconnaissance satellites, *Av Wk* reported. Mainland Chinese had confirmed earlier speculation that reentry packages returned from three PRC spacecraft were of that type. *Av Wk* also said that *Cosmos 1031* launched Sept. 8 from Plesetsk by the USSR, appeared to be a close-look reconnaissance satellite to report on NATO's Northern Wedding exercise to repel a simulated invasion of the Shetland Is. north of Britain. The satellite had an unusual perigee location of 60°N over the Shetlands; its inclination was 62.8°. (*AvWk*, Sept. 25/78, 9)

September 26: Tass reported that, in an interview marking the first anniversary of the *Salyut 6* launch, Pilot-Cosmonaut Dr. Konstantin Feoktistov had recommended designing a new orbiting scientific Salyut laboratory with a minimum of 7 or 8 docking units. The presence of only 2 docking units on *Salyut 6* meant that, after a cargo vessel had arrived and its equipment was transferred, it had to undock to make room for another vehicle. This was irrational, Feoktistov said. When a cargo ship departed, the Salyut would lose one complete "room" of the cosmic house (with a volume of about 6m^3) that could accommodate equipment and scientific instruments. To add even one more component to a Salyut was not simple, as virtually its entire surface was covered with solar batteries, windows, television cameras, and various sensors. The nose fairing covering the station in the active phase of the launch also had restricted designers. Feoktistov stressed, however, the need to equip future space stations with several mooring ports to handle units for specialized research: technological, astronomical, and geophysical. (FBIS, Tass Intl Svc in Russian, Sept 26/78)

September 27: NASA announced that LaRC physicist Dr. Joseph Heyman had received his third IR-100 award given by *Industrial Research* magazine for one of the 100 most significant new technical products of 1978. Presentation of the award to Dr. Heyman at a banquet in Chicago Sept. 21 marked the first time the magazine had selected one person three times for the honor. Dr. Heyman's previous awards had been in 1974 and 1976. His 1978 device, jointly developed at LaRC and Washington Univ., St. Louis, was an acoustoelectric transducer that used very high-pitched sound waves to improve accuracy in measuring deviations in structural properties. Unlike conventional transducers, the new device was insensitive to the shape of sound waves, and would allow researchers to identify and evaluate more accurately material flaws and to characterize biomedical specimens and tissue. Heyman's first IR-100 award had been for a continuous-wave ultrasonic microemboli monitor, which traced impurities in a patient's blood during open-heart surgery. His 1976 award had been for an ultrasonic bolt stress monitor to indicate changes in bolt strain of less than one part in 10 000. (NASA Release 78-147; LaRC Release 78-48; *Langley Researcher,* Sept 22/78, 5)

• NASA reported it had launched a 2-stage Nike-Apache sounding rocket at 1:30 am EDT Sept. 27 from Wallops Island to investigate ionization sources in the upper atmosphere during geomagnetic activity. The experiment had been postponed day to day since June 2, awaiting an intense magnetic storm that usually followed solar flares. The rocket payload had reached a peak altitude of 183km (114 statute mi), and preliminary results indicated that all instrumentation performed satisfactorily and obtained good data. The launch was the last in a series for the Joint American-Soviet Particle Intercalibration (JASPIC) Project to compare U.S. and Soviet particle-measuring instrumentation [see May 15]. The two nations had jointly conducted eight rocket-borne experiments in June from the Wallops range and the Soviet research ship, Professor Vize, located off Va.'s eastern shore. (WFC Release 78-1)

• MSFC announced it had test-fired Sept. 6 at the White Sands test facility a version of the rocket engine that would maneuver the Space Shuttle orbiter. The 10sec-duration burn was the first firing of the 6000-lb thrust engine in combination with propellant tanks, feedlines, and other components of the orbital maneuvering subsystem (OMS) pod. OMS engines using nitrogen tetroxide and monomethyl hydrazine as oxidizer and fuel had been designed for reuse in up to 100 space missions, with 1000 starts for a total firing lifetime of 15hr. The engines would provide thrust for final orbital insertion at the end of launch phase, orbit circularization, orbit transfer, changing and rendezvous maneuvers, and deorbit burns. (*Marshall Star*, Sept 27/78, 2)

• The *Congressional Record* reported that Sen. Adlai Stevenson III (D-Ill) had submitted S. 3530 proposing specific goals for the U.S. space program in the next 10yr, including large space structures, solar-power satellites, public service communications, space processing, and various space science initiatives, with procedures for the president to translate the goals into specific budget actions.

"The United States stands today on the threshold of a 'new era in space,' . . . inaugurated with the launching of the space shuttle late next year, will be characterized by . . . 'routine utility.' The prospect of conducting frequent trips to near earth orbit—up to 60 per year according to current projections—aboard the shuttle opens a range of applied and scientific activities not feasible with expendable rockets.

". . . . a commitment by the Federal Government is needed to initiate this new era in space. . . Decisions during the next year or so will more than likely set the course of U.S. space activities for the balance of this century. A balanced and reasonable program, as set forth in this legislation, will insure this Nation's leadership in developing the space environment for the benefit of people in this country and around the world. . . " (*CR*, Sept 27/78, S 16354)

• NASA, which had dropped plans in 1977 for a 1982 mission to rendezvous with Halley's comet, had decided on a 1985 mission to rendezvous with comet Tempel 2 that would include a flyby of Halley's comet and deployment of a probe into its coma, *Defense/Space Business Daily* reported. Powered by a solar-electric (ion) propulsion system, the spacecraft would deploy its probe toward Halley in 1985 and would rendezvous with Tempel 2 in 1988. Until mid-1978 NASA had been considering as a new start for FY81 a 1985 mission using ion propulsion to comet Encke. In testimony before the House Space Subcommittee, Associate Administrator Dr. Noel Hinners had said NASA would make systems-definition studies in FY79 of the spacecraft, science, and ion-drive element of the Tempel 2 mission. (*D/SBD*, Sept 27/78, 108)

September 28: NASA announced that it and the U.S. Army Aviation Research and Development Command at WFC had accepted the first rotor systems research aircraft (RSRA) to provide the U.S. with a sophisticated and much needed research facility for continued development of rotary wing technology. NASA and the Army would obtain data from the RSRA helicopters to verify and to investigate technology of existing and advanced rotor systems. RSRA would make unnecessary costly and extensive modification of helicopter aircraft for flight investigations of each new rotor design, and could produce precise measurements of a variety of rotors under repeatable test conditions.

RSRA featured the first helicopter-crew emergency-escape system (an active isolator/balance system) and could be configured as a helicopter, compound aircraft (helicopter with fixed wings), or fixed-wing aircraft. The active isolator/balance system's main rotor vibratory forces and moments could provide wide band attenuation of vibrations to the fuselage, permitting installation of different rotors without retuning the fuselage. The emergency-escape system allowed upward extraction of the crew, using tractor rocket motors to sever the main rotor blades and separate and fragment the canopies. The system, consisting solely of mechanical and pyrotechnic components, was insensitive to electromagnetically induced radiation and lightning.

Acceptance marked completion of the design, fabrication, and flight test of the second RSRA delivered in helicopter configuration. The first RSRA had been scheduled for acceptance after completion of flight tests in the compound configuration. Sikorsky Aircraft Div. of United Technologies Corp. had built the RSRAs for NASA and the Army. (NASA Release 78-148)

• NASA announced it had rescheduled launches of the TIROS-N third-generation operational meteorological satellite and the Nimbus-G pollution-monitoring satellite, originally set for mid-Sept. to Oct. 14 and 23, respectively. NASA had postponed the WTR launches because of malfunction of an onboard computer in TIROS-N, and subsequent discovery of reworking required for tape recorders on both TIROS-N and Nimbus-G [see Sept. 18]. NASA had returned the tape recorders to the manufacturer for correction of the suspected problem (loose solder particles) and would reinstall the instruments in the spacecraft over the next 2wk. (NASA Release 78-149)

• NASA announced it would award Rep. Olin Teague (D-Tex.) its Distinguished Public Service Medal Oct. 3 during a 20th anniversary observance. Chairman Teague, who headed the House Committee on Science and Technology, had announced he would retire from Congress after 2 decades of service. NASA had declared his unswerving support of the agency and the U.S. space program an inspiration to all NASA employees. (NASA anno Sept 28/78)

• LaRC announced it had selected Dr. Larry Pinson, aerospace engineer in its structures and dynamics division, to participate in the President's executive interchange program, designed to foster improved relations between the government and private industry by yearly having 50 to 75 middle management executives from both sectors spend a yr in the opposite sector gaining in-depth on-the-job experience. The program had trained more than 450 executives since its inception under President Lyndon Johnson.

Pinson had begun his term with the Aerospace Corporation, Los Angeles, in August. He had received awards for work on vibration of space vehicles, and had been a consultant during the Apollo program on various studies of light problems. He had received a Ph.D. in engineering mechanics from the Va. Polytechnic Inst. and had begun his NASA career at LaRC in 1963. (LaRC Release 78-51)

September 29: NASA Administrator Robert Frosch announced that, upon signature by the President of the Inspector General Bill (or on October 1, 1978, whichever was later), NASA would establish an Office of Inspector General consisting of the present NASA management audit office and the NASA office of inspections and security. Pending nomination by the President and confirmation by the Senate of a NASA inspector general, Robert Allnutt would act in that position. (NASA anno Sept 29/78)

• LaRC announced that Dr. Donald "Deke" Slayton would present a portrait of Virgil (Gus) Grissom on loan from NASA to the Grissom Library, Newport News, Va. The library, opened in 1961, had been named in honor of Grissom in Aug. 1961 after his historic flight as pilot of the Mercury-Redstone 4 (Liberty Bell 7) suborbital mission. Grissom, who had lived in nearby Stoneybrook Estates from 1959 to 1962, had kept in touch with the library by working with the children's summer reading program. He had been killed in Jan. 1967 with Edward White II and Roger Chaffee in an Apollo spacecraft fire during tests at KSC. Grissom and Slayton, who were among the original seven astronauts selected by NASA in 1959, had trained and worked at LaRC during the Mercury program. (LaRC Release 78-52)

• In an interview on NASA's 20th anniversary, JSC director Christopher Kraft, Jr., said that speeding up the space program would require private capital. "Production in space will become an important part of our competitive position and have a great deal to do with the balance of trade." Speculating on the future at JSC, Kraft said he believed it "highly probable that we'll see a well developed space industry in the next ten years. I have an intuitive feeling that the thing is going to mushroom on us."

First major project for JSC would be to make the space transportation system operational: "If we're going to make this thing a routine operation, it's got to go in a more matter-of-fact manner, more like an airline operation, than we've seen in the past." For the first 6 or 10 Space Shuttle flights, center employees could expect a level of activity similar to that during Apollo and Skylab "but after that point we're going to have to phase into this new mode." He emphasized he wasn't implying airline pilots would fly the Space Shuttle: "It will still require highly trained astronauts because it is a very complex vehicle. With development of the transport system, we will see a new kind of astronaut—techniques will go into space very rapidly. The need for specific talent and skills will develop."

Kraft predicted that, in the next few years, the U.S. would "recognize a need for permanency of man in space. The best way to do that is with a space station. We will probably develop a permanent space station that would be carried up as modules in the Space Shuttle." (*JSC Roundup*, Sept 29/78, 4)

- JSC announced that technicians had installed four radiator panels and a flash evaporator system (FES) in a chamber of Building 32 to prepare for Nov. tests of the Space Shuttle heat rejection system. Radiator panels that would reject heat and maintain a livable temperature for Space Shuttle crews would contain Freon 21 flowing continuously through system loops, picking up heat throughout the ship and depositing cold; heat would be pumped to a central location and vented into space. From 14 000ft to orbit, the FES would cool the ship; in orbit, it would supplement panels that would open and, using ship's maneuvers, reflect heat. The Nov. tests would close off the 120-ft-high chamber and pump liquid nitrogen through tubes in the black walls, dropping temperatures to $-320°$; infrared lamps would then increase temperatures to the level required in space. This would be first testing of all system components as a unit. (*JSC Roundup*, Sept 29/78, 1)

September 30: The *Congressional Record* reported that former astronaut Sen. Harrison Schmitt (R-N.M.) had introduced S. 3541, the NASA Solar-Power Satellite Program Act of 1978, authorizing research, development, and demonstration of obtaining electrical power from space through solar-power satellite systems or other energy sources. Successful demonstration of the technical and economic feasibility of this concept would supply an inexhaustible future energy both for domestic purposes and for export. "Either way," said Sen. Schmitt, "such a demonstration could have a significant impact on our nation's balance of trade."

The bill would instruct NASA to begin a program aimed at resolving economic and technical problems surrounding the concept, including systems definitions; health, safety, environmental and socioeconomic studies; and technology assessments of various space energy sources compared with terrestrial energy sources. Favorable evaluations "would accelerate the design of a program to make this promising concept operational at the earliest possible date," Harrison concluded. (*CR*, Sept 30/78, S 16715)

During September: NASA announced that scientists had confirmed detection by *Pioneer Venus 1*, on the first leg of its 7mo voyage to Venus, of a powerful gamma-ray burst from somewhere in the universe. So-called gamma-ray bursts, unknown until 1973, had enormous energies and occurred about once a month from random points in earth's galaxy or even beyond. Discovery of their origin—black holes, brilliant supernova, neutron stars, or some totally unexpected source—would be a difficult task for astronomers. *Pioneer Venus* measurements of these bursts during its 482million-km mission should enable scientists for the first time to track the origins accurately.

Two other American spacecraft (*Vela*, a DOE satellite orbiting the earth, and *Helios B*, a NASA-European scientific satellite orbiting the sun) had detected the gamma-ray burst; scientists would correlate observations of all the satellites to get a rough fix on the burst. Locating the origin of the burst would give scientists an idea of the extraordinary physical events that could produce such high-energy explosions; no current hypotheses could account for them. (*NASA Actv*, Sept/78, 19)

• The USAF reported it had mapped and measured on the most detailed Shuttle models tested so far at AEDC the air pressures and loads to be encountered by the Space Shuttle during launch. The 6wk program had been the first to use mockups of the complete Shuttle orbiter, external fuel tank, and solid-fuel rocket boosters in AEDC's 16-ft transonic wind tunnel. NASA would compare results of these comprehensive tests with data obtained at AEDC and elsewhere on smaller, less detailed models.

NASA would follow up on the AEDC transonic tests with supersonic tests of the same models at Ames Research Center. AEDC had recorded surface pressures on a 3% scale model with 1646 measuring points scattered over its surface, one of the most comprehensive pressure models ever used there. It had tested a 2% scale model in the aerodynamic-force part of the study using 9 measuring instruments, the most ever installed on an AEDC model; and had measured air pressure at 25 points around the base of the Shuttle model. (AFSC *Newsreview*, Sept 78, 4)

• Using the space environment for materials processing had limited application, and should be considered only on a case-by-case basis, according to a report by the Natl. Research Council's committee on materials processing in space, which had studied NASA's program. The committee said the early NASA program had used poorly conceived and designed experiments, often with crude apparatus, from which weak conclusions had been drawn and in some cases overpublicized. However, meaningful science and technology could result from experiments in space if the problems to be investigated in space were soundly based in terrestrial science or technology, and if the proposed experiments addressed specific problems instead of being scheduled to take advantage of flight opportunities or space-facility capabilities. The committee said it had not found economically justifiable reasons for producing materials in space, and recommended that NASA deemphasize this area of technology. (NRC *News Report*, Sept 78, 7)

October

October 1: President Jimmy Carter attended the Congressional Space Medal awards [see Sept. 22] ceremony Oct. 1 at Kennedy Space Center's Shuttle landing facility. The White House press release of his remarks included the following:

" . . . We are here today to recognize and honor six American pioneers of the farthest and highest of all frontiers, the frontier of space. We honor them for individual human qualities, dedication, skill, extraordinary courage. But we do more than that. What those men have done is the most visible part of a vast and continuing collective accomplishment of many people. Tens of thousands of Americans, many of you, including scientists, engineers, administrators, skilled workers, others, have contributed directly to the success of the American space enterprise. They in turn have had the support of an entire nation and the good wishes of an entire planet. The glory that belongs to the six recipients this afternoon of the Space Medal of Honor belongs equally to those who helped them, and in a real sense, to all humanity who prayed for them and who supported them.

"It is fitting that these ceremonies take place today on the 20th anniversary of the founding of the National Aeronautics and Space Administration. And it is, of course, fitting that they take place here where the ships that took men to the moon were launched and where we will take our next great step into space with the first flight of the Space Shuttle, which I sincerely hope will be before my next birthday. . . .

" . . . In the last analysis, the challenge of space takes us very close to the heart of things. It brings us face to face with the mysteries of creation, of matter, of energy, of life itself. The men we honor today met that challenge and they were equal to it. Our nation met that challenge and was equal to it. And in the final two decades of the 20th century, Americans will reach out once more to the beauty and to the mystery of space, and once again, America, you and I, will be equal to that great task." (White House Release Oct 1/78)

• Wallops Flight Center announced that NASA, the Army's Atmospheric Sciences Laboratory (ASL), the Air Force Geophysical Laboratory (AFGL), and the National Research Council (NRC) of Canada would undertake a joint scientific and technical support program to study the total solar eclipse of Feb. 26, 1979, last observable from North America in this century, and its effect on earth's atmosphere and ionosphere. Aimed at improving predictions of atmospheric responses to a variety of disturbances, the program would include sounding rockets to be launched from a NRC/ASL small rocket site at Cochinour and from the NASA/ASL large rocket site on the Chucuni River, both in Canada, that would obtain measurements necessary for a quantitative description of the middle-atmosphere response. NASA would be lead agency for the U.S.; NRC would provide construction, ground- and flight-safety systems, launch coordination, and other launch-site support.

WFC had planned 7 launches during the eclipse: 3 payloads with joint experiments for the Univ. of Illinois and the Univ. of Bern, Switzerland; 1 payload with experiments for Cornell Univ. and NRL; 1 payload for the Univ. of Pittsburgh; and 2 payloads for Penn. State Univ. Site preparation had begun in June; launchers would be installed in Sept. WFC would provide mobile instrumentation systems to support launching, tracking, and data acquisition. (WFC *Newsletter*, Oct 1/78, 5)

October 2: NASA announced award of an approximately $4.3 million contract to Control Data Corporation, Minneapolis, Minn., for maintenance of the Langley Research Center computer complex, consisting of computer systems and auxiliary devices for processing scientific data to support center research programs and the scientific, engineering, and other technical personnel performing analyses, experimental data reduction, and real-time simulations. Work on the 2yr contract with three 1yr options for extension began Sept. 28. (LaRC Release 78-55)

• NOAA announced that the American Water Resources Assn. would sponsor the fifth annual William T. Pecora memorial symposium June 11-15, 1979, in Sioux Falls, S. Dak. The Dept. of Interior's U.S. Geological Survey EROS data center would host the program, to focus attention on satellite remote sensing for observing the hydrosphere as a means of averting water shortages and of water management. Participants—scientists from NOAA's Natl. Environmental Satellite Service, the U.S.G.S., NASA, the U.S. Army Corps of Engineers, universities, and private industry—would present papers on applications of space technology to all phases of hydrology and water resources. New satellites such as *Landsat 3*, *Nimbus 7*, the heat capacity mapping mission satellite, and *SeaSat 1*; new or improved sensors, including the scanning multifrequency microwave radiometer and synthetic-aperture radar; improved resolution on Landsat and HCMM thermal scanners; and new thermal bands on the advanced very high-resolution radiometer had affected hydrological technology and would be the subject of papers describing these improvements and capabilities. (NOAA Release 78-119)

October 3: LaRC announced it had selected Mercury Co., Tustin, Calif., to negotiate a contract for a pressure systems certification/recertification and configuration management program, including engineering evaluation of ground-based pressure systems to ensure a safe environment and to facilitate operation of the center's unique high-energy research facilities. The contractor would make in-service inspection and certification of these pressure systems in accordance with NASA guidelines and

In a ceremony at KSC marking NASA's 20th anniversary Oct. 1, President Jimmy Carter awards the Congressional Space Medal of Honor to six astronauts: Neil Armstrong (center), first man on the moon; Charles Conrad, Jr., commander of the second lunar landing and of the first crew to occupy the Skylab space station, and Frank Borman, commander of the first manned spacecraft to orbit the moon (seated at left); also to Alan B. Shepard, first American in space; Virgil I. Grissom (posthumously), second American in space; and John H. Glenn, Jr., first American to orbit the earth. (NASA photo 78-H-610)

would continue the existing configuration management program. LaRC would manage the approximately $1.15 million 2yr contract, which had a 1yr option for extension valued at approximately $637 000. (LaRC Release 78-56)

October 4: Two U.S. payload specialists would join three Europeans at Porz-Wahn, W. Germany, Oct. 6 to begin training for Spacelab 1, first combined manned space mission of NASA and ESA, Marshall Space Flight Center reported. They would attend sessions in W. Germany and two other European countries conducted by the scientific investigators planning European-sponsored experiments to fly on Spacelab 1, the two NASA mission specialists assigned to the flight, with MSFC's Charles Lewis and NASA training coordinator. Porz-Wahn was the site of ESA's Spacelab payload integration and coordination in Europe (SPICE), which had coordinated European crew training. After the European trip, the group would begin a tour of U.S. scientific installations in Jan. 1979. (*Marshall Star*, Oct 4/78, 1)

• Two National Academy of Sciences committees, those on space physics (of the Space Science Board) and on solar terrestrial research (of the Geophysics Research Board), would meet for the first time at MSFC, the center announced. The former was developing for NASA a strategy for a 1980s research program in solar system space physics. Committee members would include 35 top-level scientists from government, industry, and universities throughout the U.S. (*Marshall Star*, Oct 4/78, 2)

• The *Congressional Record*, at the request of Congressman Olin D. Teague (D-Tex), reported the remarks of Dr. Robert A. Frosch, administrator of the National Aeronautics and Space Administration, who introduced the President at the KSC ceremony, presenting for the first time the Congressional Space Medal of Honor to six astronauts. Dr. Frosch said in part:

"NASA people have been privileged, not only to continue and to advance the 20th century transportation revolution in aeronautics, but to participate in laying the foundations of the future human adventure in and use of space. . . . The true marks and memories of a civilization are found not only in its physical creations, and how it cares for itself and its people, but in the progress it has contributed toward liberation of the mind from the constraining horizons of the past. One major force has been the growth of scientific knowledge through the exercise of human reason.

"The spirit of man aspires to the stars. Every generation has looked out at the universe in search of faith, of hope, of beauty, of adventure, and of the power of knowledge. In our generation, a very few have taken the first brave steps out there to begin the great human adventure of using and understanding the universe beyond the earth. To honor those recent accomplishments, and to lead our way into America's next decades of space, it is my privilege to introduce the President of the United States." (CR, Oct 4/78, E 5400)

October 5: NASA announced it had presented Rep. Olin Teague (D-Tex), chairman of the House Committee on Science and Technology, with NASA's distinguished public service medal on Oct. 3 in an outdoor ceremony at NASA Hq. NASA Administrator Robert A. Frosch lauded Teague's efforts on behalf of NASA, noting Teague's "unswerving faith in the value and virtue of a dynamic and imaginative space program—a faith which he has conveyed to all the elements of the government, industry, and university team on which all progress in space depends." Teague had announced he would retire from Congress after more than 30 yr of service in the House of Representatives. (NASA Release 78-150; NASA anno Oct 3/78)

October 9: ESA announced that *Geos 2* had experienced a short-circuit between its solar cells and the spacecraft structure that had perturbed its functions. Although without major impact on mission objectives, the defect (origin of which had not been established) was affecting 3 of the 7 payloads; experiment S-300 to measure magnetospheric fields and waves, and experiments S-302 and 303 to measure low-energy particles. Perturbations for very short periods corresponding to illumination of the damaged part of the solar array had degraded the signals transmitted by the three affected experiments. The experimenters could probably reconstitute the small amount of data lost. The remaining four experiments and all subsystems were functioning normally. (ESA Release Oct 9/78)

• *AvWk* reported that the high-energy astronomy observatory *Heao 1* was each day discovering about three new x-ray sources in deep space. NASA expected that by the end of the mission in early 1979 the spacecraft would have found more than 1000 new sources. *Heao 1* data had also revealed a new black-hole candidate in the constellation Ara, bringing to four the number of such candidates identified. (*AvWk*, Oct 9/78, 13)

• *AvWk* commented on President Jimmy Carter's acknowledgment in a KSC speech that the U.S. had been operating military reconnaissance spacecraft, and on the possibility of much wider dissemination of the information in reconnaissance-satellite pictures. Although everyone had long known that such systems existed, official announcement of "the fact of existence" would allow the U.S. a number of strategies: the reconnaissance-satellite pictures would be evidence to Congress of Soviet or other hostile activities; the White House might make available to the public pictures with lower than current state-of-the-art resolution; and some pictures might be immediately usable by federal agencies such as the Interior and Agriculture Depts. The announcement would also inform the public that the U.S. could actively monitor arms agreements with the Soviet Union, instead of relying on Soviet assurances that they were honoring agreements. (*AvWk*, Oct 4/78, 22)

October 10: NASA announced that the two Pioneer spacecraft en route to Venus—the orbiter and the multiprobe—had passed major operations tests. Tests of orientation, timing, and separation systems aboard *Pioneer Venus 2* for split-second release of three probes 9600km (6000mi) apart over Venus's earth-facing hemisphere, and of systems on *Pioneer Venus 1* (orbiter) controlling retrofire and injection into orbit (to take place behind the planet and out of reach of earth communications), had been satisfactory.

Separate operations teams at Ames Research Center had sent 6400 commands to the orbiter since its May 20 launch and 3600 commands to the multiprobe craft since its Aug. 8 launch. Virtually all experiments and systems (including thermal, orientation, command, communications, and data-return and power systems) on each of the multiprobe's five entry craft had been operated and were working well. The orbiter's instruments had been checked and calibrated, including systems for insertion into Venus orbit Dec. 4.

Twenty days and 13 million km (8 million mi) from Venus, the multiprobe would split into a transporter bus and four probes, the bus serving as fifth probe. The multiprobe had covered 140 million km (87 million mi), the orbiter 325 million km (202 million mi). Both spacecraft still had about 180 million km (112 million mi) to travel to Venus. (NASA Release 78-153; ARC Release 78-48; ARC *Astrogram*, Oct 19/78, 1)

• The Jet Propulsion Laboratory reported that NASA and DOE had selected 10 companies to begin contract negotiations for projects to improve current methods of manufacturing modules of photovoltaic cells. The $4 million effort would focus on improving production methods to reduce costs in 1979–1981, and would supplement current DOE efforts to make solar energy economically competitive with conventional power sources.

Firms selected were Arco Solar, Inc., Chatsworth, Calif.; Energy Materials Corp., Ayer, Mass.; Kayex Corp., Rochester, N.Y.; Kluicke & Soffa Industries, Inc., Horsham, Pa.; MB Associates, San Ramon, Calif.; Motorola, Inc., Semiconductor Group, Phoenix, Ariz.; RCA

Corp., David Sarnoff Research Center, Princeton, N.J.; Sensor Technology, Inc., Chatsworth, Calif.; Siltec Corp., Menlo Park, Calif.; and Sollos, Inc., Los Angeles, Calif. In the specified time period, the firms would develop techniques for mass-producing large modules of silicon-based photovoltaic cells for use by industries producing flat (non-concentrating) photovoltaic modules. Techniques would include new ways to slice silicon solar-cell material and to assemble photovoltaic modules. JPL would manage the effort (designated Low-Cost Solar-Array Project) for DOE. (NASA Release 78-151)

October 11: NASA announced that *SeaSat*, an experimental ocean-monitoring satellite launched June 26 from WTR, had stopped transmitting data shortly after midnight EDT Oct. 10 while over Australia. Scientists and engineers at JPL and Goddard Space Flight Center were working with Lockheed Missiles and Space Co. engineers to determine cause of the malfunction. (NASA Release 78-156)

• NASA reported the status of Voyager spacecraft as of Oct. 5, 1978: Spacecraft distance from earth: *Voyager 1*, 704 692 000km (437 875 000mi); *Voyager 2*, 668 582 000km (415 437 000mi). Spacecraft distance to Jupiter: *Voyager 1*, 105 983 000km (93 816 000mi); *Voyager 2*, 201 354 000km (125 117 000mi). Spacecraft distance traveled since launch: *Voyager 1*, 816 533 000km (507 370 000mi); *Voyager 2*, 831 435 000km (516 630 000mi). Spacecraft velocity relative to earth: *Voyager 1*, 24.87km/sec (55 632mi/hr); *Voyager 2*, 27.10km/sec (60 620mi/hr). Spacecraft velocity relative to sun: *Voyager 1*, 15.43km/sec (34 514mi/hr); *Voyager 2*, 14.18km/sec (31 727mi/hr). Date of Jupiter encounter: *Voyager 1*, March 5, 1979; *Voyager 2*, July 9, 1979. Date of Saturn encounter: *Voyager 1*, November 12, 1980; *Voyager 2*, August 27, 1981. (NASA Release 78-154)

• On the basis of a 4mo interagency review requested in June 1978, President Carter announced a U.S. civilian space policy that would set the direction of U.S. space efforts over the next decade. The policy would reflect a balanced strategy of applications, science, and technology development that would:

1. emphasize space applications that would increase understanding of earth resources, climate, weather, pollution, and agriculture, with the private sector taking increasing responsibility for remote sensing and other applications;

2. emphasize space science and exploration in a manner that would retain the vitality of U.S. space technology, yet would provide short-term flexibility to impose fiscal constraints when conditions warranted;

3. take advantage of Space Shuttle flexibility to reduce the cost of operating in space over the next two decades to meet national needs;
4. increase benefits for resources expended through better integration and technology transfer among national space programs and through more joint projects when appropriate, thereby increasing the return on the $100 billion investment in space to the benefit of the American people;
5. assure U.S. scientific and technological leadership in space for the security and welfare of the nation, and continue R&D necessary to provide the basis for later programmatic decisions;
6. demonstrate advanced technological capabilities in open and imaginative ways that would benefit developing as well as developed countries;
7. foster space cooperation with nations by conducting joint programs;
8. confirm U.S. support of the continued development of a legal regime for space that would assure its safe and peaceful use for the benefit of mankind.

Since space was becoming a place to work, an extension of the environment, the U.S. would pursue future activities in space when it appeared that national objectives could most efficiently be met through space activities. Finally, the U.S. found it neither feasible nor necessary to commit itself to a high-challenge space-engineering initiative comparable to Apollo. As resources and manpower requirements for Space Shuttle phased down, the U.S. would be able to give greater attention to new space applications and exploration, continue programs at present levels, or contract them. To meet the specified objectives, an adequate federal budget commitment would be made.

The U.S. would maintain a position of leadership in space science and planetary exploration, said the President, continuing a vigorous program of planetary exploration and using the space telescope and free-flying satellites to usher in a new era of astronomy, space systems to develop a better understanding of the sun and its interaction with the terrestrial environment through space systems, and the Space Shuttle and Spacelab to conduct basic research. (White House Release Oct 11/78)

• MSFC reported the first-time assembly of all elements of a Space Shuttle—two solid-fuel rocket boosters (SRBs), an external tank, and an orbiter—in the dynamic-test stand at MSFC. Next step would be resumption of mated vertical ground-vibration testing, with the upcoming series of tests being in launch configuration, the *Marshall Star* reported. National and regional press representatives were present to film and report on the historic first assembly of the entire vehicle. The SRBs would undergo launch-configuration tests filled with inert propellant, replaced by empty units for configuration tests representing the vehicle just before SRB separation. (*Marshall Star*, Oct 11/78, 1)

• The *Marshall Star* reported that Dr. Noel Hinners, NASA associate administrator for space science, had praised the performance of *Heao 1* in a report to the subcommittee on space science and applications of the House Committee on Science and Technology. Dr. Hinners noted that, before launch of *Heao 1*, scientists knew of about 300 X-ray sources in the sky; *Heao 1* was discovering about 3 new X-ray sources every day [see Oct. 9].

Another major result from *Heao 1* was its observation of a hot plasma, apparently about a half billion degrees, extending beyond earth's galaxy and possibly through the entire universe. A plasma of this temperature and density distributed uniformly throughout the universe would constitute about half the mass required to close the universe. "This concept is of such major importance for cosmologists that new *Heao 1* data are being analyzed as they come from the spacecraft," Hinners said. He also noted that *Heao 1* data indicated a new black-hole candidate in the constellation Ara. Scientific results such as these had been "so significant that we made a decision to extend the operation of *Heao 1* through March 1979," Hinners concluded. (*Marshall Star*, Oct 11/78, 4)

October 13: NASA announced it had given David Johnson, director of DOC's National Environmental Satellite Service (NESS), the 1978 William T. Pecora award for his work in applying satellite remote sensing to monitoring weather and the environment. NASA and DOI had presented the award annually since 1974 for outstanding contributions by individuals and organizations to remote sensing.

The 1978 award citation, presented to Johnson Oct. 11 in Sioux Falls, S.D., as part of the fourth annual Pecora memorial symposium on remote sensing, said that Johnson "has devoted a major portion of his scientific and managerial life to the development and implementation of remote sensing systems designed to observe weather patterns, impacts of weather and man on our environment, the climatological trends of the earth, and in the process eliminated or alleviated loss of property and life as a result of natural disasters." Dr. Pecora, who died in 1972, had been undersecretary of the Interior and a motivating force in NASA's Landsat earth-resources survey satellite program and in DOI's EROS (earth resources observation systems) program. (NASA Release 78-155)

• *Spaceport News* reported astronomers studying data obtained by the international ultraviolet explorer *IUE* satellite might have discovered a third black hole believed to be at the center of a globular cluster (a miniature galaxy shaped like a globe) some 15 000 light-yrs away. The IUE large ultraviolet telescope had penetrated the background light coming from a globular cluster and reached its core, where researchers found radiation from a group of 10 to 20 bright blue stars, apparently orbiting a black hole possibly small in size but with a mass equal to 1000 stars like the sun. Researchers would study movements and interrelationships of the stars to define the rotation about the center of the million-star cluster. (*Spaceport News*, Oct 13/78, 7)

• JPL reported it had awarded CalTech a 5-yr extension, effective Oct. 1, 1978, of its contract in force since 1962 for R&D work performed at JPL. Cost of the 5-yr extension, in contrast to previous 2 or 3-yr extensions, was estimated at more than $250 million annually. JPL would mark the 20th anniversary of its affiliation with NASA in Dec. 1978; before 1958, JPL had worked primarily for the Army Ordnance Corps. Running concurrently but separately with the R&D contract was an ongoing contract to provide buildings, equipment, and other facilities necessary for work performance. A separate memorandum of understanding (MOU) negotiated in Dec. 1968, stating terms of agreement between NASA and CalTech for JPL activities, had been updated and renegotiated and was due for completion by the end of 1978. (JPL *Universe*, Oct 13/78, 1)

• NASA and the state of Texas had signed a 3-yr agreement that the Texas natural-resources information system (TNRIS) would use Landsat computer-processed data in managing the state's natural resources, Johnson Space Center reported. Project costs were estimated at $600 000 for NASA and $750 000 for Texas. Day-to-day operations would use a mix of manual and computer-assisted remote-sensing methods to monitor Texas natural resources; photographic and electronic data from aircraft, ground-truth measurements, and weather-station environmental data would supplement satellite data during early phases of the project. JSC and a TNRIS task force would design and implement the data system; NASA would provide technical expertise. (JSC *Roundup*, Oct 13/78, 1)

• Sixty-five percent of the food carried on Shuttle missions would be cooked, packaged, and tested in a new JSC food lab, JSC reported. Orbiter menus would be standardized on a 6-day cycle. About 15% of Shuttle food—mainly flexpacks and thermostabilized food items—would be supplied by the U.S. Army's Natick Research and Development Command in Mass. Another 20% would be off-the-shelf items, from chewing gum and snap-top puddings to peanut butter.

The orbiter's lower deck would contain a galley with an oven, serving trays, water dispensers, and other serving equipment. When the galley was installed and operational (about mission SS-7), an hr before meal time the crewperson cooking would spend about 5 min rehydrating dehydrated food items, loading meal trays, and placing them in the galley oven. After meals, trays and silverware would be cleaned and sterilized for reuse.

JSC nutritionist Rita Rapp, who has seen space-food technology evolve from bland squeeze-tube pastes carried on Mercury spacecraft to the electrically-heated food trays in the Skylab wardroom, said, "We try to present foods in a normal way, more like people are accustomed to eating on earth. Operational experience permitted us to add new food items—wider selection and better packaging—as we learned weightlessness did not cause the problems everyone anticipated." (JSC *Roundup*, Oct 13/78, 4)

• Scientists and marine biologists trying to preserve the Great Lakes from pollution would receive vast amounts of new and more accurate information from a Nimbus-G satellite scheduled for launch Oct. 23, Lewis Research Center announced. Among its seven sophisticated instruments to measure parameters ranging from stratospheric aerosols and ozone to polar sea ice extent would be a coastal zone color scanner (CZCS) to measure water color and temperature, two characteristics affected by amounts of pollution and algae in water. The scanner would operate exclusively over water, taking pictures of an area encompassing the entire Great Lakes system in 2min, or of any particular area every 4 out of 5da, and would transmit the Great Lakes data to GSFC for recording and processing. Information in the form of photographs and digital-computer tapes would be archived and disseminated to assigned study groups.

A Great Lakes experiment team with scientists from LeRC, EPA, NOAA, the Canadian National Water Research Inst. and Center for Remote Sensing, the Environmental Research Inst. of Michigan, the Univ. of Minnesota, Case Western Reserve Univ., Ohio State Univ., and State Univ. of New York would assess the regional value of CZCS water-quality measurements. LeRC would be responsible for interpreting Great Lakes CZCS data and comparing it with coordinated ship measurements. (*Lewis News*, Oct 13/78, 1)

• The *Washington Post* reported the Carter Administration had decided to allow the Peoples Republic of China to buy an American communications satellite launched by the U.S. The administration was waiting for the Chinese to reconfirm interest in the deal, the most sophisticated technology transfer so far between two countries. Several yr ago the Chinese had reserved a slot for a communications satellite with the U.N. agency monitoring the field. Discussions were still tentative, as the Chinese would have to negotiate with the few U.S. comsat manufacturers and would have to buy an undetermined number of ground stations. A satellite would cost $15 to $17 million; placing it in orbit would cost $13 to $23 million, depending on the rocket used. Regardless of any military implications, the satellite sale would probably annoy the USSR, which had repeatedly warned of danger in Washington's "playing the China card" (cultivating closer Chinese ties while Peking-Moscow relations were strained). (*W Post,* Oct 13/78, C-14)

October 16: NASA announced that, as of Oct. 1, it had transferred the USNS Vanguard, which had supported Apollo, Skylab, and Apollo-Soyuz manned-spaceflight missions for 12yr and was last of an original 5-ship tracking and reentry-coverage flotilla, back to the U.S. Navy for navigation and ocean survey work. Vanguard, built by General Dynamics Corp.'s Quincy, Mass., shipyards and commissioned Oct. 15, 1966, had been the last of three identical ships (Mercury, Redstone, Vanguard) specifically designed for and dedicated to manned spaceflight support. The ships had demonstrated their versatility by supporting automated scientific-satellite missions and their extensive capabilities, functioning as well as any ground tracking station. The Mission Control Center at Houston had transmitted and received voice commands, biomedical data, spacecraft-environment reports, and command/control functions directly through the ship, which had formed part of a tracking network to ensure communications with spacecraft out of contact with land tracking stations. Space Tracking and Data Network (STDN) veterans estimated that fully half the Vanguard's NASA lifetime had been spent in a mission-support status, the rest being in port, yard repair, or transit time. (NASA Release 78-157)

• ESA announced its *Ots 2* satellite had successfully completed its first transmission tests of television programs beyond continental Europe. During a conference on "The Role of Space Technology for Development" in Cairo Oct. 7–12, ESA had demonstrated TV-signal transmission from the British Post Office's Goonhilly Downs station via *Ots 2* to a transportable receiving station in Cairo. From Oct. 7 to 11, nearly 8hr of BBC-1 TV programs were transmitted directly from Britain to Egypt at a rate of about 1.5hr per day. Ferranti Microwave Division of Ferranti Electronics Ltd., in association with the U.K. Dept. of Industry, had developed the transportable station.

After 5mo of preliminary tests, the *Ots 2* communications-test program had reached an operational level that would last at least 3yr, to prepare for subsequent exploitation of the operational European Communications Satellites (ECS) that would begin service in 1981. Tests had included telephone as well as TV routing and propagation experiments and new applications using much simpler terminals. Some 50 institutes, universities, and telecommunications entities would participate in the second category of tests, using more than 30 small terminals with 3m-diameter antennas. (ESA Release Oct 16/78)

October 17: NASA announced it had awarded the Douglas Aircraft Co.'s McDonnell Douglas Corp. a modification of its contract for engineering investigations, analyses, design studies, and component testing to evaluate alternatives in designing laminar-flow control subsonic commercial transports for the early 1990s. The modification for work to be completed by late 1979 would permit more complete exploration of promising design alternatives and evaluation of structural concepts, and would include an option to conduct a feasibility study on use of a representative transport aircraft for leading-edge systems testing and full-wing chord laminar-flow control flight testing. (LaRC Release 78-62)

• *Defense/Space Business Daily* reported that the United Nations had presented its Peace Medal to Dr. Robert Frosch, NASA administrator, at NASA Hq in recognition of NASA assistance to a recent conference on technical cooperation among developing countries in Buenos Aires, Argentina [see Aug. 30]. UN Undersecretary-General Bohdan Lewandowski presented the medal and a letter of appreciation from Secretary-General Kurt Waldheim for NASA's offer of the *Cts* satellite to the UN for communications between UN Hq in NYC and the conference. (NASA photo 78-H-625; *D/SBD*, Oct 17/78, 200)

October 18: NASA announced that Dr. Robert Frosch, NASA administrator, and Roy Gibson, ESA director general, had met in Paris Oct. 7 at ESA's annual Spacelab program meeting, during which NASA and ESA had signed three memoranda of understanding on the use of data transmitted by *Landsat 2* and *3, Nimbus G,* and *SeaSat.* The memoranda covered acquisition, preprocessing, and dissemination to national Earthnet stations (the European ESA network) of data transmitted by the spacecraft. They also dealt with acquisition and transmission of oceanographic data from microwave systems onboard SeaSat by the SeaSat Users Research Group Europe, under auspices of the European Association of Remote Sensing Laboratories (Earsel), to study possible inclusion of microwave detectors on future European satellites. (NASA Release 78-158; ESA Release Oct 9/78)

• Two experiments for HEAO-C had been delivered to the prime contractor, TRW, Redondo Beach, Calif., and the third was expected the following wk, MSFC reported. Scheduled for launch in late 1979, HEAO-C would survey gamma-ray emissions and study cosmic-ray particles from space. Like the first high-energy astronomy observatory launched in 1977, HEAO-C would be a scanning mission to perform sky surveys. The HEAO-B scheduled for launch in November 1978 would be a pointing mission, focusing on selected X-ray sources.

The two delivered experiments were a high spectral-resolution gamma-ray spectrometer (experiment C-1) and a heavy-nuclei experiment (C-3). Arriving Oct. 23 would be C-2, the European-produced experiment on isotopic composition of primary cosmic rays. Mechanical integration of the experiments would begin as soon as KSC returned mechanical ground-support equipment it was using to prepare for HEAO-B's launch. Meanwhile, TRW would check out individual experiments to assure the devices had arrived in good working condition. Following mechanical integration, TRW would make individual data runs through late Nov. Electrical integration would begin after KSC returned the computer equipment now supporting the HEAO-B launch. (*Marshall Star,* Oct 18/78, 1)

• MSFC reported that NASA had awarded Edward Buckbee, director of the Alabama Space and Rocket Center, its distinguished public service medal, highest award to persons outside the federal government, in a ceremony Oct. 17 in Washington, D.C. NASA Administrator Robert Frosch presented the medal to Buckbee for informing the public through the Alabama Space and Rocket Center of NASA's programs and activities, and for contributing to public understanding of the complexity of space and rocket technology. Buckbee had been a MSFC employee before joining ASRC. (*Marshall Star,* Oct 18/78, 1)

October 19: NASA reported its scientists and engineers were continuing attempts to revive *SeaSat,* the ocean-monitoring satellite that had suddenly gone mute before an orbital pass over the Santiago, Chile, tracking station [see Oct. 11]. Gene Giberson, SeaSat project manager at JPL, said that engineers there and at GSFC were assembling and analyzing tracking-station data from the spacecraft to determine possible cause of data interruption. NASA engineers would then work backward to reconstruct a "failure model" matching the data received just before and up to the instant the spacecraft ceased transmitting. If such a match could be made, the next step would be to find an alternate method of reestablishing contact with the spacecraft. The steps could take up to 2wk.

SeaSat had not had full benefit of its solar panels because they had received full sun during only half the spacecraft's orbits. In December, when the spacecraft would be in full sun during 100% of its earth orbits, its batteries might be charged enough to overcome a problem caused by insufficient electrical power. (NASA Release 78-159; *JPL Universe,* Oct 13/78, 1)

• NASA Deputy Administrator Dr. Alan Lovelace announced the convening of a review board to determine the cause of the *SeaSat* failure [see Oct. 11]. Lovelace named Dr. Bruce Lundin, recently retired director of LeRC, to head the review board, with other members to be named shortly. In 1973, Dr. Lundin had chaired the board that had investigated failure of *Skylab I*'s solar panels to deploy in orbit. (NASA Release 78-160)

October 20: NASA announced that JPL, acting for the Dept. of Energy, had awarded study contracts to Acurex, Inc., Mountain View, Calif.; Boeing Engineering & Construction Co., Seattle, Wash.; and General Electric Space Division, Valley Forge, Pa., to develop a solar-energy concentrator for generating electric power. The contracts, each valued at about $240 000, would cover the first part of a two-phase program to develop a low-cost point-focusing solar concentrator. The concentrator would direct mirror-reflected sunlight to a heat absorber and heat-driven engine turning a generator to produce electricity to supplement electrical power for small communities and rural areas.

The solar concentrator to be developed under the contracts would include the optics (mirrors) and the mechanism that allowed the concentrator to track the sun. Heat absorbers and engine generators would be developed under separate contracts. Prime objective of the contracts would be to obtain the best thermal performance at the least cost per concentrator. As many as six would be tested and evaluated at JPL's Solar Thermal Test Site at Edwards, Calif. JPL would manage the study for DOE's Small Power Systems Program under an interagency agreement with NASA. (NASA Release 78-161)

• Dryden Flight Research Center commemorated the final flight 10yr ago of the X-15, piloted by Bill Dana, that reached in its 11min free flight a top speed of 3716mph (5.38 times the speed of sound) and peak altitude of 255 000ft. This chapter in aeronautical history would be celebrated by the Edwards, Calif., AIAA section at a dinner meeting featuring Bill Dana. The article in *DFRC X-Press* concluded: "Recent press reports call the Space Shuttle engines the first reusable liquid rocket engines. How soon they forget." (*DFRC X-Press,* Oct 20/78, 2)

• President Carter at a press conference with editors and news directors was asked: "There was a big spread in the local (Cape Canaveral) newspapers yesterday that you were ordering cutbacks in the space program. And there are also rumors going around the Kennedy Space Center that as soon as the shuttle becomes operational, you will order even more cutbacks in an austerity program. So my question is this: What kind of space policy can we expect from your administration?"

The President said: "I think a very aggressive space policy. Anyone who reads the documents that have been prepared very carefully, very thoroughly by the Defense Department, the CIA, NSC, all those who will use them, including Agriculture, Commerce, and finally approved by me, would say that it's a very sound program based on scientific need and actually capitalizing now upon the great exploratory efforts that have been made in space. We look upon the Space Shuttle as a way to change dramatic, very costly initiatives into a sound, progressive, and innovative program to utilize the technology that we have available to us.

"We'll continue interplanetary space exploration. We'll have a greatly expanded effort concerning astronomy assessments of the earth, weather, communications. We'll expand our effort to bring into the space program now, both foreign countries and also private firms in our nation. And I think it is accurate to say that the Space Shuttle, which is approaching completion—we hope the first orbital flights will be less than a year from now—will open up a broad vista of new uses for our technology.

"So, we're not going to minimize or decrease our commitment to space at all. I think the spectacular efforts to send men to the moon and to make the first orbital flights, and so forth, have been just a precursor to now the more practical and consistent and effective use of our space technology.

"So, it's not a matter of playing down the importance of space; it's a matter of using what we've already learned in the most effective way." (Press Doc, Oct 20/78, 1976)

October 23: NASA's General Management Review Report presented data on the status of NASA program resources.

FY 1978 Obligations (in millions of dollars)

R&D:	Annual Plan	Actual (through Sept.)	%
Space Shuttle	1349.2	1346.7	100
Space Flight Operations	263.8	261.0	99
Expendable Launch Vehicles	134.5	120.3	89
Physics/Astronomy	224.2	192.9	86
Lunar/Planetary	147.2	128.0	87
Life Sciences	33.3	19.8	59
Applications	234.8	214.3	91
Aeronautics	228.0	197.3	87
Space R&D	88.7	83.9	95
Tracking/Data Acquisition	278.3	265.1	95
Technology Utilization	9.1	6.3	69
Energy	7.5	6.8	91
Planning/Program Integration	4.0	1.3	33
Low-Cost Systems	9.0	6.4	71
Total R&D	3011.6	2850.1	95
Construction of Facilities	162.3	97.0	60
Research & Program Mgt.	890.2	889.7	100
TOTAL NASA	4064.1	3836.8	94

(NASA GMRR, Oct 23/78, 15)

- A Senate operational remote-sensing bill introduced by Sen. Harrison Schmitt (R-N.M.) and a House space industrialization bill introduced by Rep. Don Fuqua (D-Fla.) had increased pending legislation to carry civilian space policy beyond that outlined by President Carter, *AvWk* reported. Both bills called for formation of government corporations to expand benefits of space exploration beyond the R&D stage. All the legislation would be reintroduced in the next Congress where, as a bloc, they would constitute the largest recent push for an expanded federal space effort.

  Other bills included a Space Policy Act introduced by Sen. Adlai Stevenson (D-Ill.), for extensive design and application of large space structures in orbit over the next 10yr; National Space and Aeronautics Act introduced by Sen. Schmitt, for a 30-yr planning process in U.S. space and aeronautics research; Earth Data and Information Service Act introduced by Sen. Stevenson, for an operational Landsat remote-sensing system as a new division of NASA; Solar-Power Satellite Research Development and Demonstration Program Act introduced by Rep. Ronnie Flippo (D-Ala.), for $25 million in near-term funding to research technologies required for a solar-power satellite; NASA Solar-Power Satellite Program Act introduced by Sen. Schmitt, for greatly expanded work on solar-power satellite technology. (*AvWk,* Oct 23/78, 19; *CR,* Oct 13/78, S 19075; *D/SBD,* Oct 24/78, 234; Oct 27/78, 257)

- Europe's space industry was looking beyond Spacelab to future programs that would maintain technological advances and employment as well as joint development activities with the U.S. in manned spaceflight, *AvWk* reported. With delivery of the first Spacelab scheduled for next yr, European industry said U.S. delay in ordering a second flight unit might disrupt skilled manpower employment and management performance. Also, European industry officials had told NASA that its charges for Space Shuttle flights were too high, and that Europe should have special treatment in Space Shuttle/Spacelab use. W. Germany's ERNO, Spacelab prime contractor, had proposed to NASA and ESA a second Spacelab as the basis for a barter agreement [see Sept. 12] under which costs of the second Spacelab would offset future Space Shuttle services for Europe.

  An ERNO spokesman had warned that NASA should not increase Spacelab 2 technical requirements to a point where the cost to Europe's space industry was unreasonable. On the other hand, the barter agreement should not give the U.S. full autonomy in operating Spacelab or producing future Spacelabs. Commercialization of the Space Shuttle was essential, because "twenty million dollars (1975 price basis) for a standard Shuttle launch and possibly the same amount for a standard Spacelab flight service are hardly suited to arouse the interest" of European users, particularly industry, the spokesman said. (*AvWk,* Oct 23/78, 17)

October 24: NASA announced that fabrication and testing of components was continuing at various locations throughout the U.S. in preparation for a scheduled first manned orbital flight of the Space Shuttle September 28, 1979. NASA had accelerated production and installation of thermal protection system (TPS) tile for Orbiter 102, had increased staffs at both the Rockwell facility and the Lockheed plant where the tiles were made, and had obtained new tile-inspection equipment.

Testing of the Space Shuttle main engine (SSME) continued at the Natl. Space Technology Laboratories; between Sept. 10 and Oct. 12, two engines had been test-fired for a total of 3794sec. Ten of the tests reached rated power level (RPL), accumulating 3096sec. Engine firings through Oct. 11 totaled 350 for 26 530sec total time, including 8909sec at RPL. NASA had scheduled full-duration testing of the three-engine cluster forming the main propulsion system for early in 1979, when the first manned orbital-flight configuration engines would become available.

The third static-test firing of a solid-fuel rocket booster (SRB) motor occurred Oct. 19 at Thiokol Chemical Corp.'s test site near Brigham City, Utah. Early data indicated the test firing and gimbaling of the motor nozzle were satisfactory.

MSFC had mated all the Space Shuttle elements—two SRBs, an external tank, and an orbiter (101)—for the first time in vertical vibration testing to verify that the structure would perform as predicted during various stages of flight. (NASA Release 78-162)

• NASA Deputy Administrator Dr. Alan Lovelace announced appointment of John Boyd as deputy director of DFRC effective Jan. 1, 1979. Since 1970 Boyd had served as deputy director of aeronautics and flight systems at ARC, where he had worked since 1947. Author of many technical reports, Boyd had received several honors for his work, including NASA's exceptional service award. (NASA Release 78-163; DFRC Release 20-78)

• NASA announced that its *Tiros-N* third-generation weather satellite launched from WTR Oct. 13 had returned excellent quality data and pictures from a near-perfect circular polar orbit at 870km (543mi) by 860km (537mi) altitude. Its advanced very high-resolution radiometer, the automatic picture transmission system, and the high resolution picture transmission system designed to provide local area coverage to hundreds of weather forecasters around the world were all working. After NASA completed a checkout in 3wk, it would turn over the satellite to NOAA for operation. (NASA Release 78-164)

October 25: NASA announced launch of Nimbus-G (*Nimbus 7* in orbit) on Oct. 24, 1978, at 4:14am EDT (08:14:00.72Z). Orbital values were: apogee, 954.6km; perigee, 953.6km; inclination, 99.2°; and period, 104.1min.

Onboard recorded data dumped during the first Madrid pass showed an apparent reduction in indicated telemetry level, starting at approximately "plus 4 min" and continuing for 30min, at which time the telemetry levels returned to normal. Analysis was under way. Solar arrays had extended, and were normal to and tracking the sun; spacecraft attitude was stable. All spacecraft systems parameters were normal, and activation of the experiment systems had begun. (MOR E-604-78-08 [postlaunch], Oct 25/78)

• LaRC announced selection of the Riggins Company, Inc., Hampton, Va., for negotiations leading to award of a contract to provide materials and services for construction of the Natl. Transonic Facility (NTF) at LaRC. The NTF would embody a unique wind-tunnel concept permitting use of temperatures as low as $-300°F$ to improve simulation of airflow over planes in full-scale flight conditions. The contract, to be managed by LaRC, would include provision of liquid nitrogen, gaseous nitrogen, cooling water, and air and vacuum piping systems required for the wind-tunnel test environment. Riggins Co. estimated the contract value at $2.9 million, with work to be completed by late 1981. (LaRC Release 78-64)

• The third of four Space Shuttle solid-propellant booster motors (DM-3) to be tested during the development phase had been fired successfully the previous week at Thiokol Corp.'s Wasatch Division in Brigham City, Utah, the *Marshall Star* reported. First of a prototype solid-fuel rocket motor (SRM) to be tested in flight configuration, the DM-3 included several features not tested in earlier firings: The motor aft skirt, containing a thrust-vector (directional) control-actuation system for the nozzle was on the motor for the first time. Also part of the overall test was recording data on the pressure inside the exhaust plume (flame), showing how much force would impact the launch platform at liftoff, that JSC and KSC would use in launch-facility studies. An onboard hydraulic system powered the thrust-vector control, moving the nozzle to change vehicle direction, for the first time in an SRM development test. Final event was a test of the linear-shaped explosive charge that would sever the nozzle aft-exit cone from the motor about 25mi above earth during Space Shuttle flights. (*Marshall Star*, Oct 25/78, 2; JSC *Roundup*, Oct 27/78, 1)

• The *Marshall Star* reported that MSFC had begun ground-vibration tests of the Space Shuttle in its liftoff configuration in the dynamic-test stand, first test of the Space Shuttle with all elements mated in flight configuration. The test phase would run through November. Test-preparation crews from Bendix worked around the clock to prepare the stand, the Shuttle, and the shaker systems for the second phase of ground-vibration testing. The Space Shuttle's weight rested on the aft skirts of the two solid-fuel rocket boosters bolted to hydrodynamic support systems at the base of the test tower. When pressurized, the support systems would allow adequate vertical, lateral, and rotational movement of the entire Space Shuttle.

Test engineers had predetermined a total of 83 shaker positions for the combination of Space Shuttle elements; using a spacecraft computers system to drive the shakers, the engineers could select different positions to determine vehicle response. The computerized system would automatically control up to 24 shaker channels from the 61 that were available, rated at 1000lb, 150lb, and 50lb force. (*Marshall Star*, Oct 25/78, 1)

October 29: FBIS carried from Tass a *Pravda* report that cosmonauts Vladimir Kovalenok and Alexander Ivanchenkov were concluding the program of scientific explorations in their 4.5mo orbital flight and preparing for return to earth, by reactivating the transport ship *Soyuz 31* and closing down *Salyut 6*. The mission had produced new information for biologists, medical personnel, glaciologists, volcanologists, oceanologists, and firemen. One example was the previously unknown regularity of glacial movement in Latin American mountains; the cosmonauts also had observed seven forest fires in Australia. Visiting crews had already returned major portions of experimental results to earth. (FBIS, Tass in English, Oct 29/78)

October 30: NASA declared the launch of ESA geostationary operational environmental satellite *Geos 2* on July 14, 1978 successful. Launch-vehicle performance was nominal and had placed the spacecraft and its apogee-boost motor (ABM) into the desired transfer orbit. Satellite performance was satisfactory during the transfer orbit, and the ABM was fired successfully July 16, 1978. The satellite was being maneuvered over the South Atlantic to a position at 6°E above the equator. All subsystem functional checks were complete and satellite status was satisfactory. The mission had replaced *Geos 1*, whose Delta vehicle had put it into an incorrect orbit in April 1977. (MOR M-492-302-78-02 [postlaunch] Oct 30/78)

• NASA announced it had appointed a board to investigate the SeaSat spacecraft failure, in addition to Dr. Bruce Lundin, previously named chairman [see Oct. 19]. Those appointed were Parker Counts, science and engineering department, MSFC; James Hannigan, flight control division, JSC; Maj. James Mannen, U.S. Air Force Space and Missile Systems Organization; T. Bland Norris, director of the astrophysics division, NASA Office of Space Science; Daniel Shramo, director of space systems and technology, LeRC; and James Stitt, director of electronics, LaRC. Board affiliates were Merland Moseson, director of flight assurance, GSFC liaison; Charles Terhune, deputy director of JPL, JPL liaison; Robert Kinberg, NASA legal counsel; and Dell Williams, NASA Office of Aeronautics and Space Technology, executive secretary.

In a letter to Dr. Lundin reporting the appointments, Dr. Alan Lovelace, NASA deputy administrator, urged that the board move as expeditiously as possible, "consistent with a thorough analysis of the facts, to determine the actual or probable causes of this mission failure." Dr. Lovelace assured Dr. Lundin that appropriate NASA resources and all background information would be available to board members. (NASA Release 78-168)

• NASA announced it had planned, weather permitting, to have its Project Cameo release large bluish-white luminescent clouds of barium in the skies over Alaska beginning Saturday, October 28, and lithium over northern Scandinavia at a later date [see Sept. 19]. The agency said Project Cameo clouds should be visible to residents of Alaska, eastern Russia, western Canada, and the northwestern U.S. The chemicals would have been released from the orbiting second stage of a Delta rocket that would have launched the *Nimbus 6* satellite into polar orbit Oct. 24. Of the many high-altitude sounding-rocket chemical-cloud launches conducted by NASA throughout the world, this would be the first from an orbiting vehicle; the Delta second stage was in a circular near-polar orbit at about 950km (590mi) altitude. (NASA Release 78-169)

• NASA announced award of a contract to Boeing Commercial Airplane Company, Seattle, Wash., for development and evaluation of selected advanced aerodynamics and active-control concepts for commercial transport aircraft. The contract, to be managed by LaRC, would be part of the second phase of energy-efficient transport (EET) work that had begun in early 1977 under two NASA contracts. The second phase would concentrate on ways to improve energy efficiency and operational economy for subsonic long-haul commercial transport aircraft, including wingtip extension or winglets, surface coatings for drag reduction, wing-load alleviation systems, and active controls/guidance systems. Boeing estimated the contract value at $17 million. (LaRC Release 78-65)

During October: NASA issued a press kit in observance of its 20th anniversary, having opened for business October 1, 1958. To commemorate the anniversary, NASA Administrator Dr. Robert Frosch wrote:

"By the time a man or a woman has lived for 20 years, a new individuality has been established, physical maturity has been achieved, important experiences have been undergone and digested, there have been some triumphs and some tragedies, the outlines of the individual's role in society are beginning to come into focus, and it becomes possible to do some sensible planning and make some educated guesses about the future. Most importantly, a unique new entity with largely undetermined potential has come to exist in the world.

"On its 20th anniversary, NASA—composed of some 23,000 men and women—is just such an entity. We are a unique agency, mandated by the people of our country through their elected representatives to develop and utilize space technology both for immediate practical application and to expand our knowledge of the earth, its environment, the solar system and the universe. We are charged with assisting the Department of Defense in the use of space to maintain the security of our nation, and with the promotion of international cooperation in space for peaceful purposes. An important part of our charter calls for research and development to maintain U.S. leadership in aeronautics, and to improve civil and military aeronautical vehicles while minimizing their energy consumption and environmental degradation. We are further charged with the dissemination to all potential users of new knowledge and technology acquired in the course of all these activities, and, alone among Federal agencies, we are required by law to 'provide for the widest practicable and appropriate dissemination of information concerning (our) activities and the results thereof.'. . .

"We have, after all, in just two decades, for the first time ever, put life from the planet earth on another body of our solar system, and in our explorations of the moon learned more about the nature and origins of that system than humanity was able to determine in all the centuries that went before.

"In those two eventful decades we have landed extensions of our intelligence on Mars, begun an automated investigation that will eventually extend to all the planets orbiting the sun, achieved significant increase in our knowledge of sun-earth interactions and relationships, and through remote sensing satellites made order of magnitude improvements in how we view the natural and manmade phenomena of the whole earth, as a first step on the way to better management of all our resources. And we have put astronomy observatories into orbit above the obscuring atmosphere, which are beginning to supply data that may well change our conception of the universe.

"Closer to home we have initiated a communications revolution. Though barely begun, the communications satellite program has tied the nations and peoples of our planet together in a way never before possible, and it promises the benefits of intercommunity contact to the most remote and isolated areas. . .

"In our 20th year, we are deeply involved in solving the primary problems of our time—through remote sensing, helping to locate new sources of fossil fuels while working with the Department of Energy to develop alternate energy resources.

"With passenger cars alone burning almost a third of all petroleum products used in this country each year and causing most of the air pollution that blights our metropolitan areas, we are applying our scientific and engineering expertise to improving auto efficiency, economy, and environmental acceptability and to developing advanced auto propulsion systems.

"Finally, exactly as is true of a man or a woman, the future from the perspective of 20 years is full of challenge, excitement, and opportunity. Consider, for example, that the 12 months following our 20th anniversary will see, among other events:
—launch of the second HEAO to follow up the major astronomical discoveries of its predecessor;
—arrival of the Pioneer orbiter and probes at Venus;
—exploration of Jupiter and its moons by both Voyagers; and
—first orbital test flight of the Shuttle.

"Beyond our 21st year, we can see the Shuttle evolving as the major factor in all our operations in space, facilitating and accelerating progress toward nearly all our goals...

"In short, on our 20th anniversary we at NASA can take pride in what we have already accomplished, satisfaction from the knowledge that we are doing useful work on the frontiers of science and technology for the benefit of our people and our country, and exhilaration at the prospects which lie ahead for the continued expansion of human knowledge and the improvement of the condition of mankind on the planet earth.

"Truly a unique new entity has come to exist in the world, an entity of which each one of us can be proud to be a part."

Also included in the press kit was a chronology of major NASA milestones, descriptions of all NASA centers, and NASA's major launch record. (NASA Release 78-133)

November

November 1: NASA announced it would rotate the Skylab space station 180° in its orbit Nov. 3 because of low temperatures encountered by one of the control moment gyros (CMG) as a result of periodic long-term shading from the sun. The sun angle was a function of the orbit inclination (in Skylab's case, 50° relative to the equator), the position of the earth around the sun, and certain other seasonal factors. When the CMG, essential to Skylab's holding a stable attitude, was in shadow for excessive periods, the bearing temperature dropped, interfering with lubrication and increasing friction on the bearings that could lead to possible CMG failure. Turning the Skylab in its orbit before mid-November would expose the CMG to more sun and maintain normal bearing temperatures. The reversed position was not expected to change Skylab's orbital-life predictions significantly. The maneuver would use none of the remaining nitrogen thruster fuel now on reserve for docking with the teleoperator-retrieval system, planned for the second Space Shuttle flight. (NASA Release 78-172; *Marshall Star*, Nov 8/78, 2; JSC *Roundup*, Nov 10/78, 1; *W Post*, Nov 4/78, A-6; *AvWk*, Nov 13/78, 65)

• NASA Administrator Dr. Robert Frosch had announced establishment of a $250 000 Center Director's Discretionary Fund to support innovative ideas in research and technology at NASA centers, the *Marshall Star* reported. Funds would be used exclusively for pursuit of research and technology tasks, not for facility, personnel, travel, or R&D project problems. Priority would be given to support of in-house activities as opposed to studies or research by contractors.

Dr. William Lucas, director of Marshall Space Flight Center, said he considered establishment of the fund a significant and favorable step, and he intended to be personally involved in all facets of the use of the fund and to retain direct control of all fund allocations. A small advisory panel of center personnel would assist the director in evaluating proposed fund uses and the progress of approved activities. (*Marshall Star*, Nov 1/78, 1)

• The *Marshall Star* reported that industry was being invited to bid on a cost-sharing contract to commercialize an electricity-saving device invented by MSFC engineer Frank Nola [see May 2, June 14]. The inexpensive power-factor controller could reduce markedly the power consumed by millions of electric motors in use throughout the U.S. MSFC's RFP specified that the contractor selected "shall have capability of manufacturing 30,000 units per month, and should provide as a part of this effort the estimated production costs based on production rates of 10,000, 20,000 and 30,000 units per month." This type of cost-sharing contract had been one method used by MSFC's Technology Utilization (TU) Office to make new technology available to the public, as required by the Space Act of 1958, and to demonstrate the utility and reliability of the technology. "By joining with industry in this type of project," Aubrey Smith, TU director, said, "we encourage the transfer of new space technology to the private sector." (*Marshall Star,* Nov 1/78, 1)

• Wallops Flight Center announced it had launched a two-stage solid-propellant sounding rocket, the Taurus-Tomahawk, at 2:47p.m. Oct. 31. Both the Taurus and Tomahawk rocket motors were currently in use, but this was the first test flight of the combined configuration, which would be used for future NASA sounding-rocket launches. Peak altitude was 554km (344 statute mi), and the successful test flight qualified the vehicle to launch scientific payloads. First operational launches would be in March 1979 from Poker Flat Research Range in Fairbanks, Alaska, to carry a 27.4kg (60lb) payload to an altitude of 575km to conduct scientific experiments in the near-earth environment. (WFC Release 78-21)

• Despite tight new restraints on federal spending, President Carter had instructed his aides to give special protection to R&D programs—currently scheduled to get $28 billion—in the FY80 federal budget, the *Washington Post* reported. This came a day after the President had announced an anti-inflation program that included spending cuts plus reduction of the FY80 budget deficit to $30 billion or less, a goal at least $3 billion tighter than his previously announced target.

In a handwritten note Carter instructed his aides: "I want to maintain our strong support for R&D as (percentage) of budget." This meant high priority for spending consideration in a field usually considered a prime target for any budget-cutting program. One White House aide indicated Carter's instructions meant that R&D would be virtually exempt from new federal spending constraints; another said that R&D might still face some restraints if inflation did not ease.

The President had previously expressed special concern that basic research (which accounted for $3.5 billion of the total R&D budget and had been dwindling until the Ford Administration reversed the trend) be protected from any budget-cutting considerations. Frank Press, Carter's science adviser, said Carter's decision constituted the nation's "first presidential expression of an overall science and technology policy." (*W Post,* Nov 1/78, A-6)

November 2: NASA announced plans to launch NATO-IIIC, third and final comsat in a new series to serve the North Atlantic Treaty Organization (NATO), on Nov. 15 from Kennedy Space Center. A U.S.–NATO agreement had specified launch on a Delta rocket, and NATO would reimburse NASA $8.9 million for the launch vehicle and services. A Delta had previously successfully launched *Nato IIA* and *IIB* in March 1970 and February 1971, respectively; *Nato IIIA* in April 1976; and *Nato IIIB* in January 1977.

NATO-IIIC would provide in-orbit backup to *Nato IIIA* and *IIIB* in the NATO Integrated Communications System (NICS). At a geosynchronous orbital altitude of 35 900km (22 300mi), NATO-IIIC would orbit earth once every 24hr synchronized with earth's 24-hr rotation period, keeping the satellite on station over the same spot above the equator. Once in its transfer orbit, the satellite would be controlled by U.S. Air Force Satellite Control Facility network on behalf of NATO. The U.S. Air Force Space and Missile Systems Organization (SAMSO) served as satellite contracting agency for NICS. Ford Aeronautic and Communications Corp., Palo Alto, Calif., built NATO III. (NASA Release 78-171; MOR M-492-207-78-03 [prelaunch], Nov 15/78; *Spaceport News*, Nov 10/78, 1)

• NASA reported successful static firing of a Space Shuttle main engine Oct. 30 at the National Space Technology Laboratories, Bay St. Louis, Miss. The firing lasted for more than 13min, testing its ability to return the Shuttle orbiter to a landing site in case of mission abort during launch. The main engine operated continuously for 823sec, longest burn time an engine should require during an actual mission. A Shuttle orbiter would use 3 main engines having a normal burn time of about 8min. (NASA Release 78-173)

• FBIS reported the longest manned flight in history—139da 15hr—had ended Nov. 2 at 14hr 5min Moscow time, after completing a program of scientific-technical researches and experiments on board the Salyut-Soyuz orbital complex. The *Soyuz 31* module landed in a preset area of the Soviet Union 180km southeast of Dzezkazgan (Kazakhstan). A preliminary medical checkup at the landing site indicated the cosmonauts had withstood the long flight well.

Vladimir Kovalenok and Alexander Ivanchenkov's flight to *Salyut 6* had begun June 15 in *Soyuz 29*. The international crew of *Soyuz 30*, including Poland's Miroslaw Hermaszewski, had joined the occupants of the orbital complex June 28 and returned to earth July 5. *Soyuz 31*, whose crew included the GDR's Sigmund Jähn and returned in *Soyuz 29*, 29, docked there Aug. 27 and later returned the *Soyuz 29* crew. Fuel for the joint engine installations, equipment, life-sustaining materials, and materials for studies and experiments had been regularly delivered to *Salyut 6* by automatic transports *Progress 2, 3*, and *4*.

On Nov. 16 a Soviet medical expert said the cosmonauts had readapted quickly to earth's gravity and had suffered no significant physical disabilities, the *NY Times* reported. These cosmonauts had exercised more during their flight than had previous crews, and the extra activity apparently helped them to readjust after their return. Only on their first day back had the cosmonauts been seriously affected by body weight, causing some uncoordinated movements. Red corpuscles and hemoglobin in the blood had dropped insignificantly, and white blood cells had increased moderately. Kovalenok had been underweight by 5lb, regained in 3 da; Ivanchenkov by 8.5lb, regained in 12. On their second morning back on earth, they had asked to walk in the park, 4da earlier than the medical experts had planned; their pulse rates increased by only 15 to 20 beats a min after 1400 steps. Soviet medical experts were quoted as saying that it took the cosmonauts about 10da to readjust completely to earth's gravity. (FBIS, Tass in English, Nov 2/78; *NYT*, Nov 17/78, A27; *W Post*, Nov 3/78, A 14; *AvWk*, Nov 6/78, 21, 13/78, 21; *D/SBD*, Nov 3/78, 15; *A/D*, Nov 3/78, 17; *Langley Researcher*, Nov 17/78, 1)

November 3: NASA announced preparations to launch its second high energy astronomy observatory HEAO-B, continuing a three-mission program to study some of the most intriguing mysteries of the universe: pulsars, quasars, exploding galaxies, and black holes. Launch would be from KSC on an Atlas Centaur about Nov. 13. Carrying a focusing x-ray telescope and a variety of sensitive instruments, HEAO-B would maneuver and point for long periods of time at selected x-ray sources previously identified by *Heao 1*. Pictures returned by HEAO-B would be the first spacecraft-generated x-ray images of objects other than the sun. HEAO-B's images, acquired by x-ray telescope, would be converted to telemetry received and taped by ground stations and eventually reconstructed as photographs showing size, structure, and detail of objects viewed by the spacecraft. HEAO-B, with a design 1yr mission lifetime, would be placed in low circular orbit far enough above the atmosphere to detect radiation normally unable to reach earth's surface.

MSFC had managed HEAO for NASA's Office of Space Science. KSC would manage launch operations including prelaunch checkout, launch, and flight through observatory separation in orbit. Lewis Research Center managed launch-vehicle procurement and related activities for the HEAO program. MSFC would control the in-orbit HEAO observatories, in conjunction with TRW flight-control engineers operating from Goddard Space Flight Center. Cost of the three-mission HEAO program was about $248 million, HEAO-B's approximately $87 million. (NASA Release 78-165; MOR S-382-78-02 [prelaunch] Oct 30/78)

• NASA announced it had awarded Vought Corporation, Dallas, Tex., a modification of its contract to provide Scout launch vehicles. Under the modification, Vought would provide NASA with 12 additional guidance systems for use on Phase 8 Scouts (107 vehicles had constituted Phases 1 through 7), the same system used on the launch vehicle since the Scout program began. Vought had been the Scout project's prime contractor since 1959; the first Scout launched in 1960 was designed to put a 130-lb payload into a 300-nautical-mi circular earth orbit. The latest Scouts were designed to place a 404-lb payload in that orbit. Scout vehicles had been launched from Wallops Island, Va., Vandenberg AFB, Calif., and San Marco, Kenya. Vought valued the modification at approximately $6.5 million. Langley Research Center would manage the 2.5-yr contract. (LaRC Release 78-67)

• FBIS reported that the USSR and India had signed a protocol for launch of a second India satellite. The two countries were working on a flying model of the spacecraft and preparing ground stations. Representatives of the USSR Academy of Sciences and India's outer-space exploration organization had discussed satellite plans for several yr. The second India spacecraft, like the first, would be launched from Soviet territory, and the carrier rocket and launching complex would be provided gratis by the USSR. (FBIS, Tass in English, Nov 3/78)

• FBIS reported the USSR announced it had fired a geophysical rocket, Vertikal 7, to a height of 1500km for research on the ionosphere and on shortwave radiation. The instrument package, which separated at 175km on the ascent trajectory, contained equipment prepared by scientists from Bulgaria, Czechoslovakia, Hungary, Romania, and the Soviet Union. Launch apparently took place from the facilities at Kapustin Yar, site of previous Vertikal launches. (FBIS, Tass in English, Oct 3/78)

November 6: Ames Research Center Director C.A. Syvertson announced appointment of A. Thomas Young, director of the planetary program in NASA's Office of Space Science, as deputy director of ARC effective Feb. 1, 1979. Young, who had held his current position since Nov. 1976, had served as missions operations manager and missions director of Project Viking as well as of other projects at LaRC, which he had joined in 1961. Young had received LaRC's sustained superior performance award in 1967 for contributions to the lunar orbiter project and in 1977 for his work on Viking. In 1977 he received NASA's Distinguished Service Medal for contributions to Viking. (ARC Release 78-53)

November 7: A new gallery, "Exploring the Planets," opened Nov. 7 at the National Air and Space Museum, the *Washington Star* reported. Melvin Zisfein, acting museum director, said the gallery was "jammed full of information and contains more facts and figures and concepts than any other gallery in the museum. Yet the information is easy to digest. We tried hard to make sure you didn't need a degree in astronomy to understand it."

Keeping up with new information on the solar system (the gallery would be updated continuously) was one of the challenges. The gallery, designed by Lucius Lomas, had succeeded in teaching without neglecting imaginative features visitors had come to expect at the museum. Visitors could make a "Descent to Venus" in a simulated cockpit of an interplanetary spaceship circa 2150. A "Flight over Mars" from an altitude of 100 000 to 200 000ft (actually relief maps of the Martian landscape sculpted on to rotating cylinders and viewed through special windows) was based on photos taken from *Mariner 9* and *Vikings 1* and *2*. A full-scale replica of the Voyager spacecraft was the largest single object in the gallery. The relative sizes of Jupiter and 21 other planetary bodies was conveyed in a corner by brightly colored spheres made to scale: Jupiter was 10ft in diameter; Ceres, the largest asteroid, 1inch; the earth, 1ft. Six computer terminals offered "Space Academy exams" where visitors could test their knowledge of the solar system. (*W Star*, Nov 7/78, F 3)

November 8: INTELSAT announced award of a contract to improve satellite communications in regions plagued by frequent heavy rainfall that caused satellite signals to fade at higher frequencies. Under the $39 827 contract, the Technische Universitat Graz, Austria, would conduct yr-long experiments for simultaneous comparison of fading caused by rainfall in four different locations. Results would determine whether alternative earth stations for use during rainfall would solve the fading problem. (INTELSAT Release 78-31-M)

• INTELSAT announced it was seeking bids from U.S. domestic communications-satellite operators to provide television services between the U.S. and Mexico. Mexico had asked INTELSAT help in obtaining international television between the Tulancingo earth station near Mexico City and a number of points within the U.S. Although INTELSAT had already established international comsat service over the Atlantic, Indian, and Pacific oceans between a total of 192 earth stations and 94 countries, its satellites (configured primarily for long-haul transoceanic communications) could not provide the coverage Mexico required. Consequently, INTELSAT's board of governors had authorized it to seek alternative ways of supplying Mexico's need until INTELSAT could do so with its own facilities. This was the first time INTELSAT had sought to lease long-term space-segment communications capacity outside its own resources. (INTELSAT Release 78-32-I)

• ESA's Spacelab program board had turned down NASA's proposal to barter Space Shuttle launch services for the second Spacelab NASA was committed to acquire, and the action was seen as final, *Defense/Space Business Daily* reported. NASA had hoped to exchange launch service in return for the Spacelab in order to reduce costs [see Sept. 12]. Although ESA itself favored the barter plan, the European governments had refused to commit themselves in advance to Space Shuttle flights. In lieu of the barter deal or any increased NASA funding, ESA might be asked to prefinance work on the second Spacelab until NASA could get the money. (*D/SBD,* Nov 8/78, 39)

November 9: LeRC would host a major conference for U.S. industry Jan. 31, 1979, to review the marketing potential and technology status of large wind-turbine electric generators, the *Lewis News* reported. NASA had supported DOE in developing, with the aid of U.S. industry, large wind-turbine electric generators as a source of supplemental power for utility networks. LeRC was conducting both a technology-development program and an experimental program with the utility industry. LeRC engineers believed that large wind turbines could become a cost-competitive source of nondepletable electrical power, and that the time was right for additional industrial-equipment designers and producers to become involved in the nation's wind-energy program. Those attending the meeting would discuss two advanced-design wind turbines in the planning stage: a 200kw wind turbine for operation at wind speeds lower than those used by current wind turbines, and one in the 1000 to 4000-kw range to produce electricity at a reduced cost per kw hr over smaller-scale units.

The January conference would review elements of the U.S. wind-energy program including objectives, present status of the technology, and current cost projections, emphasizing design of presently operating wind turbines, designs in progress for advanced machines, and plans to procure new lower-cost wind-turbine designs from industry. (*Lewis News*, Nov 9/78, 1)

• The U.S. reconnaissance-satellite program had apparently moved into a new phase emphasizing fewer satellites and longer orbital stay-time, *Aerospace Daily* reported, despite the fact that the USSR had an operational antisatellite program that might have influenced President Carter's backing of a U.S. ASAT project. An *Aerospace Daily* study of worldwide satellite launches since the early 1960s indicated the U.S. had stopped using satellites for a close look at specific ground targets. These satellites, launched by the Titan 3B/Agena booster combination, had previously flown simultaneously with an area-surveillance class launched by Thor/Agena. The U.S. had apparently withdrawn its area-surveillance models from service in 1972 and substituted a type able to do both area and close-look surveillance—Lockheed's so-called Big

Bird—and dropped the original close-look type. It had retained Big Bird, with 14 launches to date, complemented by an even newer type that might have had the same dual mission but performed better.

Only two of the latest models had been launched, one in December 1976 and one on June 14, 1978. This model could have been the subject of a controversy over alleged theft by a former CIA employee of an operating manual for a satellite code-named KH-11, which had an orbital life at least three times that of Big Bird, which had grown to an impressive 179da. The new model might have had a nominal lifetime of 540da; the first one launched was approaching 24mo in orbit. This extended orbital life meant that the USSR and other countries were under constant surveillance; however, as the U.S. had only one or two satellites operating at any one time, the vulnerability of its monitoring effort had probably increased. (*A/D*, Nov 9/78, 44)

November 10: After each Space Shuttle flight, parachutes used to slow the descent of the Shuttle's solid-fuel rocket booster (SRBs) would have to be washed, dried, and repaired before reuse; and KSC had modified for this purpose a facility formerly used as a news center and to refurbish Gemini parachutes, *Spaceport News* reported. Recovery vessels at sea would pull the parachutes from the water and wind them like fishing line on large deck reels for delivery to the facility, where they would be unwound, defouled, and hosed down. A chute arranged on hooks on an overhead monorail would proceed through the world's largest automatic washer and dryer [see March 8] to remove any traces of corrosive sea water. Next the chute would be inspected for damage, with rips mended on special sewing machines. The chute then would be stretched on a worktable almost the length of the 220-ft building, and each individual panel would be systematically folded into place resulting in a tubular shape. A monorail would guide the bundle to a wedge-shaped packing canister. Depending on conditions, one parachute could move through the entire operation in less than 50hr. (*Spaceport News*, Nov 10/78, 3)

• Despite the wealth of data on Mars provided by *Mariner 9* and the Viking missions, some mysteries remained, including the "White Rock," a unique topographic feature nestling inside a crater not far from the Martian equator, JPL *Universe* reported. White Rock's discovery by *Mariner 9* had intrigued scientists at Jet Propulsion Laboratory because of the reflectivity of its unknown basic material, and the fact that no other formation on Mars resembled it.

At first glance White Rock looked like the paw-print of some many-toed animal, as it was rounded on two-thirds of its perimeter and on the remaining sides had great serrated formations like claws, referred to as "reentrants" and "dunes," reaching into the floor of the surrounding Martian crater. Scientists believed the ridged area had been eroded by

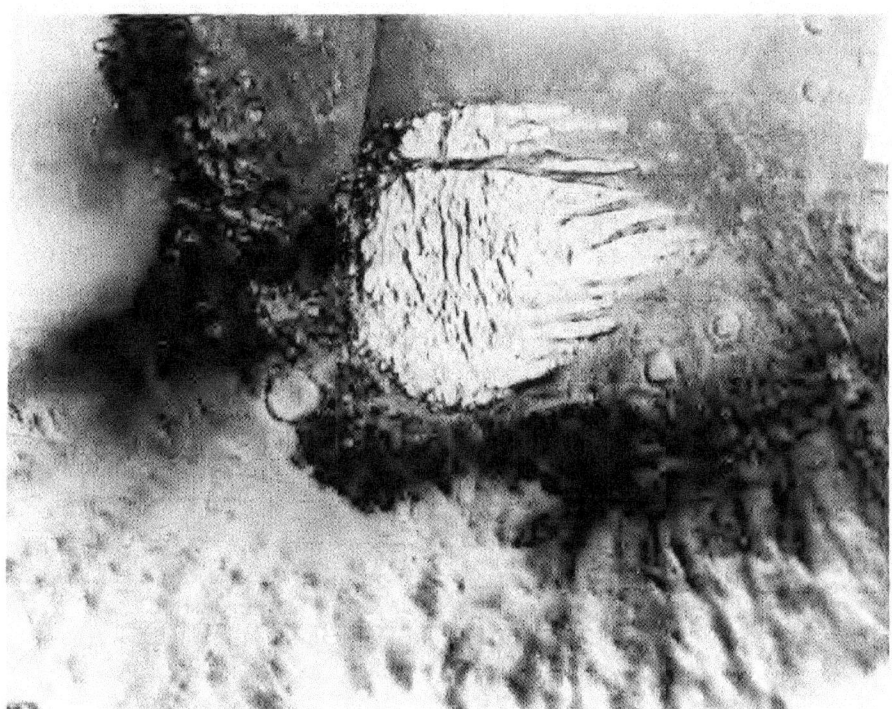

White Rock, strange topographic feature inside a crater on Mars whose composition had puzzled scientists: not snow or ice, because of its proximity to the equator, the formation exhibited high reflectivity first noticed by Mariner 9 *in 1972. This photomosaic came from the Viking orbiter in Sept. 1978 (straight lines on the image resulted from assembly of the photographs into a mosaic). (NASA photo 78-H-634)*

ever-present Martian surface winds. The possibility of ice cover on its surface had been rejected early on because of its proximity to the equator. Speculation on its mineral composition had included salt and/or sand theories.

Roughly 18km across, White Rock appeared to rise from the floor of an impact crater some 93km in diameter. Very small impact craters on the floor of the larger crater and on the feature itself might suggest the age of the formation in relation to its surroundings. Although labelled "white," general opinion was that the feature had a very light orange or pink tint. After *Mariner 9* discovered the unique feature, Viking had been programmed for a closer look; Viking's more sophisticated vidicon would offer five times better resolution at a similar altitude. Computer processing of Viking data could produce information to help solve the mystery of White Rock. (JPL *Universe*, Nov 10/78, 2)

• Disagreement about the age of the newly discovered rings around Uranus was expressed at the Nov. 9 meeting of the American Astronomical Society in Pasadena, Calif., the *Pasadena Star-News* reported. Discovered by astronomer Jim Elliot of Cornell Univ. in March 1977, the rings had sparked many scientific theories. Elliot originally had observed five rings; subsequent observations had disclosed four additional rings. Uranus's rings did not spread into broad bands circling the planet like those of Saturn, but were narrow and confined to tight belts.

Attracting most attention was the fifth ring, Epsilon, which was thicker and wider than the other rings, although its width varied from 20 to 100km. The inner edge of the Epsilon ring did not move around the planet at a rate different from that of the outer edge, contrary to what might have been expected. Dr. Richard Goldreich of CalTech theorized that the Epsilon ring was bound together by its own gravity, which, in turn, meant it had substantial mass (possibly as much as 25grams per cm^3, extremely dense by ring standards), causing him to conclude the rings around Uranus were very old. However, Judd Boynton, director of Geophysics Labs and Research in Berkeley, who had predicted that rings would be found around Uranus, had concluded the rings were of relatively recent origin, perhaps as young as 10 million yr. (JPL newsclips, Nov 10/78)

November 13: NASA launched from Cape Canaveral the high energy astronomy observatory *Heao 2* on an Atlas-Centaur launch vehicle [see Nov. 3]. This was the second in a series of three NASA high energy astronomy observatories designed to survey the entire sky for x-ray sources, make measurements of the gamma-ray flux, determine source locations and line spectra, and examine the composition and synthesis of cosmic-ray nuclei. The 20ft 7000lb *Heao 2* was equipped to provide precise pointing capability of 1 arc-minute or better, and would focus on specific x-ray sources for detailed examination. Orbiting the earth at 290mi altitude, *Heao 2* had a design life of 1yr. Nicknamed "Einstein" by project scientists, the mission would operate under a consortium representing the Harvard-Smithsonian Center for Astrophysics, Mass. Inst. of Technology, Columbia Univ., and GSFC. The spacecraft was built by TRW, Inc., under NASA direction. (NASA MOR S-832-78-02 [prelaunch] date?; *NYT,* Nov 14/78, C-3)

• NASA reported that Cornell Univ. scientists had for the first time turned xenon, rarest of stable rare gases, into a metal by applying tremendous pressure on solid xenon. David Nelson, Jr., and Professor Arthur Ruloff of Cornell's department of materials science and engineering, in work sponsored by LeRC, reported pressures of 320 000 atmospheres applied to a tiny amount of solid xenon at $-241°C$ ($-402°F$)

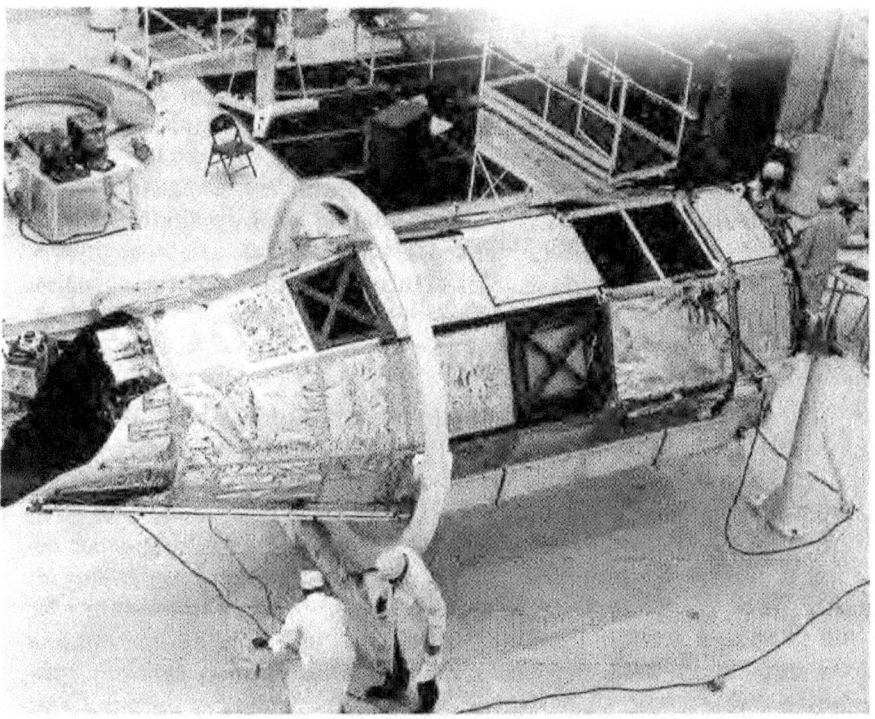

NASA's second high-energy astronomy observatory, heaviest manned earth-orbiting spacecraft, undergoing checkout before launch on a mission of general sky survey, mapping x-ray sources and measuring low-energy gamma-ray flux. (NASA photo 78-H-623)

had produced the metal. Xenon frozen solid would not conduct electricity; however, at the pressures applied in the Cornell experiment, the electrical conductivity of the frozen xenon increased more than a hundred billion times—that is, it behaved like a metal.

Because the xenon returned to the insulating (nonconductive) state when the pressure was removed, no engineering applications were apparent. However, Ruloff noted, "The production of metallic xenon is of considerable scientific interest. Furthermore, the development of these techniques is a step forward in high-pressure research." Ruloff believed that his group could make oxygen and krypton metallic, and possibly nitrogen, argon, and even hydrogen. If frozen hydrogen could be made metallic and remain in that state when the pressure was removed, it might have practical use as an extremely powerful rocket propellant, a superconductor, or an improved source for fusion energy. NASA would explore the possibility of making metallic hydrogen. (NASA Release 78-174)

• NASA announced appointment of Arnold Frutkin as NASA associate administrator for external relations, effective immediately. Frutkin had served as deputy associate administrator for external relations since March 2, 1978 [see March 1], and acting associate administrator since June 4. Previously he had headed NASA's Office of International Affairs for 18yr. Frutkin would be responsible to the administrator for developing external policy and for coordinating NASA relations with the public, the international community, universities, state and local governments, and the Department of Defense and other federal agencies. Before joining NASA, Frutkin had been deputy director of the U.S. Natl. Committee for the Intl. Geophysical Year. (NASA Release 78-175)

• The U.S. Navy was assessing its current and future space systems with three studies completed and a fourth under way, *Av Wk* reported. The studies were "Soviet Threat," assessments of how Soviet spacecraft affected U.S. naval operations and how the Soviet systems could be used in war; "Current U.S. Navy Status," assessment of Navy dependence on current U.S. spacecraft systems and the implications for the future; "Mid-Term Look," study of Navy exploitation of space in the next 15yr and what new systems should be considered; and "Far-Term Look," a new study by the Natl. Science Foundation under a $200 000 Navy contract to forecast where Navy space systems were headed by the late 1990s and into the 21st century.

In a move with immediate impact on management of Navy-associated space systems, the Navy had put ocean-surveillance programs under the Chief of Naval Operations because of the importance of command, control, and communications in ocean-surveillance activities and to coordinate these elements more efficiently. Ocean surveillance previously was managed along with antisubmarine warfare. (*AvWk*, Nov 13/78, 67)

November 14: NASA announced it was putting *Pioneer Venus 2* through maneuvers in preparation for release of four probes to study the atmosphere of Venus. A large probe would be released Nov. 15, three smaller probes Nov. 19. The probes and their transporter bus would arrive at Venus and begin descent into the atmosphere Dec. 9. On Nov. 9, while *Pioneer Venus 2* was 11 265 400km (7 000 000mi) from Venus, controllers at ARC had made a small trajectory correction to ensure the desired entry point at Venus for the large probe. They also commanded a 90° rotation of the spin axis to point the aft-mounted medium-gain horn antenna toward earth, obtaining a fourfold increase in communications capability. On Nov. 15, just before large-probe release, ARC would maneuver the spacecraft to proper entry angle at Venus, followed Nov. 16 and 19 by maneuvers to reach the desired small-probe entry points spread 9660km (6000mi) apart over Venus's earth-facing hemisphere.

Each of the four probes had its own command, communications, power, and other systems. Heat shields and titanium pressure vessels would help the probes withstand Venus's 480°C (900°F) heat, corrosive atmosphere, and crushing atmospheric pressure, 100 times that of earth. The multiprobe and its sister spacecraft, the orbiter, would conduct 30 experiments to improve understanding of Venus's weather, which in turn would help scientists understand the forces driving earth's weather. (NASA Release 78-177)

November 15: NASA announced that, after nearly a mo of attempts to reestablish contact with SeaSat, its experimental oceanographic satellite, project officials had formally terminated the mission. JPL would process the data collected during the spacecraft's 106-da lifetime; project officials expected the task to occupy more than 1.5yr. SeaSat's synthetic-aperture radar had completed some 300 data-gathering passes, during which it collected about 60hr of data including images of sea ice, waves, coastal conditions, and various land forms. The spacecraft's scatterometer (measuring sea-surface wind speeds) and scanning multifrequency microwave radiometer (measuring sea-surface temperature and wind speed) had collected data for 99da. SeaSat's altimeter and visual and infrared radiometer returned data for 70 and 52da, respectively. The instruments also completed a series of sea-surface fact-finding experiments.

The SeaSat program had three objectives: to demonstrate techniques to monitor earth's oceanographic phenomena and features from space on a global scale; to provide timely oceanographic data to scientists studying marine phenomena, and to users of the oceans as a resource (ocean shippers, fishermen, marine geologists, etc.); and to determine the key features of an operational full-time ocean-monitoring system. Analysis of SeaSat-collected data by teams of scientists, engineers, and user representatives indicated the first of the objectives would be largely met, as well as a high probability of accomplishing most of the other objectives. A failure review board [see Oct. 30] was working to determine the causes of the premature end of SeaSat operation. SeaSat had been launched June 26 from Vandenberg AFB; contact was lost Oct. 9 on its 1502nd earth orbit when a so-far-unexplained short circuit drained all power from the batteries. (NASA Release 78-178)

• Thiokol Corp.'s Wasatch Div. had delivered the last of eight empty solid-fuel rocket motor-case segments to MSFC for use in the final burnout phase of Space Shuttle ground-vibration tests scheduled for early 1979, the *Marshall Star* reported. Current vibration testing was in the "liftoff" phase using motor cases filled with inert propellant; this phase would be completed and the Space Shuttle elements removed from the dynamic-test stand by the end of 1978. All the elements would be returned to the stand, including the empty motor-case segments, for final vibration-test phase simulating the actual Space Shuttle situation just before solid-fuel rocket-booster separation after expending the solid propellant. (*Marshall Star,* Nov 15/78, 1)

• Almost all major structural elements and systems of the first two flight-model solid-fuel rocket boosters for the Space Shuttle had either been delivered to KSC or were scheduled to arrive shortly, the *Marshall Star* reported. Delivered elements included eight booster-separation motors for one SRB, with eight for the other SRB due within the wk; all major components of the thrust-vector control (TVC) subsystems except the actuators, which were to arrive on Nov. 23; frustums and nosecaps for both boosters; the forward and aft skirts for the left-hand SRB (the aft skirt for the right-hand booster was en route by rail, and the forward skirt was scheduled for delivery Nov. 20); and most of the electrical-systems components. The parachute-recovery subsystem for the left-hand booster was at KSC; the system for the right-hand booster was due Nov. 22. (*Marshall Star*, Nov 15/78, 4)

• After a flawless launch from KSC on an Atlas-Centaur at 12:24a.m. EST Nov. 13; *Heao 2* was functioning satisfactorily in a near-perfect earth orbit, the *Marshall Star* reported. The observatory was expected to become fully operational by Nov. 16. First steps in orbit were to open the solar arrays and to separate the launch vehicle, both of which happened exactly on time. Ground stations were receiving and evaluating engineering data, as one by one each instrument was turned on to low power and checked out. All leftover gas had to be purged from the spacecraft before high voltage could be turned on to avoid electrical arcing or damage to delicate onboard instruments. MSFC was in control of *Heao 2* together with TRW flight-control engineers operating at GSFC. (*Marshall Star*, Nov 15/78, 1)

November 17: NASA reported completion of a 3yr Large Area Crop Inventory Experiment (LACIE) using space-age technology to monitor global wheat production. A symposium at Johnson Space Center attended by more than 700 conferees from 22 nations representing federal agencies, private companies, universities, and foreign governments, presented results of the experiment.

The experiment had begun in 1974 to determine if data from a Landsat orbiting some 805km (500mi) above earth could be used with surface observations and weather information derived from U.S. operational environmental satellites to predict the size of the world's wheat crop. Major foreign study areas were Canada and the Soviet Union, with preliminary examination of wheat-growing areas of Australia, People's Republic of

China, Brazil, India, and Argentina. The U.S. Great Plains area served for extensive test and evaluation of techniques, since it was the best source of statistical data with known reliability. LACIE had been designed to improve global wheat-production forecasts by making foreign-production predictions as early as possible in the crop season. The goal of the project was to develop estimates accurate within 10% of the true production at harvest time in 9yr out of 10; tests over the winter-wheat area of the Great Plains showed the goal could be met. Using LACIE techniques to monitor Soviet wheat crops harvested in 1977, the system estimated a production of 91.4 million metric tons, less than 1% below the Soviets' official announcement of 92 million tons. However, the system's ability to achieve such accuracy from yr to yr had yet to be determined. Findings were presented at the four-day symposium by the LACIE participants: NASA, USDA, NOAA, and cooperating universities and industry.

A peer-evaluation team composed of prominent scientists, chaired by Dr. Don Paarlberg, former director of economics for USDA, reviewed the LACIE techniques and reported on them to the symposium. Dr. Paarlberg stated that "LACIE results to date clearly demonstrate that present remote-sensing capabilities can be combined with or substituted for conventional methods of information collection in order to improve foreign crop production estimates." He concluded that " . . . for global wheat regions such as the USSR the LACIE technology can be made operational . . . for regions where technology requires improvement, funding for further research and development should be continued." (NASA Release 78-179)

• NASA announced it had asked scientists to propose experiments for a 1984 mission to obtain a first global view of the Venus surface. Launched by the Space Shuttle, the Venus-orbiting imaging radar (VOIR) spacecraft would circle the planet for at least 7mo taking radar pictures and measuring the atmosphere and the surface. It would be the most detailed scientific examination to date of the surface of Venus, perpetually covered by clouds.

Announcing the solicitation, Dr. Noel Hinners, NASA associate administrator for Space Science, said: "NASA is engaged in a continuing program of space exploration, a major part of which is directed at the solar system. The VOIR program is a key step in that endeavor . . . While considerable knowledge of Venus's atmospheric properties will be forthcoming from the Pioneer Venus mission, and while the Soviet Venera missions have provided valuable data on surface properties at three localities, information regarding the surface morphology on a global scale and all of the insights that this scientific information provides is lacking. A major technical goal of the VOIR mission, therefore, is to obtain global radar imagery with sufficient resolution to address fundamental questions regarding the origin and evolution of the planet, and to lay the groundwork for subsequent more detailed investigations of the surface and interior." Hinners emphasized that VOIR had not been approved but said that early selection of scientific participants and investigations would assure a well-defined program upon approval.

Under present plans, in 1984 NASA would launch one spacecraft from the Space Shuttle to Venus with a typical trajectory beginning in Dec. 1984 to arrive at Venus in May 1985. The spacecraft would remain in an elliptical parking orbit around Venus through May and June 1985, while the Deep Space Network was occupied with the Galileo project [see Nov. 10]. In late June or early July a propulsion maneuver would put VOIR into a near-polar orbit at an altitude of 300km (180mi). Radar mapping and other science gathering would begin in July and continue for 5mo.

VOIR imagery should verify the presence or absence of continents, ocean basins, mountain belts, rift valleys, fault belts, or volcanoes, and might reveal the nature and time sequence of plate tectonic activity, as well as any relationship between this and volcanic episodes in the history of the planet. If impact craters were present (as suggested by earth-based radar observations), the mission might determine their size and frequency. The VOIR orbiter, weighing approximately 5000kg (11 000lb) at launch, would consist of a bus with a synthetic aperture radar and other science instruments. (NASA Release 78-180)

• NASA was proceeding cautiously toward its prospective 30/20GHz satellite communications demonstration program, with development starting in FY82 instead of the faster pace envisioned earlier, *Aerospace Daily* reported. Several questions of policy and organization remained unresolved, one of them being administration guidelines that appeared to contradict President Carter's recently announced civilian space policy [see Oct. 11]. Major issues in the 30/20GHz program included whether a dedicated satellite was to be flown and, if so, whether and how it should be used, apart from demonstrating NASA technology objectives. Another issue concerned the relationship between NASA and the Commerce Dept.'s National Telecommunications and Information Administration (NTIA), set up to take over support formerly provided by NASA to public service satellite-communications users.

The 30/20 GHz program had been presented in May to the NASA Advisory Council as a prospective FY80 development program, leading to 1984 launch of a satellite using a Space Shuttle-compatible antenna less than 15ft in diameter, about 20 spot beams, and fast onboard switching. However, in November, an ad hoc satellite-communications subcommittee of the NAC learned that the current prospect was for multiple-contractor Phase B definition studies in FY80, with technology work to precede FY82 development and a 1985 launch. (*A/D*, Nov 17/78, 78)

- Launches of USSR military space missions continued a marked decline, with total flights in 1978 23% fewer than in 1977, a possible Soviet effort to influence the atmosphere for a SALT II agreement, *Defense/Space Business Daily* reported. Notable omissions from the Soviet military-spaceflight schedule were satellites related to antisatellite interceptor development and nuclear-powered ocean surveillance. These systems had come under intense public scrutiny and criticism in the past yr, as President Carter emphasized that they were major instruments of instability and needed to be banned.

No antisatellite system launch had occurred for several months, since the Soviets had agreed to discuss a ban of such weapons and just before the convening of the first such meeting in May. The last Soviet ocean-surveillance satellite was *Cosmos 954*, which had reentered over Canada Jan. 24 trailing fragments of its nuclear reactor [see Jan. 28, Feb. 20]. At that point in 1978, the USSR had sent 50 military missions into space, a significant reduction from the 65 completed in the same time period in 1977. However, the USSR had not reduced photographic reconnaissance/surveillance activity; launch of the high-resolution observer *Cosmos 1047* Nov. 15 was the 28th photographic recon/survey mission in 1978, launched at the same rate as in 1977. (*D/SBD,* Nov 17/78, 79)

November 18: NASA announced launch of *Nato 3C*, third and last in a series of defense-related comsats, at 7:46pm EST from ETR on a Delta into a synchronous transfer orbit, where it would gradually approach station at 50°W over the equator above the Atlantic. Early orbit parameters were 35 891km apogee, 184.5km perigee, and 27.2° inclination. The Ford Aerospace-built drum-shaped craft covered with a cylindrical solar array and weighing 346kg without its apogee kick motor would transmit voice, data, facsimile, and telex messages among military ground stations, serving as backup for *Nato 3A* and *3B* launched in Apr. 1976 and Jan. 1977, respectively. NATO would reimburse NASA $8.9 million for launch services, including the vehicle. Mission life was 7yr. (MOR M-492-207-78-03 [prelaunch] Nov 15/78, [postlaunch] Feb 5/79; *NASA Actv,* Nov 20/78; Spacewarn Spx-301 [gives date as Nov 19/78, 0046UT])

November 19: Heao 2 had radioed back its first picture Nov. 18, an image of x-rays coming from the edge of what may be a black hole, the *Washington Star* reported. When the picture first appeared on a GSFC television monitor as a collection of bright dots against a black background, scientists and engineers were described as "ecstatic." Dr. Riccardo Giacconi of the Smithsonian Astrophysical Observatory said: "The significance of today's event is that for the first time x-ray astronomy has become of age, and it has achieved the kind of resolution and sensitivity which has been traditional in the fields of radio and optical astronomy." He said the x-ray telescope on *Heao 2* should be able to detect x-rays from objects two-thirds of the way toward the edge of the universe, meaning that the satellite would be seeing radiation created roughly 10 billion yr ago. Giacconi added, "The scientific significance is what it tells us we are going to get in the future," and noted the telescope was 100 to 1000 times more sensitive than previous x-ray detectors placed above earth's atmosphere.

The picture reconstructed by computer from data radioed back by the satellite was a test to demonstrate the capability of the telescope, which had a 23-in-wide mirror and an 11-ft focal length. Target for the test was a star system in the Milky Way galaxy called Cygnus X-1, selected because of its well known location and its brightness. Cygnus X-1, 6000 light-yrs away, also happened to be the best candidate for a black hole; observations by a smaller x-ray satellite had shown that its x-ray emissions were unlike those seen from other objects. (*W Star*, Nov 19/78, A-13)

November 21: GSFC announced that the third international sun-earth explorer *Isee 3* had been injected into a halo orbit, most unusual ever proposed for a NASA space mission, to increase understanding of solar-terrestrial relationships. *Isee 3* was orbiting around the L-1 libration point, a point between the earth and the sun about 1.5 million km (1 million mi) from earth at which the centrifugal and gravitational forces acting on a satellite were exactly counterbalanced. In this orbit, *Isee 3* would be able to monitor characteristics of the solar wind and other solar-induced phenomena such as solar flares about an hr before they disturbed earth space.

GSFC's Dr. Robert Farquhar, who devised the halo orbit, said it had been designed "to pass slightly above and below the ecliptic plane so that it will avoid excessive solar interference with spacecraft communications back to earth stations. To a person observing *Isee 3* from our planet, it will appear to be orbiting the sun, but it actually traces a halo above earth." On entering the halo orbit, *Isee 3* would accomplish three spaceflight firsts: a libration-point satellite; a spacecraft stationed in a halo orbit; and a spacecraft orbiting a point in space rather than a body such as the sun, earth, or moon.

Libration-point orbits had aroused interest as possible ideal locations for future space colonies; they would also offer advantages in lunar-farside communications and as staging locations for lunar and interplanetary transportation systems. At least five of the libration points had been mentioned as possible sites for space colonies; one of them (the earth-moon L-5 point) was even the namesake for the leading group of space-colony enthusiasts, the L-5 Society. However, research concerned with space colonization had been mainly theoretical. *Isee 3* should provide the first real flight experience for orbital operations of future space colonies. (GSFC Release G-78-85)

• Fourteen satellites, four of them NASA's and two starting the Intelsat V series of international communications satellites, were on the preliminary schedule of NASA launches for 1979, *Aerospace Daily* reported. These missions would use expendable launch vehicles, and NASA hoped that its most significant mission of 1979 would be number 15: first flight of the Space Shuttle. The tentative 1979 schedule would begin the transition of payloads from expendable launchers to the Space Shuttle. The Delta launch vehicle, workhorse of the 1970s, would carry only 4 of 1979's 14 missions. Only three of the payloads were commercial, and half the 1979 launches would be from Vandenberg AFB and Wallops Island rather than KSC. The three launches scheduled in the first quarter of 1979 would use all three launch sites and three different launchers. Second-quarter payloads, all sponsored by organizations outside NASA, would also use all three launch sites. NASA planned five launches in the third quarter, which it hoped to close with the Space Shuttle's first flight. (*A/D* Nov 21/78, 95)

• FBIS reported that scientists and staff members of the Prague Institute of Geophysics were working around the clock as they received and processed data coming from the first Czechoslovak artificial satellite, *Magion*. Beginning Nov. 14, when *Magion* separated from *Intercosmos 18* and went into an independent orbit, it had returned data on the earth's ionosphere and magnetosphere four times a day to the Czechoslovakian Panska Ves Observatory, where incoming information was computer-processed and transmitted to the Flight Control Centre and to stations in the Soviet Union. P. Triska, in charge of the *Magion* experiment, said that onboard systems were working efficiently and that he and his staff were cooperating closely with their Soviet colleagues: "We draw up together working programs for *Intercosmos 18* and *Magion* and exchange information." (FBIS, Tass In English, Nov 21/78)

November 22: LaRC announced it had awarded General Electric Company, Newport News, Va., a contract to provide rigid closed-cell foam-insulation material for use in the National Transonic Facility (NTF), a cryogenic wind tunnel under construction at LaRC. The insulation would be installed on the internal surface of the tunnel's cylindrical pressure shell. The contract, managed by LaRC and estimated by GE at approximately $1.4 million, covered just over 1yr; work would be performed at GE's Thermal Systems Programs plant in Tacoma, Wash. (LaRC Release 78-81)

November 23: ESA reported results of an attempt to quantify economic returns from its expenditures on space research, by cataloging and quantifying the positive benefits to industrial firms participating in the European space program and by identifying economic spinoffs outside the space effort generated in European industry from ESA contracts. ESA wanted to know what benefits its payments had procured for European firms and what amounts they represented in financial terms. It had asked the Theoretical and Applied Economics Department of Louis Pasteur Univ., Strasbourg, France, to conduct the study.

Information collected from ESA contractors had identified and quantified 171 specific cases of benefit: for example, sales of new products, increased sales, maintenance of a production team, reduction of production and R&D costs, and amortization of equipment. The study classified the benefits under four major headings: technology, commerce, organization and methods, and work force efficiency. Benefits obtained outside ESA-member states accounted for 28% of the total, and 68% of the total benefit related to nonspace activities. Analysis showed the benefit/contracts ratio for each country was fairly uniform: for example, Germany 3.2, United Kingdom 2.9, Italy 2.6, France 2.5, and other member states, 2.2.

This study of secondary economic effects generated by ESA contracts confirmed the agency's success in promoting European industry worldwide. Moreover, the showing of economic benefits to individual countries demonstrated that they exceeded the amounts governments had invested in space research. A second phase of the study would produce a more precise analysis of the benefits by including economic factors not previously taken into account. (ESA Release Nov 23/78)

November 24: If research now in advanced stages went well, Space Shuttle orbiter crews would enjoy tasty and nutritious meals served in an attractive manner, NASA announced. Menus would be well balanced (3000 calories a day) with hot entrees, "reasonably cold" fruit beverages or dessert, and plenty of condiments. A typical daily menu on the Space Shuttle might be: breakfast—orange drink, peaches, scrambled eggs, sausage, sweet roll, cocoa; lunch—cream of mushroom soup, ham and cheese sandwich, stewed tomatoes, banana, cookies, and tea; dinner—shrimp cocktail with sauce, beefsteak, broccoli au gratin, strawberries, pudding, cookie, and cocoa. Many of the foods (breakfast bars, for example) would come off supermarket shelves. All meals would be prepared with variety in mind, so that a crew would have a different menu for six successive days; on the seventh day the cycle would begin again. Dining would be from a food tray at a table in the galley, or the tray could be held in the lap or affixed to the wall. Ordinary eating utensils would be used.

Foods planned for Shuttle crews included (thermostabilized) thermal/heat-processed foods canned or packaged in laminated foil pouches, such as cheese spread, tuna, beef with barbecue sauce; (intermediate moisture) foods preserved by controlling the available moisture, such as dried apricots, peaches, breakfast bars; (rehydratable) dehydrated foods reconstituted with water, such as scrambled eggs, beef patty, chicken and noodles, all beverages; (irradiated) foods exposed to ionizing radiation to effect preservation, such as bread, rolls, and beefsteak; (freeze-dried) foods with all water removed to be rehydrated, or eaten as is with saliva acting as the moistening agent, such as strawberries, shrimp, bananas; (natural form) foods such as nuts and cookies. All this was far from the space meals of an earlier era, when Mercury and Gemini astronauts had to squeeze various kinds of food into their mouths from tubes, or when floating crumbs had to be recaptured before they got into the instruments. Some of the same people who had pioneered space food preparation were still working at JSC putting together the Space Shuttle food systems.

From the beginning of the national space effort, the topic of food had captured the attention of both observers and participants. On short spaceflights, food was not too important; it became a more serious consideration, then a vital one, as crews began to range farther and stay longer. First menus had included only the tubes of paste-like food served at room temperature, lukewarm dehydrated cubes, or a few freeze-dried items. By the time of the long-duration Apollo flights, the types of and systems of preparing space food had become more sophisticated, and new means of heating and cooling had brought improved palatability. A breakthough came during *Apollo 8* when the crew was surprised with a home-cooked Christmas dinner: pieces of turkey with most of the trimmings, eaten with a spoon rather than squeezed from a container. The turkey and gravy was a thermostabilized product packaged in a laminated foil pouch, leading to the development of a "spoon-bowl" plastic pack for rehydratable foods into which water was injected through a valve. Contents were spooned out through a zippered opening. Main differences between Space Shuttle food and that served on earlier flights would be greater variety and larger amounts, more efficient handling of food, and more comfortable natural dining arrangements. (NASA Release 78-176)

November 27: The U.S. Postal Service would soon begin a 1yr test of an electronic message service system that would reduce letters to electronic impulses, relay them to their destination by satellite, reconvert them to letter form using high-speed printers, and deliver them by regular carriers, *Defense/Space Daily* reported. If the test was successful, the agency could offer electronic mail services in up to 10 cities within 3yr. The initial test would send messages by satellite from one room in a Postal Service facility in Rockville, Md., to another room in the facility.

A $2.3 million RCA study of the electronic message service system had concluded that it would cost the Postal Service 10 to 11 cents to deliver this kind of mail, a price equal to or less than present costs. RCA suggested that the system could be used within 6 to 8yr for up to one-fourth of the 100 billion letters sent through the U.S. mails annually and estimated it would cost the Postal Service $2 billion to install the system in 87 cities around the U.S. (D/SD, Nov 27/78, 120)

• *Av Wk* reported that Europe's aerospace industry would probably get a sizable slice of procurements for the emerging global maritime satellite-communications system (variously designated as the second-generation system, the Marisat 2 system, or the pre-Inmarsat system). Great Britain's ambassador Peter Jay and France's ambassador Francois de Laboulaye had made strong representations on Europe's behalf to U.S. Deputy Secretary of State Warren Christopher. The fact that all prime contracts (and most of the dollar volume of procurement) for the first global satellite communications system, owned and operated by INTELSAT, had gone to U.S. firms had been a sore point in U.S.-Europe relations for more than a decade. ESA's Marecs satellite scheduled for launch on an Ariane booster would figure in both plans for the new global system that had received most-favored consideration, as U.S. government agencies (led by the State Dept.) and ComSatCorp struggled to formulate a U.S. position. One plan favored by the U.S. for the interim system was to combine maritime communications packages on three Ford Aerospace-built Intelsat 5 satellites with two or three Marecs satellites. The other favored plan was to combine commercial communications packages on the Leasat system being established by Hughes Communications Services, Inc. for the U.S. Navy with two or three Marecs satellites. (*AvWk*, Nov 27/78, 13)

November 28: The most extensive study ever made of Venus would begin in December when *Pioneer Venus 1* and *2* arrived at the planet, NASA announced. *Pioneer Venus 1* would swing into orbit around the planet Dec. 4, taking pictures and making measurements for one Venusian yr (225 earth days) or more. On Dec. 9 the four probes and transporter bus constituting *Pioneer Venus 2* would plunge into the Venusian clouds at widely separated points to make detailed measurements of the dense atmosphere from top to bottom. These flights, the first devoted to studying the atmosphere and weather of another planet on a global scale, would use the largest number of vehicles so far for such a study, and would make measurements at the greatest number of locations. They also would seek information on characteristics of Venus's upper atmosphere and ionosphere, as well as the lower atmosphere, and would study the interactions of these regions with the solar wind.

At periapsis (orbital low point) the Pioneer orbiter would come as close as 150km (90mi) to the planet's surface, dipping into the thin

Venus upper atmosphere, sampling its composition, and making radar measurements of surface elevations and roughness. It would also take daily ultraviolet and infrared pictures of the thick clouds and atmosphere. The orbiter's primary mission was designed to cover the time it took Venus to make one revolution on its axis; the slow rotation of the planet under the periapsis of the orbit would permit closeup radar mapping of the entire circumference.

The four probes of *Pioneer Venus 2* would enter the atmosphere at points 10 000km (6000mi) apart over the planet's earth-facing hemisphere: the large probe would enter Venus's atmosphere near the equator on the day side; the day probe would enter Venus's southern hemisphere

Artist's conception of Pioneer Venus orbiter and multiprobe investigating the cloud-covered planet on their arrival in late 1978. The orbiter (left) would circle Venus for 8mo, relaying images of the clouds and measurements of the atmosphere, while making radar maps of the surface. The probe (right) would split into five entry craft, four to sample the atmosphere down to the surface at widely separated points; the transport bus would continue to sample the upper atmosphere until it burned up from friction. (NASA photo 78-H-208)

on the planet's day side; the night probe, the night southern hemisphere on the night side; and the north probe, the north polar vortex on the night side. The transporter bus would enter in the southern hemisphere on the day side.

Critical operations on arrival near the planet would be to inject *Pioneer Venus 1* into orbit while it was out of radio contact behind Venus, and the entry and descent through the atmosphere of the four probes and bus. On Dec. 2, controllers at ARC would orient the orbiter for orbit insertion, and the next day would load command-memory sequences. On Dec. 9, the probes would be in position to make their hr-long descent through Venus's atmosphere. Major operations problem would be retrieving probe data; during the critical 1hr 38min of probe and bus operations at Venus, all the Pioneer spacecrafts would be transmitting simultaneously. Radio frequencies of the probe signals would shift rapidly because of huge entry decelerations—from 42 000km/hr (26 000mph) to about 65km/hr (40mph)—and signal distortions caused by Venus's dense atmosphere. One consequence would be that, after the 10- to 15-sec communications blackout during entry, signals would reappear again at new frequencies. To ensure recovery of the data, NASA had installed special equipment in DSN stations at Goldstone, Calif., and Canberra, Australia. To avoid loss of data, both DSN stations would use newly developed receivers and recorders with a frequency band wide enough to ensure receipt of the four shifting signals. The real-time system would use automatic tuners, but would lose some data.

Scientists believed that the coordinated Pioneer Venus atmospheric data, combined with similar data from Mars, Jupiter, and other planets, would aid understanding of atmospheres in general. Studies of the interactions of temperatures, pressures, composition, clouds, and atmospheric dynamics different from those of earth should define mechanisms on another planet that would improve scientists' understanding of earth's complex weather processes. (NASA Release 78-181)

• ESA's meteorological satellite, *Meteosat 1*, had been in orbit for a yr carrying out its image-taking, dissemination, and data-collection missions, ESA announced. Ground facilities at the European Space Operations Center (ESOC) in Darmstadt, Germany, for receiving, processing, archiving, and disseminating meteorological data from the satellite were ready to fulfill all requirements for the first GARP experiment, scheduled to begin Dec. 1. *Meteosat 1*, which transmitted every half hr visible and infrared images of the earth, had been retransmitting to primary and secondary data-user stations some 300 formats per day in various digital and analog forms. It also acted as a relay for image data from the U.S. satellite *Goes 1* over the Atlantic, transmitted through the Centre de Meteorologie Spatiale at Lannion at the rate of about 20 formats per day. *Meteosat 1* had also participated in Nov. in the first transmission tests of image data from the U.S. *Goes 3* satellite over the Indian Ocean, so that users had received data from three geostationary satellites covering approximately two-thirds of the globe.

Data-collection platform experiments had demonstrated the high quality of the links between *Meteosat 1* and the mobile platforms; about 15 DCPs were operational, and about 10 were in the process of being admitted and would come into service in early 1979.

Another mission of *Meteosat 1* was to supply immediately usable meteorological products (wind speed, cloud height, sea-surface temperature, etc.). Meteosat image quality should maintain the serviceability of these products; results already achieved were at least comparable to those obtained by other geostationary satellites participating in the GARP. Use of the water-vapor channel, an original feature on *Meteosat 1*, was considered by the meteorological community as a major advance in refinement of interpretation techniques. (ESA Release Nov 28/78)

November 29: Heao 2 continued to operate satisfactorily in its third wk in orbit, and plans were to rotate a new instrument into the focal plane of the observatory's x-ray telescope within several days, the *Marshall Star* reported. Four of the five instruments aboard had been mounted on a revolving platform at the focus of the telescope so that ground controllers could rotate each into the focal plane on command. A high-resolution imager located in the focal plane at launch had transmitted to earth Nov. 18 the first picture ever of an x-ray star and had sent many such pictures since then.

Replacing the high-resolution imager would be a solid-state spectrometer able to observe the entire spectrum at once, measuring the energy (and therefore the wavelength) of each photon that struck the experiment's silicon crystal. This would permit a better understanding of recently discovered astrophysical phenomena such as black holes, pulsars, and bursters. The solid-state spectrometer, the only life-limited instrument aboard, would have priority in the observatory's initial viewing program. (*Marshall Star*, Nov 29/78, 1)

• MSFC announced it had awarded RCA Government Systems Division an $880 000 contract to develop and deliver three video-instrumentation tape recorders for use on Spacelab 1 and subsequent missions. Under the contract, RCA would deliver three STAR-4 units before April 1980 for integration into Spacelab program equipment before the first launch planned for 1981. Spacelab crew members would use the units to record visual and analog data during communications blind spots, while the Space Shuttle was out of contact with NASA ground stations. The information would be played back for transmission when communications were reestablished.

The units, capable of about 500 record-playback operations between tape changes, could record for 45min simultaneously on one video and two audio channels under computer control. Video bandwidth was 5MHz, audio bandwidths 50 to 8000hz. The instruments would undergo rigorous testing for high performance and reliability, and to ensure equipment compatibility with the manned-spacecraft environment. NASA would install two STAR-4s as part of the Spacelab 1 payload and keep the third as a spare. (*Marshall Star*, Nov 29/78, 4)

• U.S. and Chinese technical officials had begun discussion Nov. 28 of a plan for the People's Republic of China to buy an American communications satellite, the *Washington Post* reported. The deal would constitute the most sophisticated transfer of technology yet between the two countries. A 15-member Chinese delegation led by Jen Hsinmin, director of the PRC's Space Technology Research Institute, had met with officials from the State and Commerce Depts. and NASA at the start of a 23-day visit to the U.S. The *Baltimore Sun* quoted a State Dept. official as saying the U.S. would require, as a condition of purchase, that the PRC not use the satellite for military purposes, meaning that the satellite's signals could be transmitted only within China's borders. A White House official said it was understood the Chinese planned to use the satellite to transmit educational television. The communications system could cost as much as $50 million, including the price of sending the satellite into orbit. (*W Post*, Nov 29/78, A-19; *B Sun*, Nov 27/78, 13)

November 30: NASA announced that an experimental oceanographic satellite had achieved an unexpected technological breakthrough: for the first time, earth-contour measurements by a satellite microwave sensor had exhibited an accuracy comparable to that of conventional ground and aerial surveys used for small-scale mapping. An experiment sponsored by WFC had produced terrain contouring from *Geos 3* (geodynamics experimental ocean satellite) radar-altimeter measurements at a height of 840km (520mi) above earth. As *Geos 3* circled the globe at 7km/sec (4mps), the radar altimeter pointed vertically down and measured the distance between the satellite and the earth's surface with a precision of 50cm (20in). Nineteen *Geos-3* passes over the San Joaquin valley of southern Calif. had recorded terrain elevations for 40 000km^2(15 500mi^2), and comparison of the satellite-derived contours with U.S. Geological Survey small-scale topographic maps showed agreement within 1m (3ft).

Since the altimeter had been designed as an ocean-sensing system, the acquisition of meaningful land data was unexpected. The ocean-optimized design of the altimeter and the nature of radar backscatter was expected to limit measurement of terrain variations to smooth or gently rolling topography such as a coastal plain, interior plain, or valley. The *Geos 3* overland results showed that satellite-borne altimeters could measure terrain elevations for more rugged areas with accuracy sufficient to satisfy many small-scale mapping requirements, at a cost competitive with conventional methods. This technique would be particularly useful in remote areas of the world, such as arctic and desert regions where conventional surveying was difficult and expensive. (NASA Release 78-183)

• NASA announced that a frame structure for mounting experiments inside the Space Shuttle orbiter's payload bay had arrived by ship in Savannah, and would be transported by tug to the vehicle assembly building at KSC. Called an OFT (orbital flight test) pallet, the structure, fitted with NASA-developed subsystems, would provide electrical and other support to the experiments. This was the first of two pallet systems destined for use on two of the first six Space Shuttle flight tests.

Designed and assembled by British Aerospace Corp. for ESA as part of the joint ESA/NASA Spacelab program, the pallet structure was nearly identical to that of experiment pallets to be used in Spacelab missions beginning in 1981. It was 3m (10ft) long and constructed in a U-shape to fit into the 4.5m (15ft)-diameter orbiter payload bay. Later in the Spacelab program, five pallets would be available for space missions. MSFC was responsible for delivery and system integration of the OFT pallets. (NASA Release 78-182)

• NASA's Office of Space Science (OSS) had become less optimistic about how much money it could expect from Congress for Spacelab in the 1980s, and was trying to reduce the estimated cost of such missions, *Aerospace Daily* reported. The cost of Spacelab operations, together with inherent limitations on the work it could support, was working against previous forecasts of heavy use.

Two yrs previously, OSS had viewed Spacelab as an opportunity for growth of NASA's science program and had projected annual Spacelab budgets of $250 million or so, excluding launch costs, by the mid-1980s. Dr. Noel Hinners, NASA's Associate Administrator for Space Science, said, "Now we don't have the same optimism," given the expected budget environment, Spacelab orbit staytime and schedule limits, and competition within OSS for continued development of free-flyers. A more realistic budget prospect now would be $100 to $125 million per yr, essentially permitting three missions per yr (one for physics and astronomy, one for solar terrestrial programs, and one for life sciences), and development of one facility-class instrument per yr once steady-state operations were reached about 1984. That level of funding would be about the same as that for one major flight of planetary spacecraft per yr. (*A/D*, Nov 30/78, 127)

During November: Two Space and Missile Test Center (SAMTEC) test ranges had provided NASA with radar tracking and data-relay support during the reentry of *Pegasus 1*, AFSC *Newsreview* reported. The micrometeoroid-detector satellite was NASA's largest unmanned spacecraft when it was launched in February 1965. The Eastern Test Range at Patrick AFB, Fla., had served as lead range in satellite tracking, furnishing data from eastern sensors to GSFC; the Western Test Range at Vandenberg AFB had tracked the satellite after receiving acquisition data from NORAD.

Employees of Federal Electric Corp., SAMTEC's range technical-services contractor, had worked from early Sept. 12 until late Sept. 16 when reentry occurred off the coast of Angola. The company had used the TPQ-18 radar on south Vandenberg AFB, the FPQ-14 radar at Kaena Point, Hawaii, and the FPQ-6 radar at Pillar Point, Calif. In addition, SAMTEC Operations Control personnel at Vandenberg had supported the tracking and data relay provided to ETR from SAMTEC's Kwajalein missile range in the Marshall Islands. (AFSC *Newsreview*, Nov 78, 1)

• The AFSC *Newsreview* reported that the Air Force Avionics Laboratory (AFAL) was preparing critical electronic equipment for the NavStar global-positioning system (GPS) to undergo initial flight tests with GPS satellites. A generalized development model (GDM) designed by AFAL and built by Rockwell Intl. Corp.'s Collins Avionics and Missile Systems Division, Cedar Rapids, Ia., would demonstrate operation of GPS navigation equipment in a hostile electromagnetic-jamming environment. One GDM delivered to ASD's 4950th Test Wing had been installed on a C-141 aircraft for testing.

Evaluation of a GDM installed in two satellites had begun in August at the Naval Air Station, El Centro, Calif., and flight testing with three satellites against a ground jammer would begin in late Nov. GDM tests with four satellites in March 1979 would determine the equipment's full antijam performance, using ground-based jammers and possibly an airborne jammer. A second GDM, to be delivered to AFAL in Feb. 1979, would be integrated with the laboratory's GPS evaluator, after which AFAL engineers would conduct jamming tests in various simulated environments. (AFSC *Newsreview*, Nov 78, 12)

• The USAF's Space and Missile Systems Organization (SAMSO) was conducting initial ground tests of a prototype laser-communications (LaserCom) system to transfer data from satellite to satellite as well as to and from ground and airborne users, the AFSC *Newsreview* reported. Over a 2yr period SAMSO would test the system at the White Sands Missile Range, N.M., in three phases: ground-to-ground transmission and reception of the laser beam, by reflecting it off a 24in steerable mirror 1 to 6mi from the laser; interim air-to-ground tests, evaluating aircraft ability to acquire and point the laser transmitter toward a ground-based beacon laser; and full demonstration of the air-to-ground system, using prototype equipment.

Although most of the tests had been successful in laboratory simulations, a few unknowns remained, including laser-beam communications through the atmosphere. LaserCom had some inherent advantages over existing satellite communications systems: for example, its narrow beam width enhanced security by allowing only the intended user to receive the beam. It also offered an unprecedented data-transmission capability because its 1-gigabit-per-sec rate would accommodate a great number of spacecraft, airborne and surface vessels, and ground users that would be relaying information to a synchronous-altitude LaserCom satellite for subsequent real-time or delayed transmission to other users. After completing its tests in 1980, DOD would evaluate the system for technical capability, requirements, and cost; should LaserCom application prove feasible, in-orbit operation of the system could begin as early as 1986. (AFSC *Newsreview,* Nov 78, 1)

December

December 1: *Aerospace Daily* reported that John Yardley, NASA's associate administrator for Space Transportation Systems, had addressed the National Space Club in Washington, D.C. Nov. 28 on the importance of Congress's role in an aggressive space program, contrasting the cautious approach of the Carter Administration. He cited the 1978 introduction of national space-policy bills by Sen. Adlai Stevenson (D-Ill.) and Sen. Harrison Schmitt (R-N.M.). On the House side, Rep. Ronnie Flippo (D-Ala.) had introduced, and the House overwhelmingly passed, a bill authorizing an unrequested $25 million to broaden and accelerate R&D for a solar power satellite system, including a hardware flight-demonstration phase that the administration had opposed as "premature."

The staff of the House Science and Technology subcommittee on space science and applications, chaired by Rep. Don Fuqua (D-Fla.), had been working on ways for Congress to facilitate "the space industrial revolution," Yardley noted. He added that "if Congress does all these things, it will be in their tradition," including taking the lead in developing the nation's railroad system and passing the Homestead Act in the 19th century. "The same kind of action will be needed soon for the space industrial revolution," he said; the industrial activity required will "be brought to bear if the climate is correct. It's up to the government to provide the environment to stimulate growth." Whether this occurred, he concluded, "may be the difference in the next 50 years as to whether the United States continues to be world leader not only in space but economically—or becomes a second-class power." (*A/D,* Dec 1/78, 137)

December 4: NASA announced plans to launch the Canadian comsat Telesat-D (to be called Anik-B in orbit) from ETR on a Delta no earlier than Dec. 15, 1978. Telesat/Canada would reimburse NASA $19.2 million for launch support under a March 11, 1978, contract. Telesat-D would be the fourth domestic comsat launched by NASA for Telesat/Canada, which owned and operated the satellites. Telesat-D would replace *Anik A1,* launched in Nov. 1972 and positioned at 109°W over the equator south of California. With a 7-yr design life, Telesat-D would provide point-to-point voice, TV, and data communications traffic to Canada's 10 provinces.

In addition to 12 commercial channels in the 6 and 4GHz frequency bands, Telesat-D would offer 4 channels in the 12 and 14GHz frequencies; the entire capacity of the higher frequency bands would be leased to

Canada's Department of Communications for 2yr with an option for a 2yr extension. CDC would devote these channels to social experiments such as telemedicine, tele-education, teleconferencing, and Eskimo broadcasting to demonstrate how satellites can reach remote locations with small ground stations (a use begun by the communications technology satellite *Cts*, a joint U.S.-Canadian experiment). These channels would also carry out purely technical assignments to gather data on signal propagation and on power generation and usage.

NASA planned to launch Anik-C1 and -C2 spacecraft (Telesat-E and -F) on the Space Shuttle in 1981, to replace *Anik A2* and *A3* (*Telesat B* and *C*) launched in April 1973 and May 1975, respectively. (NASA Release 78-184; MOR M-492-201-78-04 [prelaunch], Dec. 12/78)

• Space platform studies by NASA's Office of Space Science (OSS) had raised questions about the extent to which OSS would use Spacelab during the 1980s, *Aerospace Daily* reported. During the next yr or so, users would evaluate cost-effectiveness of the modular Spacelab being developed by ESA to fly in the Space Shuttle orbiter cargo bay against that of free-flying satellites and larger undefined space platforms. In the meantime, experiments and instruments for Spacelab would be selected partly on the basis of their adaptability to later missions aboard free flyers.

Operating-cost estimates and the outlook for funding had tempered earlier OSS optimism about using Spacelab [see Nov. 30]; the platform studies would likewise curb optimism if they substantially described activity that might be carried out better or cheaper on free flyers. A member of the NASA Advisory Council's space science advisory committee noted at a Spacelab review that NASA could develop two Explorer-class free flyers at about the same cost as a few Spacelab flights.

The space platform studies also had some political dangers: NASA had sold Spacelab previously as an integral part of the space transportation system, and would lose credibility if it backed off. However, Dr. Noel Hinners, NASA's associate administrator for space science, pointed out to the committee that the platform studies were intended to avoid the Spacelab situation, which had handed science and applications researchers a facility they had little if any role in defining. If NASA proceeded with a platform, Hinners said, it would be because mission requirements had been identified, "not because somebody who likes to build things says, 'Here it is.'" (*A/D,* Nov 4/78, 144)

• U.S. analysts believed the USSR would continue to fly manned military space station reconnaissance missions even though the Soviets had not launched any such Salyuts since June 1976, *Av Wk* reported. *Salyut 3,* launched in June 1974, and *Salyut 5,* launched in June 1976, were successful military missions; the *Salyut 2* spacecraft, which failed in orbit in April 1973, was also apparently a military Salyut. Significant design differences between the military and scientific versions included docking configuration, overall structure, and solar-panel placement. Introduction of the military configuration as early as the second Salyut flight was considered to be evidence of USSR emphasis on military-man-in-space planning and engineering. Photo reconnaissance was the prime objective of Salyut military missions; other military activities such as high energy research for possible weapons application were secondary. (*AvWk*, Dec 4/78, 17)

December 5: Lewis Research Center and Dryden Flight Research Center would conduct flight tests of advanced models of propellers, some with 8 or 10 blades, intended for new high-speed fuel-conserving aircraft, DFRC announced. LeRC had already measured aerodynamic performance of the approximately 1/7 scale (2ft-diameter) models in its wind tunnel, at a simulated cruise speed of 855kph (530mph) and altitude of 10 668m (35 000ft). NASA estimated that, in those cruise conditions, an advanced turboprop engine with the new-design propeller could save 20 to 40% in fuel over current turbofan engines and 10 to 20% over advanced turbofan engines. (DFRC Release 23-78)

• RCA Americom Communications, Inc., announced it would launch its Satcom-3 domestic comsat in Dec. 1979, more than a yr earlier than planned, and would study procurement of a fourth comsat either based on current design or the first of a new generation of spacecraft, *Aerospace Daily* reported. Satcom-3 previously appeared on a Space Shuttle schedule for launch in mid-1981. Latest plans called for a Delta launch like *Satcoms 1* and *2*; the Shuttle reservation would be assigned to Satcom-4.

RCA Americom said it was advancing the Satcom-3 launch because traffic on the first 2 satellites was nearing capacity particularly in cable television, now using 900 earth stations, with 1400 to 1500 expected by the end of 1979. Satcom-3 would have the same 24-transponder capacity as the first two satellites, but would carry spare transponders and additional stationkeeping fuel to exceed its planned 8-yr lifetime. The company had asked the FCC for permission to station Satcom-3 at synchronous altitude over the equator at 132°W to serve the 50 states, mainly for cable-TV relays. RCA's Astro-Electronics Div., which had built the three Satcoms, estimated the cost of the third spacecraft and its launch at $40 million. (*A/D*, Nov 5/78, 150)

• The Aerospace Safety Advisory Panel had recommended elimination of the Space Shuttle solid-fuel rocket booster's thrust-vector control (TVC) system and modification or replacement of the Shuttle orbiter's auxiliary power unit (APU), *Aerospace Daily* reported. The panel said its proposals should make the Shuttle a safer system during its operational phase. With "emphasis shifting to the problems of reusable operation from those of technical breakthrough," it said in a summary report, ". . . it is not too soon for NASA to review the Space Shuttle design from this point of view." The report concluded that the recommendations were "not reservation on the panel's part as to the readiness for SS-1," but rather said the Shuttle "will fly well, and probably on the scheduled date."

Modifications could be expensive, and panel members noted that such work might absorb the expected "Shuttle dividend," funds from declining Shuttle development costs that might be devoted to other programs. NASA Administrator Robert Frosch responded that NASA might apply the panel's proposals in designing its next space transportation system, instead of modifying the present system. (*A/D,* Dec 5/78, 148)

December 6: Early analysis of Space Shuttle liftoff-configuration vibration-test results at Marshall Space Flight Center had revealed two primary deviations, the *Marshall Star* reported: excessive movement in a forward section of the solid-fuel rocket boosters that contained guidance gyros, and a problem with transfer of loads (or forces) at the attach points between the boosters and external tank. After installing brackets to strengthen the boosters at the gyros location, NASA would conduct individual tests on that booster section at a test facility used for smaller components. Additional studies of load distribution at the attach points would determine if additional tests were necessary. After the most recent test phase, engineers found that about 80% of the test modes had verified predictions of hardware behavior during flight. The final test phase, beginning early in 1979, would simulate the Space Shuttle configuration just before burnout and separation of the solid-fuel rocket boosters at an altitude of about 43.5km (27mi). (*Marshall Star,* Dec 6/78, 1)

• Harold Berger of the Natl. Bureau of Standards would give the colloquium lecture Dec. 11 on "Nondestructive Testing—Its Growing Importance for the 1980s," Langley Research Center announced. The aerospace and nuclear industries, which required unusually high reliability and quality in products and operations, had used nondestructive testing extensively since World War II. Berger would describe the early use of NDT, including radiographic, ultrasonic, magnetic, penetrant, and eddy-current techniques. Although early NDT had increased safety and extended product life, it had the potential for significantly improving industrial productivity and conserving energy and materials.

NDT was now facing growing problems such as an aging fleet of military airplanes; demand for a clean environment with NDT implications for nuclear plants, pipelines, and offshore facilities; and needs associated with product liability. These challenges had stimulated new developments; as an example, Berger would discuss radiography, including real-time methods, image enhancement, and unusual radiation techniques using neutrons and protons. (LaRC Release 78-83)

• The *Marshall Star* announced that the most extensive study ever made of Venus had begun with the arrival at the planet of *Pioneer Venus 1,* the orbiter half of the two-spacecraft Pioneer Venus mission. It went into orbit Dec. 4 and was scheduled to take pictures and make measurements for one Venusian yr (225 earth days) or more. On Dec. 9, the four probes and transporter bus of *Pioneer Venus 2* would plunge into the Venusian clouds at widely separated points to measure the dense atmosphere in detail from top to bottom. A spokesman at Ames Research Center, manager of the project for NASA's Office of Space Science, reported Dec. 4 that *Pioneer Venus 2* was on course and on schedule. (*Marshall Star,* Dec 6/78, 2)

• On Dec. 12, the U.S. Coast Guard would receive its first lightweight portable firefighting module, developed under direction of MSFC's Technology Utilization Office, the *Marshall Star* reported. The manufacturer, Northern Research and Engineering Corp. of Cambridge, Mass., would deliver it to NASA representatives in Boston for transfer to the Coast Guard on completion of an acceptance test.

Since MSFC had developed technology for lightweight highly efficient pumps in rocket engines, the Coast Guard had requested help in designing a unit small enough to be lowered onto the deck of small craft but capable of pumping hundreds of gallons of water per min. The module was 4ft wide, 5ft high, and 6ft long, and weighed 2500 lb, including fuel for 3hr of operation, and all equipment stored internally. Two people could set it up rapidly without tools, and it could deliver full output flow within 10min.; a special feature was a pair of fiberglass water cannons, or monitors, weighing 40lb. The module could draw water from the sea through a lift up to 20ft long and pump at least 1000gal per min through each monitor.

Commercial applications included use on light trucks by local fire departments, on docks and ships for harbor fires, and for forest, offshore oil rig, and high-rise building fires, as well as an emergency booster pump for municipal water and for flood control and military operations. Aubrey Smith, head of TU activities at MSFC, said the firefighting module showed "how NASA-developed technology and expertise can be adapted for down-to-earth uses." (*Marshall Star,* Dec 6/78, 1)

• The *Marshall Star* reported delivery to Cape Canaveral of a light ion-mass spectrometer (LIMS) developed by MSFC scientists, for integration with other instruments on the USAF's SCATHA (satellite charging at high altitudes) scheduled for launch on a Delta Jan. 25, 1979. SCATHA instruments would collect data on electrically charged particles in the upper ionosphere and magnetosphere to help solve the problem of static-electricity buildup on satellite surfaces at geosynchronous altitude. The LIMS would measure density, temperature, and speed of low-energy ions from all directions, as sensors separated hydrogen, helium, and oxygen ions and recorded their speed and numbers. Martin Marietta Co., integration contractor, had delivered the instrument to MSFC 3wk previously, and MSFC personnel had installed flight sensors and completed final calibration before delivering it to the Air Force at the Cape. Designed for a nominal 1yr lifetime, SCATHA might return data for possibly 3yr or more. (*Marshall Star,* Dec 6/78, 2)

• ESA announced successful completion Dec. 5 of the fourth development test of the Ariane launcher's first-stage inflight configuration by the Société Européenne de Propulsion, Vernon, France. The 142.9sec test (a propulsion time corresponding to depletion of the N204 propellant) confirmed that the propulsion system would function for the nominal duration of the stage's flight, 142.5sec. Also tested was the silicon and phenolic-resin (SEPHEN 301) material to be used for Viking-engine nozzles, replacing other materials exhibiting inadequate resistance in previous tests.

The second test of the third-stage flight configuration, scheduled for Nov. 28 at Vernon, had been interrupted when failure of the cryogenic engine (H8) to ignite caused a small explosion that damaged the propulsion bay, but not the test stand. ESA was investigating the incident and its effect on the development timetable; it would issue a new timetable in early January 1979, with the target of an operational Ariane in 1981 unaffected. (ESA Release Dec 6/78)

December 7: ESA announced that its council had unanimously authorized Roy Gibson, director general, to sign an agreement with Canada on Canadian participation in ESA's basic activities. Canada would participate in meetings of ESA's delegate bodies, and would have two delegates on the Council who could vote on questions relating to activities covered by the agreement. Canada's financial contribution supporting its participation in general studies of future projects would be 1% of the net fixed common costs of ESA's general budget. The agreement, a step toward closer relations between Canada and ESA, confirmed the Canadian government's intention to expand collaboration in European space activities and to establish permanent institutional relations with ESA. Cooperation between Canada and ESA had begun in 1972, when ESA agreed to provide systems for Canada's experimental comsat CTS.

A March 1977 agreement defined the terms of cooperation in remote sensing. (ESA Release Dec 7/78)

December 8: NASA announced that 8 women volunteers aged 45 to 55 had emerged in November hale and hearty after nearly a month of tests at ARC's Human Research Center during which they were isolated from friends and family. Like 40 other men and women aged 25 to 55 who had volunteered for space program tests since 1973, the ARC group had participated in a study of the effects of weightlessness on the human body.

Researchers from ARC's biomedical research division were interested in the tolerances and responses of a body reentering earth's gravity after experiencing zero gravity in space. The long-term NASA experiment, the Bed Rest Program, would extend over the next 12mo with 2 more groups—8 men, then 8 women, all aged 55 to 65—undergoing the same tests. Prolonged horizontal bed rest (a physiological approximation on earth of zero-gravity) would cause bodily changes within 24 to 48hr and would maintain them throughout the program. Having approximated the effects of zero gravity, researchers could study human responses to the introduction of both normal and abnormal Shuttle-reentry gravity forces. Other tests in cooperation with Johnson Space Center had evaluated the protection offered by antigravity suits against reentry stresses.

Those volunteering for the program in response to newspaper advertisements had undergone a series of exacting physical and psychological tests. The 3-phase study began with the 8 subjects moving into the 12-bed ARC research facility. After a 9-day control period came 10da of total bed rest, followed by a 5-day recovery period. During the bed rest experiment, participants had to remain flat except to raise their heads slightly to eat. The subjects even showered in a horizontal position, being moved on a wheeled gurney to a specially designed shower. They had no visitors, but could have incoming and outgoing telephone calls. The routine did not deviate from three meals a day, lights off at 11pm, on at 7pm. At times the lower body underwent negative-pressure tests in a box. Researchers took daily blood samples and deep body temperatures. Electrocardiograms constantly recorded heart rates, and a "biobelt" around the waist constantly sent data to recorders. Before and during bed rest, the subjects took centrifuge rides designed to measure reflexes and coordination as the body made adjustments. The researchers tried to keep the subjects cheerful and to relieve natural boredom with books and games; each bed had stereo and color television. Medical attendants (doctors, nurses, and technicians) were on hand at all times. (NASA Release 78-185)

• NASA announced it would physically integrate all Space Shuttle/Spacelab experiments at Kennedy Space Center. Extensive analysis and cost studies had confirmed an earlier tentative decision to do the work at KSC rather than at centers managing the Spacelab mission. "Physical integration" consisted of installing experiments on Spacelab's racks and pallets. After Spacelab operations became routine in the mid-1980s, physical integration should require about 100 civil service and contractor man-yrs per yr at KSC. The analyses required to integrate experiments for each mission would continue at assigned Spacelab management centers. (NASA Release 78-186)

• MSFC announced it had awarded 28 contracts valued at $50 000 or more during October 1978. Organizations receiving contracts were United Technologies Corp.'s United Space Boosters, $250 888, modification of a facility for Shuttle assembly, checkout, launch operations, and refurbishment; McDonnell Douglas Co.'s Spacelab Integration Div., $2 400 540, engineering changes for Spacelab integration; Sperry Rand Corp., $335 329, increased funds for delivering Shuttle SRB multiplexer/demultiplexer; Garrett Corp., $469 143, engineering changes in design, development, manufacture, test, and delivery of solar heating and solar heating/cooling systems for operational testing; Martin Marietta Corp., $600 000, cost adjustments for designing structural attachments for large space structures; Univ. of Chicago, $3 460 230, addition of cosmic-ray investigation provisions to contract for Skylab 2 experiment; Univ. of Iowa, $2 125 198, ejectable plasma-diagnosis experiment for Spacelab 1 and similar experiment for Spacelab 2; Smithsonian Inst., $806 192, work on HEAO-B x-ray telescope; Bendix Corp., $1 000 000, changes to contract for Shuttle SRB electronics assembly and associated equipment; and Naval Research Laboratory, $1 275 300, increase in funds for solar-ultraviolet experiment on Spacelab. (MSFC Release 78-132)

• Representatives of LeRC, ARC, LaRC, other government agencies, and industry had met at Dryden Flight Research Center for what was probably the final symposium on NASA's YF-12 aircraft program begun in the late 1960s, the *Lewis News* reported. The airplanes would be mothballed by March 1979. DFRC's Dr. James Albers, symposium chairman, said the symposium was to "disseminate recent research findings and stimulate exchange of information among participants engaged in design or research related to high supersonic flight."

The YF-12 had given NASA an opportunity for research on sustained supersonic flight. Several NASA centers had investigated propulsion-system performance and the dynamics of mixed compression inlets; aeroelastic effects and thermal problems of high Mach number flight; and techniques of measuring aerodynamic data and operational parameters in extreme temperatures. Researchers had conducted a parallel series of laboratory experiments to augment the findings of the flight program.

The symposium ended with an industry-government panel on "Future Perspectives in Supersonic Flight." (*Lewis News,* Dec 8/78, 7)

- INTELSAT announced it would use both the Space Shuttle and ESA's Ariane launch vehicle to orbit later spacecraft in the Intelsat V series. INTELSAT's board of governors had ordered two Shuttle launches from NASA and one Ariane launch from ESA. Worth a minimum of $82 million, the orders would cover the orbiting of the 5th, 6th, and 7th satellites in the Intelsat V series during 1981-82. The Shuttle would give INTELSAT its first experience with reusable launch vehicles; Ariane would be INTELSAT's first use of a non-U.S. launch vehicle.

INTELSAT said its decision was difficult because neither vehicle had been proven in operation. The organizations had scheduled its earlier Intelsat V satellites, like the then-current Intelsat IV and IV-A satellites, for launch on NASA's Atlas Centaur. The agreement had provided for an Atlas Centaur backup for the series, should the first Shuttle not be available. (INTELSAT Release 78-34-I)

December 11: The National Air and Space Museum would commemorate the 75th anniversary of the Wright brothers' historic Kitty Hawk, N.C., flight with the opening of a new exhibit featuring the 1903 Wright Flyer, the *Washington Star* reported. The exhibit, including pages from the 1903 patent, photographs, and other memorabilia, would open three days before the actual anniversary of the Wright brothers' flight Dec. 17, 1903. The special Wright exhibit would be on display for a yr.

Paul Garber, first curator of the Smithsonian's old National Air Museum, wrote in his book commemorating the occasion: "I was not at Kill Devil Hill to see Orville Wright make that most epochal of all pioneer flights. I wish I had been. But six years later I saw him piloting the Wright military flyer at Fort Myer, Va., and 39 years after that, I had the honor of bringing the first Wright 'aeroplane' back home and installing it in the Smithsonian Institution's Arts and Industries building. (The plane had been in England.) I also helped obtain our third Wright airplane (the Vin Fiz) . . .

"One day an acquaintance visited the museum with Orville Wright; . . . and I had the pleasure of meeting the first man to make a powered, sustained, and controlled flight . . . Later, he assisted me in preparing an exhibit illustrating some of the accomplishments of his brother and himself. I saw him on many occasions, and as I recall those times I only wish that I had asked him more questions! Regrettably he died on Jan. 30, 1948 [Wilbur Wright had died in 1912 of typhoid fever.]" In an interview before the exhibition opened, Garber said, "I can look through the window of a modern jetliner and see the Wright aircraft flying along, changing the course of history." (*W Star,* Dec 11/78, DC-1)

• *Av Wk* reported the Federal Aviation Administration prediction of an increase in the general aviation fleet from 186 600 active civilian aircraft in 1978 to 310 800 in 1990. Preliminary data indicated that hrs flown in general aviation aircraft reached 38.6 million in FY78, up 5.2% from 36.7 million in FY77. The FAA also predicted that by 1990 general aviation aircraft would be flying 67.4 million hr, increasing 4.8% yearly to 74.6% over the 1978 total. The lower increase of annual hrs flown resulted from the current emphasis on conservation of resources, FAA said. Operating costs, expected to rise more rapidly than inflation, would affect private flying for pleasure. (*AvWk,* Dec 11/78, 70)

December 12: NASA announced it had signed a $77 million contract with General Electric Space Division to build Landsat-D, most advanced earth-resources-monitoring satellite system to date, scheduled for launch in 1981 to view the earth from more than 640km (400mi) altitude. In addition to a multispectral scanner (MSS) like that on the first three Landsats, Landsat-D would carry a thematic mapper (TM) with spatial resolution three times as detailed as that of MSS. The new sensor could distinguish area features as small as 0.2-acre instantaneous field-of-view (IFOV) compared to the 1.2-acre IFOV resolution of present systems. The cost included a $5 million fee with possible additional earnings up to $4.3 million, depending on system performance once the satellite was in orbit. The contract included a backup spacecraft, data management system, operations control center at GSFC, transportable ground station, and a Landsat assessment system to demonstrate advantages of the TM over the MSS.

Landsat data had served primarily for monitoring earth resources and managing food, fiber, and water resources; mineral and petroleum explorations; and landcover and land-use mapping. The U.S. Depts. of Agriculture, Commerce, Interior, and State, and a number of states and foreign governments would participate in the Landsat-D mission. (NASA Release 78-188).

• FBIS reported that Oleg Gazenko, director of the USSR's Manned Spaceflight Institute of Medico-Biological Problems, said in an interview that cosmonauts Vladimir Kovalenok and Alexander Ivanchenkov were physiologically "as healthy as they were before the [*Soyuz 29*] spaceflight" [see Nov. 2]. The record 140-day flight had caused some functional changes: . . . "the cosmonauts lost a part of the muscle mass, which they have now restored. They returned to earth with erythrocytes that were born in space. The average life span of erythocytes is 120 days, and they are fully renewed during this period."

Gazenko said that the cosmic erythrocytes tended to be smaller, though they fully retained the main function of carrying oxygen to the tissues. "As time passes, the erythrocytes that were born in weightlessness will perish. But what will those that will replace them be like? Identical to those that were born before the space flight? Apparently yes, because our organism is a wonderful instrument of self-regulation; it excellently arranges all the processes taking place in it in such a way that they would meet the demands made by those conditions in which man is staying."

Gazenko had directed research that substantiated the possibility of manned spaceflight, and had received the 1978 USSR state prize for studies of ways to counteract unfavorable effects of weightlessness on the human organism, making lengthy manned spaceflights possible. (FBIS, Tass in English, Dec 12/78)

December 13: MSFC reported that LaRC scientists were using the Pioneer Venus mission for an atmospheric-drag experiment to measure for the first time the orbital decay of a spacecraft in orbit around another planet. Orbital decay would result from atmospheric friction as the low point of *Pioneer Venus 1*'s orbit came within 150km (93mi) of Venus's surface. From atmospheric drag, scientists could determine variations in the density of Venus's upper atmosphere and deduce composition and temperature. Changes in the frequencies of the spacecraft's radio signals has determined orbital decay with an accuracy of better than one part in a million. (*Marshall Star,* Dec 13/78, 2)

December 14: The Intl. Council of Scientific Unions and the World Meteorological Organization (WMO) announced that more than 140 nations would participate in the largest international scientific experiment ever held, beginning in January 1979. The Global Weather Experiment (GWE) would establish the practical limits of weather forecasting and design a global observation system to achieve them. During the 12mo experiment, a massive array of scientific personnel and observing equipment would monitor earth's oceans and atmosphere to compile the most complete record of global weather data ever assembled. The experiment, part of the Global Atmospheric Research Program (GARP), would be a joint effort of the 2 organizations and all 147 member nations of WMO; 5 other international organizations were also contributing. U.S. participants would include the Depts. of Commerce, Defense, Energy, Interior, Transportation, and State; NASA; the Natl. Center for Atmospheric Research; the Natl. Science Foundation; the Defense Nuclear Agency; the Air Force, Army, Navy, and Coast Guard; and many academic institutions. NOAA would coordinate U.S. participation.

The experiment would supplement the existing World Weather Watch (now generating more than 40 000 observations daily) with 10 satellites, about 50 research vessels, 110 aircraft (including 10 research planes), 300 high-altitude constant-level balloons, and 300 instrumented drifting buoys. One focus of attention would be the broad belt around the equator, an enormous heat sink absorbing much of the sun's energy and generating much of the world's weather, but so far comparatively unobserved. To obtain more data on this crucial region, GWE would operate observing systems during two special periods (Jan. 5 to March 5, and May 1 to June 30) to gather vital tropical-wind information. The combined systems would generate weather data for 85% of the earth's area not covered by the World Weather Watch.

At the end of the Global Weather Experiment Nov. 30, 1979, centers in a number of nations would process the data and forward the results to World Data Centers in Moscow and Washington, D.C., and to meteorological research laboratories in Bracknell, England, and Princeton, N.J. Scientists would use computers to refine the data and construct for the first time a complete 12-mo world-weather history. They would run computer experiments using this unique record to test their results, and to determine not only the practical limits of weather prediction but also the data needed to reach these limits and the kind of system needed to gather the data. The socioeconomic impact of improved weather forecasts could be dramatic; industries, for example, could benefit from knowing what the weather would be as far as 8 to 10 (or even 14) days in advance. (GWE Release 78-1)

December 15: DFRC and the Jet Propulsion Laboratory had joined in a study of the "Mars airplane" (a derivative of the Mini-Sniffer), the DFRC *X-Press* reported. The idea of a Mars airplane resulted from a January 1977 meeting between David Scott, former director of DFRC, and JPL director Dr. Bruce Murray, who had offered JPL's assistance in perfecting the hydrazine engine of the Mini-Sniffer remotely-piloted vehicle (RPV) designed to fly at 70 000 to 100 000ft. The plane was originally a low-cost alternative to the U-2 for atmospheric research; DFRC's Dale Reed and JPL personnel had recognized the possibility that the Mini-Sniffer could cruise halfway around Mars in the planet's low gravity and thin atmosphere, using its unique nonairbreathing hydrazine engine.

Developmental Sciences, Inc. had received a $60 000 contract for designs to send four Mars airplanes folded into capsules to conduct low-altitude surveillance flights down Martian canyons and volcanoes and over icecaps. The researchers believed technology was advancing so quickly in lightweight solar cells, beamed-microwave energy, lightweight electric motors and batteries, and lightweight radioisotope electric generators, that it would soon produce a generation of high-flying long-endurance long-range aircraft, the paper noted. (DFRC *X-Press,* Dec 15/78, 3)

• A jack-in-the-box parachute had proved to be an effective spin-recovery device, DFRC *X-Press* reported. Known as jack-in-a-box because of its deployment mechanism, the parachute had been installed in the nose of a spin-research vehicle, a 23-ft model of a modern fighter air-launched from a B-52 at 50 000ft and flown into a spin at 25 000ft. On command from a pilot on the ground, the nose parachute had deployed and recovered the vehicle from the spin. After parachute jettison, the SRV ground operator flew the SRV to a near-perfect landing on the Rogers dry lakebed. Results indicated that maintenance of a relatively clean airflow around the nose of the aircraft would permit use of a significantly smaller parachute, simplifying its installation. (DFRC *X-Press,* Dec 15/78, 3)

• INTELSAT announced reduced charges for many of its satellite services for the ninth consecutive yr. Meeting in Washington, D.C., INTELSAT's board of governors had reduced the monthly charge for full-time two-way telephone circuits on an INTELSAT satellite (about 80% of the INTELSAT system) from $1140 to $960 beginning in Jan. 1979. Charges for on-demand telephone circuits (known as SPADE circuits) would drop from 16 to 14 cents/min. Original rate for satellite telephone circuits established by INTELSAT in 1965 was $5334/mo; since then, advancing technology had increased demand and improved efficiency so that INTELSAT could progressively reduce its rates. Marcel Perras of Canada, chairman of the board of governors, said INTELSAT was considering a 5-yr plan to lower the rate for satellite telephone circuits to less than $700/mo by 1983. (INTELSAT 78-35-1)

December 18: JSC announced that one of the world's most unusual freight airplanes, the Super Guppy, would deliver a high-fidelity mockup of the Space Shuttle orbiter crew compartment from the contractor in Calif. to neighboring Ellington AFB in Tex. for transfer to NASA, marking first use of the Super Guppy for delivering Space Shuttle equipment. NASA would install the mockup (a detailed model of the cockpit and living quarters on the Space Shuttle orbiter) in a JSC training area for use in crew familiarization. The nonworking mockup was built by Rockwell Intl., prime contractor for the orbiter.

The Super Guppy took aboard the mockup at Los Alamitos Naval Air Station near Los Angeles and flew it to Ellington, refueling at Davis-Monthan AFB, Tuscon, and Dyess AFB, Abilene. A modified Boeing YC-97J, the Super Guppy had the world's largest internal dimensions in an aircraft; it could accommodate the fuselage of a 747, if sectioned. Its present configuration, test-flown by late NASA pilot Joe Walker in late 1965, was first used by NASA early in 1966. The Super Guppy, equipped with 4 Pratt & Whitney T34 turboprops, had flown more than 2 million mi carrying NASA cargo and had seen extensive use in the Apollo and Skylab missions. It had last landed at Ellington AFB in June 1976. (JSC Release 78-58)

- LaRC announced selection of Telos Computing, Inc., Santa Monica, Calif., for negotiations leading to award of a contract to develop software for the Natl. Transonic Facility (NTF) wind tunnel at LaRC that would allow networking of four medium-sized computers to be the nucleus of NTF's instrumentation complex. The contract would cover collection and display of all research measurements, as well as automatic control of many wind-tunnel test parameters. Telos estimated the value of the cost-plus-fixed-fee contract at $1.2 million. Work would be performed at LaRC over a 3-yr period. (LaRC Release 78-86)

- Engineers from Rocketdyne and NASA were trying to determine why a liquid-oxygen heat exchanger on a Space Shuttle main engine had failed during a test at the Natl. Space Technology Laboratories, Bay St. Louis, Miss., *Av Wk* reported. This was the first failure of a heat exchanger (an engine component converting liquid oxygen to gaseous oxygen to pressurize the Space Shuttle's external propellant tank). Earlier in 1978 the Natl. Research Council's Assembly of Engineering had noted that, although the exchanger had not encountered any difficulties, the system "poses a potential threat to the total shuttle system" from a single-point-failure standpoint. The committee said that, whereas redesign to add systems redundancy was impractical at that point, the system warranted continued examination for potential design improvements. NASA agreed with that assessment.

Inflight heat-exchanger failure might start a serious fire or explosion that could result in loss of the spacecraft. Failure during the test had resulted in some engine damage, but none to the test-stand. Testing with different engines was continuing. (*AvWk*, Dec 18/78, 8)

December 19: NASA announced it had decided not to reboost or deorbit the Skylab space station in orbit around the earth because success was unlikely. Uncertainties that had developed in Skylab systems, Skylab lifetime, Space Shuttle schedule, and teleoperator-retrieval system (TRS) delivery had made further planning for a rescue mission impractical. Factors leading to the decision included deterioration of systems aboard Skylab (recurring control-moment gyro anomalies, marginal power conditions, and minimal attitude-control impulse gas); high sunspot activity that was accelerating decay of Skylab's orbit; and reviews showing that a Skylab-rescue mission would be difficult to arrange before April 1980.

Skylab, currently orbiting at about 426km (265mi) altitude, was being maintained in a low-drag attitude and would be continually monitored while NASA assessed the possibilities of damage on reentry, now predicted to occur between mid-1979 and mid-1980. Most of the vehicle would burn up during reentry, but some fragments probably would reach earth's surface. Skylab's flight path was 75% over the ocean; the probability of injury or damage from it was less than that from meteorites, according to estimates. (NASA Release 78-191)

At a press conference following the announcement, Bill Haines of the *Chicago Sun Times* asked: "Way back in the olden days I remember that there was a lot of talk about space rescue, and one of the philosophies seemed to be that if we got into trouble or the Russians got into trouble, the other side would come to the rescue. Have you made any approaches at all to the Russians to see if they could do anything about this? If not, why not? And do you intend to?"

John Yardley, associate administrator for Space Transportation Systems, answered: "This is a little different from the space rescues that we had talked about. As you know, part of the ASTP program was to design docking equipment that could be prototypes of that. And we still have a long-range plan with the Russians, albeit rather fuzzy, to some day have this capability.

"With respect to Skylab: After our ASTP, working with the Russians, of course, we had a fairly good working knowledge of the capabilities, at least that we knew of that were done in that program. And at least a year ago we got our own heads together and talked about whether or not they could do anything that we couldn't do. We concluded that there are many problems that we won't go into that it probably is not feasible, so we did not ask them. However, they are being informed of what we are doing here and if they make any proposals that we haven't thought of that could be useful, we will certainly entertain them." (NASA press conf Dec 19/78)

December 20: NASA announced it had selected Global Associates, Oakland, Calif., to provide operating services at the Michoud Assembly Facility, New Orleans, La. Services provided under the contract would support NASA program activity and missions of the Depts. of Agriculture, Defense, and Commerce tenanted at the Michoud facility, as well as refuse collection and telecommunications services at Slidell Computer Complex, Slidell, La. MSFC managed both facilities.

Under the 1-yr cost-plus-award-fee contract, with four 1-yr options for extension, Global Associates would provide management, personnel, and materials for plant engineering, maintenance, telecommunications, and property and supply, as well as mail, transportation, medical, and food services. Proposed cost over the 5-yr period was approximately $26.5 million. NASA's primary program at Michoud was manufacture and assembly of external tanks for the Space Shuttle, including those for service as test articles, on test flights, and on operational missions. (NASA Release 78-192)

• Boeing had awarded a $2 million contract to United Technologies'

Chemical Systems Division to build seven servoactuator systems (and two actuators for preflight testing) that would raise the inertial upper stage (IUS) to deployment position in the Space Shuttle payload bay, *Defense/Space Business Daily* reported. Each system would consist of two servoactuators (one for backup), composed of an actuator and electronic controller positioned on either side of a tilt table in the payload bay to which the IUS would be secured. On command, the system would physically tip the tilt table to a predetermined position (e.g., 58°) for IUS deployment. The actuator motor would drive an extendable shaft 41 to 57in long to rotate the tilt table. Pyrotechnic devices on the tilt table would release the IUS, and springs would launch it into space. First flight of the IUS on board the Space Shuttle was scheduled for 1980. (*D/SBD*, Dec 20/78, 239)

• ESA reported on the mechanical-qualification tests of the CAT/Apple/Meteosat-2 composite (a payload scheduled under the APEX program for the third test flight of Ariane in May 1980) taking place Dec. 12-22 in the CNES facilities at Toulouse. The tests were to check behavior of the payload structure and the subsystems under conditions arising during launch; results would be compared with mathematical models. Undergoing tests would be a dynamic mockup subjected by means of an electroacoustic exciter to longitudinal and transverse vibrations like those occurring in flight.

The payload being tested had three satellites mounted one above the other: the Ariane technological capsule (CAT), carried on all Ariane test flights; Apple, a technological satellite of the Indian Space Research Organization, which would carry a communications experiment; and ESA's meteorological satellite Meteosat-2 which would continue the Meteosat mission into 1980. Apple and Meteosat-2 would be the first satellites placed in geostationary orbit by a European launcher. These tests were the first conducted in Europe on such a heavy payload, 6.5m high and weighing 1574kg. (ESA Release Dec 20/78)

• FBIS reported the USSR had launched a communications satellite, *Horizont*, Dec. 19 under a program "for further improvement and development of communications systems and television transmissions." Satellite apogee was 48 365km; perigee, 2581km; period, 23hr 40min; inclination, 11.3°. *Horizont* carried improved multiplex-relay apparatus for transmitting telephone and telegraphic radio communications and TV programs, including those from the 22nd summer Olympic games in Moscow. It had a three-axis system for precision orientation, a power-supply system with independent guidance and alignment of the solar batteries with the sun, a thermo-regulation system, a telemetry system for transmitting data on onboard systems, a radio system for precision measurement of the orbit's parameters and control of the satellite, and an orbit-correcting system. All satellite systems were functioning normally. (FBIS, Tass in English, Dec 20/78)

December 21: NASA announced it had awarded Lockheed-California Co., Burbank, Calif., a contract to develop and evaluate in flight an augmented-stability active-control concept that would use a smaller horizontal tail on transport planes. Lockheed had estimated the total value of the cost-sharing contract, monitored by LaRC, to be $17.6 million; the government would pay approximately $15.8 million, Lockheed $1.7 million. Work would be performed over 44mo at Lockheed's Burbank facility.

An earlier NASA contract with Lockheed (Phase I of the energy-efficient transport program) was for conceiving aerodynamic and active controls that could increase energy efficiency of new and derivative civil-transport aircraft early in the 1980s. Results from that phase and from independently funded activities indicated that Lockheed was ready for flight demonstration of active controls to augment stability of a commercial air transport with a reduced horizontal-tail size. (NASA Release 78-193)

• *JPL Universe* reported NASA Administrator Robert Frosch had characterized the Pioneer Venus mission as "a superb success" in a message congratulating personnel at NASA centers and in the industrial and scientific communities. Frosch had added: "The precise execution of the multiprobe encounters December 9 and the continuing excellent performance of the orbiter could have resulted only from a well planned and competently executed program . . ."

The orbiter would continue its scientific measurements for 243da after Venus encounter; however, three small probes and a larger sounder had plunged into the Venusian atmosphere with destruction expected immediately upon impact. To the surprise of Pioneer Venus scientists, one probe had survived for 67min after impact, sending back information that would help confirm probe findings (particularly a wind experiment) before the 900° Venus temperatures silenced it. (*JPL Universe,* Dec 21/78, 1)

• INTELSAT announced it had awarded a $101 539 contract to General Electric Co., Valley Forge, Pa., for a reconfigurable-beam satellite communications antenna. Under the 15-mo contract GE would develop an antenna capable of beaming dual-polarized transmissions that could be reshaped on command while the satellite was in orbit; GE would also fabricate portions of such an antenna for demonstration. Reconfigurable-antenna developments might be incorporated into INTELSAT satellites launched during the 1980s. (INTELSAT Release 78-36-M)

• FBIS reported the USSR interplanetary spacecraft *Venera 12* had reached Venus Dec. 21 after a 98-day flight covering over 240 million km. The descent module had entered Venus's atmosphere at 11.2 km/sec and soft-landed, transmitting scientific information from the planet's surface for 110min. During descent, the module made precise chemical analyses of the composition of the atmosphere and clouds, a spectral analysis of solar radiation scattered in the atmosphere, and a study of electric charges in the atmosphere from 62m above the planet's surface. Scientific measurements continued after the module landed. The *Venera 12* station was continuing its flight in space, carrying a Soviet-French experiment for studying solar and galactic gamma flares.

Venera 11, launched 5da before *Venera 12*, would reach Venus Dec. 25. (FBIS, Moscow Dom Svc in Russian, Dec 21/78)

December 22: NASA reported *Voyager 1* was 598 975 000km (372 186 000mi) from earth; *Voyager 2*, 544 445 000km (338 303 000mi). One-way communication time for *Voyager 1* was 33.3min, for *Voyager 2* 30.2min. Spacecraft distance traveled since launch was 931 872 000km (579 038 000mi) for *Voyager 1*, 936 428 000km (581 869 000mi) for *Voyager 2*. *Voyager 1*'s velocity relative to earth was 70 830km/hr (44 013mi/hr), *Voyager 2*'s 68 537km/hr (42 586mi/hr). *Voyager 1*'s velocity relative to the sun was 49 262km/hr (30 609mi/hr); *Voyager 2*'s 44 334km/hr (27 548mi/hr). (NASA Release 78-194)

December 25: FBIS reported the Soviet Union's *Venera 11* had reached Venus Dec. 25, and its descent module had made a soft landing some 800km from where *Venera 12*'s descent module had landed. Main task of *Venera 11* and *12* was to examine the characteristics of various areas of Venus. After the descent module separated from *Venera 11* Dec. 23, the spacecraft had proceeded on a flyby trajectory some 35 000km from the planet's surface. The *Venera 11* module, during descent and while on the planet surface, had performed research similar to that done by the *Venera 12* module (scientific and technical experiments and exploration of Venus's surface at 446°C and pressure of 88 atmospheres) for 95min. Descent-module data received by *Venera 11* were relayed to earth. After leaving the zone of radio visibility and completing receipt of data, *Venera 11* would continue in a heliocentric orbit. Scientific measurement and research begun on the earth-Venus interplanetary route would continue during the flight of *Venera 11* and *12*. Data obtained from the two spacecraft would be processed and studied in the institutes of the USSR Academy of Sciences. (FBIS, Tass Intl Svc in Russian, Dec. 25/78)

December 26: NASA announced the five *Pioneer Venus* atmospheric-entry craft reaching Venus and the companion craft in orbit around

Venus were the highlights of its space-exploration activities in 1978, the agency's 20th anniversary yr. NASA had conducted 20 launches in 1978, mainly for other organizations and agencies; it was the third yr in NASA history with a 100% launch-success record. It had also seen completion of major milestones in preparing for crew-carrying Space Shuttle missions scheduled to begin late in 1979. (NASA Release 78-190)

December 27: NASA reported a delegation from the People's Republic of China had been discussing with U.S. officials since Nov. 28 the possibility of U.S.-PRC cooperation in peaceful uses of space technology. The delegation also had visited several NASA centers and U.S. aerospace industrial establishments. NASA Administrator Dr. Robert Frosch headed the U.S. delegation; Dr. Jen Hsin-min, director of the PRC Academy of Space Technology, the Chinese delegation.

Following up a visit to PRC in July by presidential science adviser Dr. Frank Press, the discussions had produced an informal agreement on developing a civilian communications system for PRC. China would purchase a U.S. satellite communications system, including ground receiving and distribution equipment; the U.S. would put the space portion of the system into geostationary orbit, and China would take over its operation. A similar informal agreement covered PRC purchase of a ground station capable of receiving earth-resources information from U.S. Landsat remote-sensing satellites including the Landsat D currently under development.

The PRC delegation would remain in the U.S. until mid-Jan. for further technical discussions. Both sides expressed satisfaction with the discussions, and agreed on further talks to work out details of the informal agreements and to consider other forms of civilian space cooperation. (NASA Release 78-197)

• NASA announced appointment of Robert Aller as director of tracking and data relay satellite system (TDRSS) program office at the NASA Hq Office of Space Tracking and Data Systems. Effective Jan. 2, 1979, Aller would be responsible for planning, direction, and evaluation of the TDRSS program. Working with NASA program and project managers, he would determine requirements and develop and implement plans for TDRSS support of NASA earth-orbital spaceflight missions. Western Union Space Communications, Inc., would provide NASA with TDRSS tracking, command, and data-relay services under a 10yr telecommunications agreement. Aller had been deputy director of the expendable launch vehicles program in the NASA Office of Space Transportation Systems, and had served in the Apollo-Soyuz Test Project and the Skylab, Apollo, and Gemini programs. He had received the NASA medal for outstanding leadership and two NASA exceptional service medals. (NASA Release 78-195)

December 28: NASA announced that 11 of its 16 launches scheduled for 1979 would be reimbursables (launched by NASA for other agencies or corporations). Paying customers included DOD, NOAA, the United Kingdom, Western Union Corp., ComSatCorp, and RCA. As in 1978, most of the 1979 launches would emphasize use of space for the direct benefit of people: communications, environmental, and meteorological information.

Activities in 1979 would include the first launch and orbital flight of the Space Shuttle; Jupiter and Saturn encounters by two Voyager spacecraft; and the *Pioneer 11* flyby of the rings of Saturn. NASA would launch the Space Shuttle from KSC into its first orbital flight Sept. 28 with astronauts John Young and Robert Crippen as crew. They would land 53hr later at DFRC.

The 1979 schedule would begin with two launches Jan. 25: DOD's spacecraft charging at high altitudes (SCATHA) from Cape Canaveral, and a stratospheric aerosol and gas experiment (SAGE-A) from Wallops Flight Center. In April NASA would launch NOAA-A for NOAA and Navy-20 for DOD from Vandenberg AFB. It would launch in May FltSatCom-B from Cape Canaveral, and UK-6 for the United Kingdom from WFC; Westar-C in July and Intelsat V-A in Aug., both from Cape Canaveral for ComSatCorp; in Sept., its third high-energy astronomy observatory HEAO-C from Cape Canaveral, and its Magsat-A from Vandenberg; in Oct., the solar maximum mission (SMM-A) from Vandenberg. NASA would launch Intelsat V-B from Cape Canaveral for ComSatCorp in Nov.; in Dec., NOAA-B for NOAA from Vandenberg, and RCA-C for RCA from Cape Canaveral. (NASA Release 78-196)

December 29: Five European and U.S. scientists named in July as experimenters on NASA/ESA's first Spacelab mission would begin training in the U.S. in Jan. 1979, NASA announced. The training tour, taking them to seven U.S. and two Canadian cities, would prepare them to operate equipment for scientific investigations that would fly aboard the Space Shuttle in 1981. Selection and training of these scientists to fly in space had been a departure from earlier NASA practice, as they were not career astronauts and had been chosen for the mission by the scientists who devised the experiments. This was also the first time that Western Europeans would fly in space, and the first time that NASA would orbit people who were not U.S. nationals. Of the five persons training for the mission, two—a European and an American—would actually fly on Spacelab 1; the others would operate ground-based experiment equipment and support the two in orbit.

The chosen five had just returned from Europe where they had been since Oct. learning to operate the experiments chosen for Spacelab by European scientists. The science payload was about equally divided between NASA and ESA experiments in terms of weight, power, and volume requirements. Areas of investigation would include stratosphere and upper-atmosphere physics, materials processing, space-plasma physics, biology, medicine, astronomy, solar physics, earth observations, and technology areas such as thermodynamics and lubrications. MSFC would manage training in the U.S. as part of its overall responsibility for the first three Spacelab missions; ESA had managed the training in Europe. (NASA Release 78-198)

During December: NASA announced it had silenced the scientific instruments aboard the orbiting solar observatory *Oso 8* by radio command sent from Goddard Space Flight Center to NASA's tracking station in Orroral, Australia, on the satellite's orbit no. 18 072. A few moments later the Australia station confirmed that the satellite's radio transmission had stopped. Last of the "sunshine" satellites that had begun investigating the sun with the launch of *Oso 1* in 1962, *Oso 8* (launched from Cape Canaveral June 21, 1975) had surpassed its 1-yr design lifetime by 2yr. It had far exceeded all prelaunch mission objectives, and had obtained the most accurate observations yet of the solar chromosphere and transition regions, giving scientists a better understanding of these.

Dr. Roger Thomas, *Oso 8* project scientist at GSFC, said that the lessons learned from 16yr of OSO satellites had been instrumental in developing new experiments for future studies of the sun such as next yr's solar maximum mission. Four of the eight experiments aboard the 1064kg *Oso 8* had been designed to measure the sun during the quiet (solar minimum) period. Solar activity had increased substantially, and the solar-science community was preparing to study a period of maximum solar activity, in part through launch of the solar maximum mission in Oct. 1979. The nonsolar high-energy astrophysics experiments on *Oso 8* also had made significant observations of the celestial sphere. These experiments had been superseded by *Sas 3, Heao 1,* and *Heao 2.* The *Oso 8* control center at GSFC would be reconfigured to support the SMM mission. (*NASA Actv*, Dec 78, 12)

• NASA reported procurements during FY78 totalled $3660 million, 3.6% more than during FY77. Approximately 81% of the net dollar value was placed directly with business firms, 5% with educational and other nonprofit institutions or organizations, 8% with the Calif. Inst. of Technology (for operations conducted by or through JPL), and 5% with or through other government agencies; 90% of the dollar value of the latter resulted in contracts with industry. About 53% of funds placed by NASA under JPL contracts resulted in subcontracts or purchases with business firms. In short, about 90% of NASA procurement was contracted to private industry. Of total direct awards to business firms, 72% were competitive procurements (either through formal advertising or negotiation) and 28% were noncompetitive.

Small business firms had received $282 million (10%) of NASA's direct awards, indicating that most agency awards to business firms were large continuing R&D contracts for major systems or items of hardware generally beyond the capability of small business firms on a prime-contract basis. However, of the $633 million in new contracts worth $10 000 or more awarded to business firms during 1978, small business received $137 million (22%); in addition to direct awards, small business received substantial subcontract awards from 75 of the NASA prime contractors participating in its small business subcontracting program. Total direct awards plus known subcontract awards to small business aggregated $513 million, 17% of NASA's total (direct plus indirect) awards to business firms. Included in awards to small business firms were $75.852 million awarded to minority business enterprises: $14.064 million in direct awards, $32.151 million under Section 8(a) of the Small Business Act, and known subcontract awards of $29.637 million.

In 1978, 49 states and Washington, D.C., participated in NASA prime-contract awards of $10 000 and over. These larger awards went to 2068 business firms in 46 states and D.C., and to 341 universities and nonprofit organizations in 47 states and D.C. Labor-surplus areas in 40 states had received 53% of the awards. (NASA ann proc rpt FY78)

• NASA announced the inception of Orbit 81, an RCA educational program to involve inner-city high school students in designing and developing a scientific experiment for launch on the Space Shuttle in 1981. The program had resulted from discussions among representatives of the Camden, N.J., public school system, NASA, and RCA. Other high schools were conducting similar programs, but NASA officials called the Camden project unique because it was the only "total involvement program covering all the disciplines, arts, and skills" taught in the curriculum. Students taking courses other than science and mathematics could be active participants in the project. Selection of an Orbit 81 experiment would take place when students had gained enough scientific insight to participate in the decision. The program would include enrichment courses in space science, summer sessions, and tours of NASA and RCA facilities. RCA had donated surplus laboratory equipment to the schools; NASA would provide literature, special programs, lectures, and demonstrations. Students of Camden and Woodrow Wilson High Schools had attended a space colloquium presented by NASA and RCA.

RCA had made an initial payment to NASA as part of the $10 000 reservation fee to ensure that the Orbit 81 payload, which could weigh up to 90kg and occupy no more than 0.14m³, would have a place aboard the Space Shuttle when launch time arrived. (*NASA Actv*, Dec 78, 11)

• NASA announced that its administrator, Dr. Robert Frosch, had accepted for NASA a document commemorating the anniversary of Charles A. Lindbergh's solo trans-Atlantic flight from New York to Paris in the "Spirit of St. Louis" on May 20, 1927. Copies of the framed Flight Philatelic Document, containing a series of American and French Lindbergh stamps and cancellations tracing the history of the aviator's achievement, had been presented to 25 persons and organizations in the U.S. and France. Friends of the Nassau County Museum, Long Island, N.Y., had produced the tribute and selected the recipients, including President Jimmy Carter, French President Valery Giscard d'Estaing, Anne Morrow Lindbergh, and museums and agencies linked to aviation. Each of the documents bore seven American and French Lindbergh stamps with key cancellations, and each document was autographed by President Carter, President d'Estaing, Paris Mayor Jacques Chirac, U.S. Postmaster General Benjamin Bailer, French Postal Secretary Norbert Segard, and Mrs. Lindbergh. (*NASA Actv,* Dec 78, 12)

• DOT's *National Transportation Report* noted that the department had been using administrative action to improve transportation. For example, while working for the Aviation Regulatory Reform Act, the department had led a federal interagency task force to develop guidelines for U.S. negotiators in reaching air service agreements with other countries. This had resulted in bilateral air service agreements with 13 nations that would increase competitive opportunities for airlines and lower passenger fares. Another major DOT aviation action was Secretary Brock Adams's decision to allow existing Concorde supersonic aircraft to fly over U.S. territory, at the same time requiring future SSTs to meet the same noise standards as subsonic aircraft. (*Natl Transp Rept* Jan 77-Dec 78, 3)

• The AFSC *Newsreview* reported that the USAF's Geophysics Laboratory (AFGL) had successfully completed a rocket test flight of equipment to cope with lightning-like discharges that could affect earth-orbiting satellites. These high-voltage discharges could interrupt transmissions of information to earth stations from geosynchronous satellites. A sudden rush of electrical current could confuse control circuits, causing a satellite to behave erratically or fail to function. Scientists were still analyzing information from the 160mi-high rocket flight, but preliminary results showed that the electrical buildup—which could reach thousands of volts—could be controlled by emitting negative or positive charges from the spacecraft.

An Astrobee F sounding rocket carried equipment that released either negative charges (electrons) or positive charges (ions) as the electrical charge built up on the rocket body. This either kept the charging to a low voltage or provided an electrically conductive path around the rocket body so the charge could leak off slowly. The rocket payload contained instruments nearly identical to those scheduled for orbit in Jan. on a SCATHA (spacecraft charging at high altitudes) satellite as part of a test program managed by SAMSO, Los Angeles. (AFSC *Newsreview*, Dec 78, 9)

• The USAF's Aeronautical Systems Division (ASD) had awarded Hughes Aircraft Co., Canoga Park, Calif., a $90.25 million contract for engineering development and test of the AGM-65D imaging infrared (IIR) Maverick missile, the AFSC *Newsreview* reported. A 35-mo engineering-development program designed to fine-tune the IIR version of Maverick would ensure that the weapon incorporated the state-of-the-art and was producible, maintainable, and reliable. ASD officials said testing, a key element in advanced development, would be equally crucial as the IIR program moved into the engineering development phase. The 16-mo test series would begin with ground-based studies and captive flight tests using helicopers and fixed-wing aircraft. Live firings of both telemetry- and warhead-equipped missiles from A-10, A7D, and F-4D and -E aircraft would complete the rigorous flight-test schedule.

The IIR guidance system (fourth and most sophisticated mate to the already-proven Maverick) would also be used with two guided bombs, the Air Force's GBU-15 and the Navy's Walleye. The IIR Maverick's digital-centroid tracker could pick up heat images generated by a target, calculate its approximate center, and guide the missile to target impact. USAF plans called for procurement of more than 31 000 IIR Maverick missiles; production would begin in mid-1981. (AFSC *Newsreview*, Dec 78, 7)

• The AFSC *Newsreview* reported that F-15 fighter pilots had come closer to better anti-G (gravity forces) protection, with airworthiness certification of a high-flow ready-pressure anti-G valve developed by the USAF School of Aerospace Medicine. Tests of the valves installed in two F-15s at Edwards AFB, Calif., had checked their reliability under flight conditions. Alar Products, Inc., Macedonia, O., was manufacturing 15 of the high-flow anti-G valves, and the ready-pressure attachment would be manufactured and added to the valves at USAFSAM. The valves would be installed in F-15s at Holloman AFB, N.M., and tested for 2 to 3mo. If test results were favorable, the valve would be further developed by the Aerospace Medical Division's life support system program office at Wright-Patterson AFB. (AFSC *Newsreview*, Dec 78, 12)

• The AFSC *Newsreview* reported that, pending arrival of the age of the

Space Shuttle, the USAF Space and Missile Systems Organization (SAMSO) was modifying a Titan III 34D booster to ensure prime and backup launch capabilities during Space Shuttle transition. The 34D was a modified Titan IIID that the USAF had put into operation in 1971; SAMSO had recommended it in 1975 to the Defense Science Board studying how DOD could manage an orderly transition from expendable to reusable launch vehicles. Instead of a variety of boosters, the 34D would use a common Titan III core for east- and west-coast launches, with the option of dedicating a 34D launch vehicle to either coast as late as 6mo before launch.

The 34D also was compatible with the new inertial upper-stage (IUS) vehicle, another SAMSO project in full-scale development by Boeing Aerospace Co. During the Space Shuttle backup era beginning in 1980, the 34D/IUS combination would be contingency booster for top-priority defense payloads. Plans called for development of seven 34D boosters to be used first for launching two defense satellite communications system (DSCS) spacecraft in July 1980 from Cape Canaveral. Four of the seven vehicles would serve east coast launches using the IUS, and three would serve polar launches from Vandenberg AFB.

Differences between the 34D and Titan IIID included solid-fuel rocket motors lengthened 5ft each to align with new common-core attach points and payload fairing modified to extend from 40 to 55ft. The new fairing would make the 34D payload environment nearly identical to that of the Shuttle orbiter cargo bay, a step toward upper-stage commonality. The 34D program was in final design stage under contracts with Martin Marietta Aerospace, McDonnell Douglas Corp., and United Technologies Corp. (AFSC *Newsreview*, Dec 78, 9)

- The Natl. Science Foundation reported that the 1979 federal budget presented in Jan. 1978 reflected greater austerity in R&D programs than during the previous 3yr, in which rate of R&D growth had exceeded the rate of inflation. R&D programs funded were those that met direct federal needs, general social and economic needs, or the need to accelerate private R&D efforts because of overriding national interest. The FY79 budget contained a rise for defense and space R&D programs of about 8%, each, with only a 3% rise for all other R&D. This would reverse a 13-yr trend in which federal R&D support had shifted steadily toward civilian programs. The 1979 change had resulted not only from increases in defense/space support at rates at least equal to anticipated inflation, but also from expected real-dollar declines in support of civilian R&D areas of health, energy, and environment.

Federal R&D obligations rose 6% to an estimated $28.0 billion in the FY79 budget, an all-time high in current dollars, but inflation would mean a drop from the 1978 total figure in constant dollars. Later congressional actions had raised the level approximately 9% above that of

1978, and action on supplemental requests might add another 2 or 3 percentage points. Defense and space programs had accounted for more than 80% of the current dollar rise in R&D support in 1979. Funding for each of these was expected to increase by about 8% in 1979, whereas all civilian functions taken together would increase by only 3%. The defense/space share of the federal total in 1979 would be an estimated 62%, more than the 1978 share but still considerably lower than the 77% share in 1969. *(Analysis of Fed R&D Funding by Function,* Dec 78, vii)

• ComSatCorp announced it had derived 57% of its 1978 operating revenues from services provided through Intelsat satellites; 37% from maritime and U.S. domestic services provided by Comsat General through the Marisat and Comstar satellite systems; and 6% from its investment share in INTELSAT and operational and technical services.

ComSatCorp had been considering expansion into the growing environmental information services field in which satellite technology could play a significant, possibly essential, role. Comsat General, in cooperation with the U.S. Geological Survey and Telesat Canada, had already demonstrated the ability of satellites and small unattended earth stations to collect water-resources data from remote areas of the U.S. and Canada. ComSatCorp would enter the field by acquiring at a price not to exceed $20 million Environmental Research & Technology, Inc. (ERT), currently engaged in a broad spectrum of environmental services. Adding ERT's expertise in environmental data collection, monitoring, analysis, and prediction to ComSatCorp's expertise in satellite communications and related technologies should make the organization better able to extend satellite-communications technology to the growing environmental-information services market.

ComSatCorp had also been working to apply advanced technology in developing and manufacturing specialized products, including earth-station components, signal-processing systems equipment, and fiber-optics communications equipment. (COMSAT annl rept 78, 1-7)

• The Aerospace Industries Assn. announced that industry sales, earnings, backlog, export performance, and contribution to the U.S. balance of trade had improved substantially in 1978. Sales, at $37.3 billion, were up almost $5 billion over 1977; the major gain was in commercial sales. The high inflation rate had distorted sales data, but 1978 sales had increased about 15% above 1977, well in excess of the inflation rate. Industry profits as a percentage of sales rose half a percentage point to 4.7%.

Aerospace exports had reached an all-time high of $9.3 billion, $1.8 billion more than in 1977. At a time when export sales were important to the U.S. economy, the aerospace industry recorded an international trade surplus of nearly $8.4 billion, leading all U.S. manufacturing industries in contributing to the nation's balance of trade. Industry backlog at year-end, estimated at $51.4 billion, was a sharp increase over $44.3 billion in 1977, resulting mostly from a surge of new orders for commercial transports. (*Aerospace*, Winter 78, 2)

Appendix A
SATELLITES, SPACE PROBES, AND MANNED SPACEFLIGHTS, 1978

World space activity in 1978, 125 launches, matched that of 1977; however, the U.S. increased launches from 24 to 32, including *Pioneer Venus 1* (an orbiter) and *2* (four probes and a transporter bus). The USSR decreased launches from 98 to 89 including a Vertikal geophysics probe, with 121 payloads, 96 in the Cosmos series. These included such specialized spacecraft as seven navigation satellites (*Cosmos 991, 994, 996, 1000, 1011, 1027,* and *1046*) and four octuple-payload launches to improve tactical communications. The Soviets also launched six spacecraft in the Molniya comsat series; one Intercosmos spacecraft carrying experiments from other countries; the magnetospheric investigator *Prognoz 7*; two Venus exploration spacecraft, *Venera 11* and *12;* and five manned Soyuz spacecraft of which *Soyuz 28* carried the first Czech cosmonaut, *Soyuz 30* the first Polish cosmonaut, and *Soyuz 31* the first GDR cosmonaut.

In 1978, other countries launched or sponsored six satellites. On Jan. 26 China launched *China 8*, recovered in Feb. In Feb., Japan launched *Exos 1* as part of the International Magnetospheric Study and *ISS 2* (*Ume*), an ionosphere sounder; in Apr. *BSE*, nicknamed *Yuri*, was launched by NASA to experiment with direct-broadcast TV transmission-and-reception technology; and in Sept. *Exos 2*, another magnetosphere-research satellite. NASA launched *Anik 4* in Dec. for Canada's domestic comsat system.

Sources of these data include the United Nations Public Registry of Space Flights; the *Satellite Situation Report* compiled by Goddard Space Flight Center's Operations Control Center; and press releases of NASA, Department of Defense, National Oceanic and Atmospheric Administration, and other government agencies, as well as the Communications Satellite Corporation. Soviet data come from statements in the Soviet press, translations from the Foreign Broadcast Information Service (FBIS), international news service reports, and announcements and briefings by Soviet officials. Data on satellites of other nations also come from the announcements of their respective governments and international news services.

SATELLITES, SPACE PROBES, AND MANNED SPACEFLIGHTS, 1978

Launch Date	Spacecraft, Country, Int'l Designation, Vehicle, Launch Site	Payload Data	Apogee (km)	Perigee (km)	Period (min)	Inclination (degrees)	Remarks
Jan. 6	Cosmos 974 USSR 1978-1A A-2 Plesetsk	Total weight: 6300kg? Objective: "Continuation of outer space investigations." Description: Unavailable.	356	188	89.6	62.8	Reentered Jan. 19. Redundant maneuvering engine separated on Jan. 18; designated 1978-01D.
Jan. 7	Intelsat IVA F-3 U.S. 1978-2A Atlas-Centaur ETR	Total weight: 1515kg. Objective: To provide 6250 two-way voice circuits plus two television channels simultaneously or a combination of telephone, TV, and other forms of communication traffic. Description: Cylinder 6.99m high, 2.38m in diameter; two main elements: rotating cylinder covered with solar cells contains power subsystem and positioning and orientation subsystems; despun earth-oriented platform atop rotating cylinder contains communications repeaters (transponders), new antenna reflectors, and telemetry and command subsystems.	35 806	35 768	1436.1	0.3	Launched by NASA for ComSatCorp.; fifth in a series of improved Intelsat IVA spacecraft with almost two-thirds greater communications capacity than Intelsat IV model.
Jan. 10	Soyuz 27 USSR 1978-3A A-2 Baykonur-Tyuratam	Total weight: 6570kg? Objective: Carry a crew to Salyut 6 orbiting laboratory. Description: Unavailable.	237 350	190 330	88.7 91.3	51.7 51.6	Carried Lt. Col. Vladimir Dzanibekhov and civilian engineer Oleg Makarov; crew returned Jan. 16 in Soyuz 26 vehicle used by occupants Yuri Romanenko and Georgy Grechko.
Jan. 10	Cosmos 975 USSR 1978-4A A1 Plesetsk	Total weight: 3800kg? Objective: "Continuation of outer space investigations." Description: Unavailable.	680	637	97.6	81.2	Probable Soviet electronic ferret.

Jan. 10	Cosmos 976 USSR 1978-5A C-1 Plesetsk	Total weight: 50kg? Objective: "Continuation of outer space investigations." Description: Unavailable.	1465	1457	115.1	74.03	Multiple launch (8 satellites) probably for military communications.
	Cosmos 977 USSR 1978-5B C-1 Plesetsk	Total weight: 50kg? Objective: "Continuation of outer space investigations." Description: Unavailable.	1465	1403	114.5	74.03	
	Cosmos 978 USSR 1978-5C C-1 Plesetsk	Total weight: 50kg? Objective: "Continuation of outer space investigations." Description: Unavailable.	1465	1421	114.7	74.03	
	Cosmos 979 USSR 1978-5D C-1 Plesetsk	Total weight: 50kg? Objective: "Continuation of outer space investigations." Description: Unavailable.	1465	1440	115.0	74.03	
	Cosmos 980 USSR 1978-5E C-1 Plesetsk	Total weight: 50kg? Objective: "Continuation of outer space investigations." Description: Unavailable.	1478	1465	115.4	74.03	
	Cosmos 981 USSR 1978-5F C-1 Plesetsk	Total weight: 50kg? Objective: "Continuation of outer space investigations." Description: Unavailable.	1498	1465	115.6	74.03	

Launch Date	Spacecraft, Country, Int'l Designation, Vehicle, Launch Site	Payload Data	Apogee (km)	Perigee (km)	Period (min)	Inclination (degrees)	Remarks
Jan. 10	Cosmos 982 USSR 1978-5G C-1 Plesetsk	Total weight: 50kg? Objective: "Continuation of outer space investigations." Description: Unavailable.	1518	1465	115.8	74.03	
	Cosmos 983 USSR 1978-5H C-1 Plesetsk	Total weight: 50kg? Objective: "Continuation of outer space investigations." Description: Unavailable.	1540	1465	116.1	74.03	
Jan. 13	Cosmos 984 USSR 1978-6A A-2 Plesetsk	Total weight: 5700kg? Objective: "Continuation of outer space investigations." Description: Unavailable.	313	215	89.5	62.8	Probable Soviet military photoreconnaissance satellite; recovered Jan. 26.
Jan. 17	Cosmos 985 USSR 1978-7A C-1 Plesetsk	Total weight: 700kg? Objective: "Continuation of outer space investigations." Description: Unavailable.	1032	960	105.0	82.9	Probable Soviet navsat.
Jan. 20	Progress 1 USSR 1978-8A A-2 Baykonur-Tyuratam	Total weight: 7020kg? Objective: Ferry fuel and materials to Salyut 6 and remove waste materials; test backup rendezvous system; and perform a small orbital adjustment on the triple spacecraft complex. Description: Unavailable.	348	329	91.3	51.6	Design, though based on the Soyuz craft, did not carry a heat shield; burned up on reentry over Pacific, Feb. 8.

Date	Satellite	Total weight / Objective / Description	Orbital params				Remarks
Jan. 24	Molniya 3J USSR 1978-9A A-2e Plesetsk	Total weight: 1890kg? Objective: Communications. Description: Unavailable.	40 631	661	736	62.8	For transmitting television programs and multichannel radio communications.
	Cosmos 986 USSR 1978-10A A-2 Baykonur-Tyuratam	Total weight: 6300kg? Objective: "Continuation of outer space investigations." Description: Unavailable.	341	179	89.4	65	Probable Soviet military photoreconnaissance satellite; recovered Feb. 7.
Jan. 26	China 8 China 1978-11A (No data on vehicle) Shuang-Cheng-Tze	Total weight: 3600kg? Objective: Unavailable. Description: Unavailable.	479	161	90.9	57.0	Reentered Feb. 7; recovered a portion Jan. 30.
Jan. 26	IUE 1 U.S. 1978-12A Delta ETR	Total weight: 669kg. Objective: To explore ultraviolet region of spectrum at wave lengths inaccessible to ground-based observers. Description: Octagonal cylinder, 4.3m long, 1.4m dia; 2 solar-cell panels with 4.3m span.	45 888	25 669	1435.7	28.6	On station at 71°W; to be used by scientists from a number of countries; orbit inclined and eccentric, to improve communications with ESA observatory near Madrid.
Jan. 31	Cosmos 987 USSR 1978-13A A-2 Baykonur-Tyuratam	Total weight: 6300kg? Objective: "Continuation of outer space investigations." Description: Unavailable.	359	183	89.6	62.8	Probable Soviet military photoreconnaissance satellite; recovered Feb. 14.
Feb. 4	Exos 1 (Kykko) Japan 1978-14A Mu-3H Kagoshima Space Center	Total weight: 103kg. Objective: Part of Japan's contribution to the Intl. Magnetospheric Study. Description: Unavailable.	3977	642	134	65.4	Designed for aurora research; launched by Institute of Space and Aeronautical Science, University of Tokyo.

APPENDIX A ASTRONAUTICS AND AERONAUTICS, 1978

Launch Date	Spacecraft, Country, Int'l Designation, Vehicle, Launch Site	Payload Data	Apogee (km)	Perigee (km)	Period (min)	Inclination (degrees)	Remarks
Feb. 8	*Cosmos 988* USSR 1978-15A A-2 Plesetsk	Total weight: 5900kg? Objective: "Continuation of outer space investigations." Description: Unavailable.	363	210	89.9	72.8	Probable Soviet military photoreconnaissance satellite; recovered.
	Pickaback USSR 1978-15F A-2 Plesetsk	Total weight: 200kg? Objective: Unannounced. Description: Unavailable.	335	201	89.9	72.5	
Feb. 9	*FltSatCom 1* U.S. 1978-16A Atlas-Centaur ETR	Total weight: 1884kg (incl. fuel). Objective: Provide communications between ships/military field units using small mobile terminals. Description: Dual hexagon composed of a payload module and spacecraft module.	35 666	35 523	1426.3	2.1	First of four planned comsats to be launched by NASA for U.S. Navy and the Dept. of Defense.
Feb. 14	*Cosmos 989* USSR 1978-17A A-2 Baykonur-Tyuratam	Total weight: 6300kg? Objective: "Continuation of outer space investigations." Description: Unavailable.	354	178	89.5	65	Probable Soviet military photoreconnaissance satellite; recovered Feb. 28.
Feb. 16	*Iss 2 (Ume)* Japan 1978-18A N rocket Tanegashima Space Center	Total weight: 140kg? Objective: Ionosphere research. Description: Unavailable.	1225	972	107	69.4	Ionosphere-sounding satellite launched by Natl. Space Development Agency of Japan.
Feb. 17	*Cosmos 990* USSR 1978-19A C-1 Plesetsk	Total weight: 750kg? Objective: "Continuation of outer space investigations." Description: Unavailable.	824	783	101	74	Probable Soviet military comsat using store/dump technique.

Date	Satellite	Details				Remarks
Feb. 22	Navstar 1 U.S. 1978-20A Atlas F WTR	Total weight: 433kg. Objective: To support USN Global Positioning System. Description: Irregular cylinder with 4 extended solar panels and complex of antennas.	20 308	20 095	718.7	63.3	Launched by USAF.
Feb. 25	Satellite Data System 6 (Sds 6) U.S. 1978-21A Titan IIIB-Agena WTR	Total weight: 700kg. Objective: "Development of spaceflight techniques and technology." Description: Unavailable.	39 377	311	703.7	63.2	Launched by USAF for communications.
Feb. 28	Cosmos 991 USSR 1978-22A C-1 Plesetsk	Total weight: 680kg? Objective: "Continuation of outer space investigations." Description: Unavailable.	1022	972	104.8	83	Probable Soviet third-generation navsat.
Mar. 2	Soyuz 28 USSR 1978-23A A-2 Baykonur-Tyuratam	Total weight: 6570kg? Objective: To carry the third crew to Salyut 6. Description: Unavailable.	246 353	192 334	88.8 91.4	51.6 51.6	Carried Col. Aleksey Gubarev and first non-Soviet, non-American to fly in space: Captain Vladimir Remek of the Czech Army; docked Mar. 3, returned to earth March 10.
	Molniya 1-39 USSR 1978-24A A-2e Plesetsk	Total weight: 1750kg? Objective: Communications. Description: Unavailable.	40 733	632	738	62.8	Carried instruments for transmitting television programs and multichannel radio communications.
Mar. 4	Cosmos 992 USSR 1978-25A A-2 Baykonur-Tyuratam	Total weight: 5700kg? Objective: "Continuation of outer space investigations." Description: Unavailable.	346	210	89.8	71.4	Probable Soviet military photoreconnaissance satellite; recovered Mar. 17.

Launch Date	Spacecraft, Country, Int'l Designation, Vehicle, Launch Site	Payload Data	Apogee (km)	Perigee (km)	Period (min)	Inclination (degrees)	Remarks
Mar. 5	Landsat 3 U.S. 1978-26A Delta WTR	Total weight: 960kg. Objective: To acquire multispectral imagery over the U.S. and foreign countries, improve remote-sensing interpretive techniques. Description: Basic structure and design as Nimbus satellites: 3.04m high, 1.52m diameter, 3.96m with solar panels deployed; carried 5-channel multispectral scanner (MSS), a 2-camera return-beam vidicon (RBV), a data-collection system (DCS), and two wideband video tape recorders (WBVTR); satellite capable of near-global coverage.	917	898	103.2	99.1	Launched into circular near-polar sun-synchronous orbit, allowing spacecraft to photograph nearly entire globe during an 18-day period.
	Oscar 8 U.S. 1978-26B Delta WTR	Total weight: 27kg. Objective: To provide continuous radio communications with small amateur ground terminals. Description: Rectangular, covered by solar cells; carried two transponders, only one to be used at a time because of power limitation.	915	904	103.2	99.0	Launched by NASA with Landsat 3 as secondary payload, replacement for Oscar 6; cooperative effort by U.S., Canada, Japan, and West Germany.
	DOD spacecraft U.S. 1978-26C Delta WTR	Total weight: 350kg. Objective: "Develop spaceflight techniques and technology." Description: Unavailable.	913	906	103.2	99.0	Plasma-interaction experiment (PIX).
Mar. 10	Cosmos 993 USSR 1978-27A A-2 Plesetsk	Total weight: 6300kg? Objective: "Continuation of outer space investigations." Description: Unavailable.	368	182	89.7	72.9	Recovered Mar. 23; maneuvering engine separated Mar. 22, 1978, designated 1978-27D.

Date	Spacecraft	Details				Description
Mar. 15	Cosmos 994 USSR 1978-28A C-1 Plesetsk	Total weight: 680kg? Objective: "Continuation of outer space investigations." Description: Unavailable.	1023	996	105	82.9	Probable Soviet second-generation "A" navsat.
Mar. 16	DOD spacecraft U.S. 1978-29A Titan IIID WTR	Total weight: 13 300kg. Objective: "Develop spaceflight techniques and technology." Description: Unavailable.	240	160	88.5	96.4	Reconnaissance satellite (Big Bird 14?) launched by USAF; decayed September 11, 1978.
	Hitchhiker 40 U.S. 1978-29B Titan IIID WTR	Total weight: 60kg. Objective: "Develop spaceflight techniques and technology." Description: Unavailable.	645	639	97.6	95.8	Supplementary payload launched with 1978-29A.
Mar. 17	Cosmos 995 USSR 1978-30A A-2 Plesetsk	Total weight: 5700kg? Objective: "Continuation of outer space investigations." Description: Unavailable.	262	221	89.1	81.4	Probable Soviet military photoreconnaissance satellite; recovered Mar. 30.
Mar. 28	Cosmos 996 USSR 1978-31A C-1 Plesetsk	Total weight: 680kg? Objective: "Continuation of outer space investigations." Description: Unavailable.	1020	970	104.8	82.9	Probable Soviet third-generation "B" navsat.
Mar. 30	Cosmos 997 USSR 1978-32A D-1 Baykonur-Tyuratam	Total weight: 9500kg? Objective: A precursor launch, possibly to test reentry. Description: Unavailable.	230	200	88.7	51.6	Two satellites launched on single vehicle, probably recovered at end of first orbit Mar. 30; flight similar to that of Cosmos 881 and 882; possible connection with Cosmos 929.
	Cosmos 998 USSR 1978-32B D-1 Baykonur-Tyuratam	Total weight: 9500kg? Objective: A precursor launch, possibly to test reentry. Description: Unavailable.	230	200	88.7	51.6	

Launch Date	Spacecraft, Country, Int'l Designation, Vehicle, Launch Site	Payload Data	Apogee (km)	Perigee (km)	Period (min)	Inclination (degrees)	Remarks
Mar. 30	Cosmos 999 USSR 1978-33A A-2 Baykonur-Tyuratam	Total weight: 6300kg? Objective: "Continuation of outer space investigations." Description: Unavailable.	376	180	89.9	71.4	Probable Soviet military photoreconnaissance satellite; recovered Apr. 12.
Mar. 31	Cosmos 1000 USSR 1978-34A C-1 Plesetsk	Total weight: 7000kg? Objective: "Continuation of outer space investigations." Description: Unavailable.	1024	978	104.9	83	Probable Soviet 4th-generation "B" (civil) navsat.
	Intelsat IVA F-6 U.S. 1978-35A Atlas-Centaur ETR	Total weight: 1500kg. Objective: U.S. commercial comsat. Description: Cylindrical, 6.99m high, 2.4m diameter, solar-panel height 2.8m; carrying 6250 two-way voice circuits plus two television channels; two-thirds increase in capacity over previous Intelsat series.	35 791	35 781	1436.1	0.0	Sixth and last in a series of Hughes-built improved comsats launched by NASA for ComSatCorp., manager of Intelsat series, for business use in geostationary orbit at 60°E over Indian Ocean.
Apr. 4	Cosmos 1001 USSR 1978-36A A-2 Baykonur-Tyuratam	Total weight: 7000kg? Objective: "Continuation of outer space investigations." Description: Unavailable.	249	205	88.7	51.6	Might be part of development of manned craft; recovered Apr. 15.
Apr. 6	Cosmos 1002 USSR 1978-37A A-2 Baykonur-Tyuratam	Total weight: 5500kg? Objective: "Continuation of outer space investigations." Description: Unavailable.	305	209	89.4	65	Probable Soviet military photoreconnaissance satellite; recovered Apr. 19.
Apr. 7	Rhyolite RH-4 U.S. 1978-38A Atlas-Agena D ETR	Total weight: 700kg. Objective: "Develop spaceflight techniques and technology." Description: Unavailable.	41 411	6528	615.5	28.4	Launched by USAF; probable U.S. early-warning satellite to monitor missile launchings.

Date	Name/Source	Description	Apogee (km)	Perigee (km)	Period (min)	Incl. (deg)	Remarks
Apr. 7	*BSE Yuri* Japan 1978-39A Delta ETR	Total weight: 678kg. Objective: Broadcast color television to remote islands of Japan. Description: Rectangular electronics module 1.3m wide, 1.2m long, 3.9m high, including apogee-boost motor; deployable solar panels 8.9m long, 1.5m wide; three-axis-stabilized.	35 797	35 774	1436.0	0.0	Broadcasting Satellite-Experimental launched by NASA for the Natl. Space Development Agency (NASDA) of Japan, in geostationary orbit at 110°E above the equator south of Japan.
Apr. 20	*Cosmos 1003* USSR 1978-40A A-2 Plesetsk	Total weight: 6300kg? Objective: "Continuation of outer space investigations." Description: Unavailable.	349	185	89.6	62.8	Probable Soviet military photoreconnaissance satellite; recovered May 4.
Apr. 26	*Hcmm (Aem 1)* U.S. 1978-41A Scout WTR	Total weight: 134kg. Objective: Produce thermal maps of earth's atmosphere; investigate use of day/night thermal imagery to generate apparent thermal-inertial and temperature-cycle data. Description: Basic module containing attitude-control, data-handling, communications, command and power subsystems; instrument module containing heat-capacity mapping radiometer (HCMR) and supporting equipment; powered by two solar panels.	626	612	97.1	97.6	Heat-capacity mapping mission also known as Applications Explorer Mission A (AEM-A); first in a series of low-cost modular spacecraft in special orbits for mission-unique experimental data acquisition, to measure earth's surface for maximum temperatures and 12 hrs later for same area's minimum temperature.
May 1	*Ams 3* U.S. 1978-42A Thor-Burner 2 WTR	Total weight: 513kg. Objective: "Develop spaceflight techniques and technology." Description: Similar to NASA's Tiros-N.	835	820	101.3	98.6	Advanced meteorological satellite; launched by USAF to support Defense Meteorological Satellite Program.
May 5	*Cosmos 1004* USSR 1978-43A A-2 Plesetsk	Total weight: 5900kg? Objective: "Continuation of outer space investigations." Description: Unavailable.	311	213	89.4	62.8	Probable Soviet military photoreconnaissance satellite; recovered May 18.
	Pickaback USSR 1978-43G A-2 Plesetsk	Total weight: 200kg? Objective: Unavailable. Description: Unavailable.	431	204	90.0	62.8	

Launch Date	Spacecraft, Country, Int'l Designation, Vehicle, Launch Site	Payload Data	Apogee (km)	Perigee (km)	Period (min)	Inclination (degrees)	Remarks
May 11	*Ots 2* U.S. 1978-44A Delta ETR	Total weight: 865kg. Objective: Test concepts for satellite links in the 1980s to provide intra-European telephone, telegraph, telex, and television in Western Europe. Description: Six-sided box containing payload module; service module; and antenna assembly, measuring 1.62m wide, 2.39m high, and 9.26m across with solar array deployed; three-axis stabilized.	35 792	35 786	1436.2	0.0	Orbital test satellite launched by NASA for ESA to replace OTS-A, destroyed Sept. 13, 1977, when its launcher exploded.
May 12	*Cosmos 1005* USSR 1978-45A A-1 Plesetsk	Total weight: 3800kg? Objective: "Continuation of outer space investigations." Description: Unavailable.	672	626	97.6	81.2	Probable Soviet military electronic ferret.
	Cosmos 1006 USSR 1978-46A C-1 Plesetsk	Total weight: 900kg? Objective: "Continuation of outer space investigations." Description: Unavailable.	417	383	92.5	65.8	Probable Soviet radar-calibration satellite.
May 13	*NavStar 2* U.S. 1978-47A Atlas F WTR	Total weight: 433kg. Objective: To support USN Global Positioning System. Description: Irregular cylinder with four extended solar panels and complex of antennas.	20 084	19 952	711.3	63.1	Launched by USAF.
May 16	*Cosmos 1007* USSR 1978-48A A-2 Plesetsk	Total weight: 6300kg? Objective: "Continuation of outer space investigations." Description: Unavailable.	384	180	89.8	72.9	Probable Soviet military photoreconnaissance satellite; recovered May 29.

Date	Satellite	Details	Weight				Notes
May 17	Cosmos 1008 USSR 1978-49A C-1 Plesetsk	Total weight: 975kg? Objective: "Continuation of outer space investigations." Description: Unavailable.	551	501	95.1	74	Probable Soviet military electronic ferret.
May 19	Cosmos 1009 USSR 1978-50A F-1m Baykonur-Tyuratam	Total weight: 3320kg? Objective: First test of new hunter-killer satellite, 15th test since program began in 1968. Description: Unavailable.	1378	971	109	66	Satellite interceptor made close approach to orbiting Cosmos 967; plunged into Pacific May 19.
May 20	Pioneer Venus 1 U.S. 1978-51A Atlas-Centaur ETR	Total weight: 549kg. Objective: Determine structure of upper atmosphere and ionosphere of Venus; observe interaction of solar wind with Venus ionosphere and with the small magnetic field; use remote sensing to observe characteristics of atmosphere and surface of Venus on a planetary scale; measure planet's gravitational field; detect gamma-ray bursts. Description: Cylinder 1.2m high, 4.5m including antenna mast, 2.5m in diameter, powered by solar cells on spacecraft exterior and 2 nickel-cadmium batteries.	(Currently in heliocentric orbit)				First of 2 NASA interplanetary s/c; scheduled to arrive at Venus Dec. 78, measure planet atmosphere and electromagnetic environment from orbit.
May 23	Cosmos 1010 USSR 1978-52A A-2 Plesetsk	Total weight: 6300kg? Objective: "Continuation of outer space investigations." Description: Unavailable.	257	218	89	81.4	Probable Soviet earth-resources observation satellite; recovered June 5.
	Cosmos 1011 USSR 1978-53A C-1 Plesetsk	Total weight: 680kg? Objective: "Continuation of outer space investigations." Description: Unavailable.	1026	978	104.9	82.9	Probable Soviet third-generation "B" navsat.
May 25	Cosmos 1012 USSR 1978-54A A-2 Plesetsk	Total weight: 5700kg? Objective: "Continuation of outer space investigations." Description: Unavailable.	280	214	89.2	62.8	Probable Soviet military photoreconnaissance satellite; recovered June 7.

Launch Date	Spacecraft, Country, Int'l Designation, Vehicle, Launch Site	Payload Data	Apogee (km)	Perigee (km)	Period (min)	Inclination (degrees)	Remarks
June 2	Molniya 1-40 USSR 1978-55A A-2e Plesetsk	Total weight: 1750kg? Objective: Transmit TV programs and multichannel radio communications. Description: Unavailable.	40 837	457	736	62.5	
June 7	Cosmos 1013 USSR 1978-56A C-1 Plesetsk	Total weight: 50kg? Objective: "Continuation of outer space investigations." Description: Unavailable.	1555	1480	116.3	74.0	Multiple launch (8 satellites) probably for Soviet military communications.
	Cosmos 1014 USSR 1978-56B C-1 Plesetsk	Total weight: 50kg? Objective: "Continuation of outer space investigations." Description: Unavailable.	1533	1480	116.1	74.0	
	Cosmos 1015 USSR 1978-56C C-1 Plesetsk	Total weight: 50kg? Objective: "Continuation of outer space investigations." Description: Unavailable.	1518	1475	115.8	74.0	
	Cosmos 1016 USSR 1978-56D C-1 Plesetsk	Total weight: 50kg? Objective: "Continuation of outer space investigations." Description: Unavailable.	1500	1472	115.6	74.0	
	Cosmos 1017 USSR 1978-56E C-1 Plesetsk	Total weight: 50kg? Objective: "Continuation of outer space investigations." Description: Unavailable.	1494	1459	115.4	74.0	

Date	Spacecraft	Details	col4	col5	col6	col7	Remarks
June 7	Cosmos 1018 USSR 1978-56F C-1 Plesetsk	Total weight: 50kg? Objective: "Continuation of outer space investigations." Description: Unavailable.	1490	1443	115.2	74.0	
	Cosmos 1019 USSR 1978-56G C-1 Plesetsk	Total weight: 50kg? Objective: "Continuation of outer space investigations." Description: Unavailable.	1491	1424	115.0	74.0	
	Cosmos 1020 USSR 1978-56H C-1 Plesetsk	Total weight: 50kg? Objective: "Continuation of outer space investigations." Description: Unavailable.	1456	1409	114.8	74.0	
June 10	Cosmos 1021 USSR 1978-57A A-2 Baykonur-Tyuratam	Total weight: 6000kg? Objective: "Continuation of outer space investigations." Description: Unavailable.	336	180	89.4	65	Probable Soviet military photoreconnaissance satellite; recovered June 23.
	Imews 8 U.S. 1978-58A Titan IIID WTR	Total weight: 820kg. Objective: "Development of spaceflight techniques and technology." Description: Unavailable.	42 039	29 929	1446.3	12.0	Launched by USAF as part of integrated missile early-warning system.
June 12	Cosmos 1022 USSR 1978-59A A-2 Plesetsk	Total weight: 6300kg? Objective: "Continuation of outer space investigations." Description: Unavailable.	374	182	89.7	72.9	Probable Soviet military photoreconnaissance satellite; recovered June 25.
June 14	KH-11-2 U.S. 1978-60A Titan IIID WTR	Total weight: 13 300kg. Objective: "Develop spaceflight techniques and technology." Description: Unavailable.	509	276	92.4	96.8	"Big Bird" reconnaissance satellite launched by USAF.

APPENDIX A

Launch Date	Spacecraft, Country, Int'l Designation, Vehicle, Launch Site	Payload Data	Apogee (km)	Perigee (km)	Period (min)	Inclination (degrees)	Remarks
June 15	*Soyuz 29* USSR 1978-61A A-2 Baykonur-Tyuratam	Total weight: 6570 kg? Objective: Carry crew to *Salyut 6*. Description: Unavailable.	355	339	91.4	51.6	*Soyuz 29* crew (Col. Vladimir Kovalenok, civilian engr. Aleksandr Ivanchenkov) docked with *Salyut 6*'s forward docking port June 16/78; returned to earth Sept. 3.
June 16	*Goes 3* U.S. 1978-62A Delta ETR	Total weight: 627kg. Objective: Continuous observation of earth's atmosphere. Description: Cylinder 190.5cm in diameter, 344cm long from top of magnetometer to bottom of apogee boost motor; primary member, thrust tube located in the center of cylinder, contains radiometer/telescope extending length of spacecraft; scanning mirror looks out through an opening in cylindrical solar arrays.	35 804	35 776	1436.2	0.7	Third of a series of geostationary operational environmental satellites launched by NASA for NOAA; on station at 135°W.
June 21	*Cosmos 1023* USSR 1978-63A C-1 Plesetsk	Total weight: 750kg? Objective: "Continuation of outer space investigations." Description: Unavailable.	822	784	100.8	74.1	Probable Soviet military comsat using data-store/dump techniques.
June 27	*SeaSat 1* U.S. 1978-64A Atlas Agena D WTR	Total weight: 2300kg. Objective: Experimental satellite to demonstrate microwave monitoring of ocean surface. Description: Two major hardware elements: Agena bus for launch function and satellite support (attitude control, power, guidance, telemetry, and command); sensor module, platform for five sensors, antennas, laser retroreflector, tracking aids, and a SAR downlink. Two modules combined length 21m, maximum diameter 1.5m without antennas.	800	776	100.6	108.0	All sensors turned on simultaneously July 7; spacecraft ceased transmitting Oct. 9; attempts to reestablish contact unsuccessful.

Date	Satellite	Description	Weight (kg)	Perigee/Apogee (km)	Period (min)	Incl. (°)	Remarks
June 27	Soyuz 30 USSR 1978-65A A-2 Baykonur-Tyuratam	Total weight: 6570kg? Objective: To carry additional crew to Salyut 6. Description: Unavailable.	244 360	194 336	88.8 91.2	51.6 51.6	Carried second international crew to join Soyuz 29 crew on Salyut 6 (flight engineer Major Miroslaw Hermaszewski of Polish Air Force and Lt. Col. Pyotr Klimuk); docked with aft port of Salyut 6 June 28/78; returned to earth July 5.
June 28	Cosmos 1024 USSR 1978-66A A-2e Plesetsk	Total weight: 1800kg? Objective: "Continuation of outer space investigations." Description: Unavailable.	40 000	630	726	62.8	Probable Soviet early-warning satellite.
	Cosmos 1025 USSR 1978-67A F-2 Plesetsk	Total weight: 6320kg? Objective: "Continuation of outer space investigations." Description: Unavailable.	680	649	97.8	82.5	Possible Soviet electronic scanner of ocean surface.
June 29	Comstar 3 U.S. 1978-68A Atlas-Centaur ETR	Total weight: 1520kg. Objective: Provide 14 000 two-way domestic high-quality voice circuits (telephone). Description: Cylinder, 610cm high, 244cm in diameter; 14 000 solar cells on exterior; nickel-cadmium batteries to provide power during launch and twice-a-year solar eclipses.	35 789	35 783	1436.1	0.0	Launched by NASA for Comsat General Corp.
July 2	Cosmos 1026 USSR 1978-69A A-2 Baykonur-Tyuratam	Total weight: 5700kg? Objective: "Continuation of outer space investigations." Description: Unavailable.	261	209	89	51.8	Probable Soviet military photoreconnaissance satellite; recovered July 6.
July 7	Progress 2 USSR 1978-70A A-2 Baykonur-Tyuratam	Total weight: 7020kg. Objective: Unmanned supply craft carrying consumables to Salyut 6. Description: Unavailable.	309	248	90.0	51.6	Docked with Salyut 6 July 9; burned up over Pacific Aug. 4.

Launch Date	Spacecraft, Country, Int'l Designation, Vehicle, Launch Site	Payload Data	Apogee (km)	Perigee (km)	Period (min)	Inclination (degrees)	Remarks
July 14	Geos 2 U.S. 1978-71A Delta ETR	Total weight: 573kg. Objective: Scientific studies of earth's magnetosphere. Description: Cylinder 1.6m dia, 1.1m high, covered with solar cells; eight booms varying in length between 1.5 and 2.0m.	35 789	35 784	1436.0	0.5	Launched by NASA for ESA to perform same experiments as Geos 1, which failed to achieve satisfactory orbit in Apr. 77.
	Molniya 1-41 USSR 1978-72A A-2e Plesetsk	Total weight: 1750kg? Objective: Provide TV and multichannel radio transmissions. Description: Unavailable.	40 660	650	737	62.8	
July 18	Raduga 4 USSR 1978-73A D-1e Baykonur-Tyuratam	Total weight: 2500kg? Objective: Provide telephone and telegraph links, TV transmissions. Description: Unavailable.	35 590	36 590	1478	0.5	In geostationary orbit at stationary position over Indian Ocean.
July 27	Cosmos 1027 USSR 1978-74A C-1 Plesetsk	Total weight: 680kg? Objective: "Continuation of outer space investigations." Description: Unavailable.	1015	979	104.8	82.9	Probable Soviet second-generation "A" navsat.
Aug. 5	Satellite Data System Sds7 U.S. 1978-75A Tital IIIB WTR	Total weight: 700kg. Objective: "Development of spaceflight techniques and technology." Description: Unavailable.	39 315	380	703.8	63.3	DOD comsat launched by USAF.
	Cosmos 1028 USSR 1978-76A A-2 Plesetsk	Total weight: 6700kg? Objective: "Continuation of outer space investigations." Description: Unavailable.	272	182	88.7	67.1	Probable Soviet military photoreconnaissance satellite; recovered Sept. 4.

Aug. 7	Progress 3 USSR 1978-77A A-2 Baykonur-Tyuratam	Total weight: 7020kg. Objective: Carry food supplies to Salyut 6 crew (Soyuz 29). Description: Unavailable.	249	195	88.7	51.6	Docked with Salyut 6 Aug. 10; reentered over the Pacific Aug. 23.
Aug. 8	Pioneer Venus 2 U.S. 1978-78A Atlas Centaur ETR	Total weight: 904kg. Objective: Carry sounders and bus to measure nature and composition of clouds, composition and structure of atmosphere and general atmosphere circulation pattern of Venus. Description: Cylindrical bus carrying four atmospheric-entry probes; height, top of main probe to bottom of bus, 2.9m; diameter, 2.5m. Four probes carried on bus by a large inverted cone structure, three small probes equally spaced around larger probe in center.	(Currently in heliocentric orbit)				Large probe separated from bus Nov. 15, three smaller probes separated Nov. 20. Four probes reached Venus surface; bus burned up in atmosphere; one probe transmitted data for an hour after impact.
Aug. 12	Isee 3 U.S. 1978-79A Delta ETR	Total weight: 469kg. Objective: Measure interaction between solar wind and earth. Description: 16-sided cylinder, 173cm dia, 161cm high; solar array located forward of equipment section, another below; instruments mounted on equipment shelf.	1 151 664	180	73 702	28.9	Third of three intl. sun-earth explorers; to be inserted into halo orbit (around a Lagrange point) in Nov.
Aug. 22	Molniya 1-42 USSR 1978-80A A-2e Plesetsk	Total weight: 1750kg. Objective: Transmit TV and multichannel radio communications. Description: Unavailable.	40 788	480	736	62.8	Forty-second operational Molniya-1 comsat.
Aug. 26	Soyuz 31 USSR 1978-81A A-2 Baykonur-Tyuratam	Total weight: 6570kg? Objective: Carry crew to Salyut 6. Description: Unavailable.	243	193	88.8	51.6	Carried third international crew to join Soyuz 29 crew on Salyut 6 (including Col. Valery Bykovsky and Sigmund Jähn of the GDR). Soyuz 31 crew returned in Soyuz 29 vehicle Sept. 3.

Launch Date	Spacecraft, Country, Int'l Designation, Vehicle, Launch Site	Payload Data	Apogee (km)	Perigee (km)	Period (min)	Inclination (degrees)	Remarks
Aug. 29	Cosmos 1029 USSR 1978-82A A-2 Plesetsk	Total weight: 6300kg? Objective: "Continuation of outer space investigations." Description: Unavailable.	353	186	89.6	62.8	Probable Soviet military photoreconnaissance satellite; fragment separated from Cosmos 1029 Sept. 7 roughly spheric, recovered Sept. 11 near Moulins, France.
Sept. 6	Cosmos 1030 USSR 1978-83A A-2e Plesetsk	Total weight: 1800kg? Objective: "Continuation of outer space investigations." Description: Unavailable.	40 100	650	725.6	62.8	Probable Soviet early-warning satellite.
Sept. 9	Venera 11 USSR 1978-84A D-1e Baykonur-Tyuratam	Total weight: 4940kg? Objective: Explore Venus. Description: Unavailable.	Heliocentric orbit				First of two Venus-exploration craft launched during 1978 window for Venus encounter (arrived Dec. 25, 1978).
	Lander USSR 1978-84E D-1e Baykonur-Tyuratam	Total weight: 1560kg? Objective: Soft landing on Venus. Description: Unavailable.					
Sept. 9	Cosmos 1031 USSR 1978-85A A-2 Plesetsk	Total weight: 6300kg? Objective: "Continuation of outer space investigations." Description: Unavailable.	351	191	89.6	62.8	Probable Soviet military photoreconnaissance satellite; recovered Sept. 22.
Sept. 14	Venera 12 USSR 1978-86A D-1e Baykonur-Tyuratam	Total weight: 4940kg? Objective: Explore Venus. Description: Unavailable.	Heliocentric orbit				Second Venus-exploration craft.

Date	Name/Designation		Total weight / Objective / Description				Notes
Sept. 14	Lander USSR 1978-86E D-1e Baykonur-Tyuratam	Total weight: 1560kg? Objective: Venus landing. Description: Unavailable.					
Sept. 16	Exos 2 Japan 1978-87A Mu-3H Kagoshima	Total weight: 100kg. Objective: Magnetosphere research. Description: Unavailable.	30 051	228	523.17	31.1	Japanese research satellite, also known as Jiki-Ken (magnetosphere).
Sept. 19	Cosmos 1032 USSR 1978-88A A-2 Plesetsk	Total weight: 5900kg? Objective: "Continuation of outer space investigations." Description: Unavailable.	249	218	88.9	81.4	Probable Soviet military photoreconnaissance satellite; recovered Oct. 3.
Oct. 3	Cosmos 1033 USSR 1978-89A A-2 Plesetsk	Total weight: 5900kg? Objective: "Continuation of outer space investigations." Description: Unavailable.	268	223	89.1	81.4	Probable Soviet photoreconnaissance satellite carrying earth-resources equipment; recovered Oct. 16.
	Progress 4 USSR 1978-90A A-2 Baykonur-Tyuratam	Total weight: 7020kg? Objective: Unmanned supply vehicle carrying fuel and other consumables to Soyuz 29 crew aboard Salyut 6. Description: Unavailable.	266 362	191 359	88.8 91.7	51.7 51.6	Docked at Salyut 6 aft port Oct. 6; refueling completed Oct. 13; on Oct. 20 Progress engines boosted Soyuz 31-Salyut 6-Progress 4 orbit; undocked Oct. 24; reentered Oct. 26.
Oct. 4	Cosmos 1034 USSR 1978-91A C-1 Plesetsk	Total weight: 50kg? Objective: "Continuation of outer space investigations." Description: Unavailable.	1479	1421	115.0	74	Multiple launch (probable military comsats), 8 on single launcher.
	Cosmos 1035 USSR 1978-91B C-1 Plesetsk	Total weight: 50kg? Objective: "Continuation of outer space investigations." Description: Unavailable.	1478	1401	114.7	74	

APPENDIX A

Launch Date	Spacecraft, Country, Int'l Designation, Vehicle, Launch Site	Payload Data	Apogee (km)	Perigee (km)	Period (min)	Inclination (degrees)	Remarks
Oct. 4	Cosmos 1036 USSR 1978-91C C-1 Plesetsk	Total weight: 50kg? Objective: "Continuation of outer space investigations." Description: Unavailable.	1479	1440	115.2	74	
	Cosmos 1037 USSR 1978-91D C-1 Plesetsk	Total weight: 50kg? Objective: "Continuation of outer space investigations." Description: Unavailable.	1480	1459	115.4	74	
	Cosmos 1038 USSR 1978-91E C-1 Plesetsk	Total weight: 50kg? Objective: "Continuation of outer space investigations." Description: Unavailable.	1485	1475	115.6	74	
	Cosmos 1039 USSR 1978-91F C-1 Plesetsk	Total weight: 50kg? Objective: "Continuation of outer space investigations." Description: Unavailable.	1550	1477	116.4	74	
	Cosmos 1040 USSR 1978-91G C-1 Plesetsk	Total weight: 50kg? Objective: "Continuation of outer space investigations." Description: Unavailable.	1525	1477	116.1	74	
	Cosmos 1041 USSR 1978-91H C-1 Plesetsk	Total weight: 50kg? Objective: "Continuation of outer space investigations." Description: Unavailable.	1507	1475	115.9	74	

Date	Satellite	Details	Col4	Col5	Col6	Notes
Oct. 6	Cosmos 1042 USSR 1978-92A A-2 Plesetsk	Total weight: 6300kg? Objective: "Continuation of outer space investigations." Description: Unavailable.	326	187	89.3	62.8	Probable Soviet military photoreconnaissance satellite; recovered Oct. 19.
Oct. 7	Navstar 3 U.S. 1978-93A Atlas F WTR	Total weight: 433kg. Objective: Support DOD's Global Positioning System. Description: Irregular cylinder, 4 extended solar panels, and complex of antennas.	20 312	20 285	722.6	62.8	Launched by USAF.
Oct. 10	Cosmos 1043 USSR 1978-94A A-1 Plesetsk	Total weight: 3800kg? Objective: "Continuation of outer space investigations." Description: Unavailable.	650	625	97.3	81.1	Probable Soviet military intelligence electronic ferret.
Oct. 13	Molniya 3-10 USSR 1978-95A A-2e Plesetsk	Total weight: 1890kg? Objective: Domestic comsat carrying radio, TV, telephone, telegraph equipment. Description: Unavailable.	40 825	467	736	62.8	Probable replacement for Molniya 3-6 (1976-127A).
	Tiros N U.S. 1978-96A Atlas F WTR	Total weight: 734kg. Objective: Provide real-time meteorological information. Description: Spacecraft, adapted from RCA's 1976 Tiros, 371cm high, 188cm dia; solar panel deployed in orbit.	866	850	102.1	98.9	Launched by NASA; turned over to NOAA Nov 6. First of 8 planned third-generation operational polar-orbiting meteosats; primary data source for Global Atmospheric Research Program's Global Weather Experiment, intl. cooperative program.
Oct. 17	Cosmos 1044 USSR 1978-97A A-2 Plesetsk	Total weight: 5700kg? Objective: "Continuation of outer space investigations." Description: Unavailable.	315	211	89.5	62.8	Probable Soviet military photoreconnaissance satellite, recovered Oct. 30.
Oct. 24	Nimbus 7 U.S. 1978-98A Delta WTR	Total weight: 987kg. Objective: Pollution monitoring. Description: Butterfly-shaped in orbit; 3m high, 2m wide; earth-oriented, three-axis-stabilized.	955	943	104.1	99.3	Last of Nimbus meteosats series first launched in 1964; first satellite designed to monitor atmosphere for manmade and natural pollutants.

APPENDIX A

Launch Date	Spacecraft, Country, Int'l Designation, Vehicle, Launch Site	Payload Data	Apogee (km)	Perigee (km)	Period (min)	Inclination (degrees)	Remarks
Oct. 24	*Cameo* U.S. 1978-98B Delta WTR	Total weight: 350kg. Objective: Study solar energy, plasma flows and electric fields in Arctic regions.	953.0	933.0	103.9	99.2	Chemically active materials ejected in orbit, piggyback payload with *Nimbus 7*, activated four barium canisters over northern Alaska, lithium canister over Scandinavia.
	Intercosmos 18 USSR 1978-99A C-1 Plesetsk	Total weight: 550kg? Objective: Study magnetosphere and ionosphere. Description: Unavailable.	644	378	96.5	82.9	International cooperative mission supported by USSR, Czechoslovakia, Hungary, Poland, and Romania.
	Magion Czechoslovakia 1978-99C C-1 Plesetsk	Total weight: 15kg? Objective: Study magnetosphere and ionosphere. Description: Unavailable.	667	383	95.1	82.9	Czechoslovakian satellite, part of *Intercosmos 18* scientific mission; separated into autonomous orbit Nov. 14.
	Cosmos 1045 USSR 1978-100A F-2 Plesetsk	Total weight: 3400kg? Objective: "Continuation of outer space investigations." Description: Unavailable.	1704	1682	120.3	82.5	Probable Soviet earth-resources satellite for sea-surface scanning.
Oct. 26	*Radio 1* USSR 1978-100B F-2 Plesetsk	Total weight: 40kg? Objective: "Continuation of outer space investigations." Description: Unavailable.	1705	1682	120.3	82.5	Carried equipment to improve amateur (ham) radio communications, as well as student experiments.
	Radio 2 USSR 1978-100C F-2 Plesetsk	Total weight: 40kg? Objective: "Continuation of outer space investigations." Description: Unavailable.	1704	1682	120.3	82.6	See *Radio 1*.

Date	Satellite	Details	Perigee (km)	Apogee (km)	Period (min)	Incl. (deg)	Remarks
Oct. 30	Prognoz 7 USSR 1978-101A A-2e Baykomur-Tyuratam	Total weight: 910kg? Objective: Scientific satellite to study magnetosphere, carrying instruments for ultraviolet and gamma-radiation research. Description: Unavailable.	202	965	483	5888.0	65.0	Onboard equipment manufactured in Hungary, Sweden, France, Czechoslovakia, and USSR.
Nov. 1	Cosmos 1046 USSR 1978-102A A-2 Plesetsk	Total weight: 5900kg? Objective: "Continuation of outer space investigations." Description: Unavailable.	353	212	89.9	72.9	Probable Soviet military photoreconnaissance satellite; recovered Nov. 14.
Nov. 3	Vertikal 7 USSR C rocket Kapustin Yar	Total weight: 1300kg? Objective: Geophysical probe; research on ionosphere and shortwave radiation. Description: Unavailable.	1500 altitude (instrument pkg separated at 175 km)					Onboard equipment made in USSR, Bulgaria, Romania, Hungary, and Czechoslovakia.
Nov. 13	Heao 2 U.S. 1978-103A Atlas-Centaur ETR	Total weight: 2720kg. Objective: High-energy astronomical observatory to detect x-rays from objects two-thirds of the way toward the edge of the universe. Description: Octagonal prism 9m long, 3m dia.	545	525	95.3	23.5	Second of three NASA high-energy astronomical observatories; this mission would concentrate on findings of Heao 1.
Nov. 15	Cosmos 1047 USSR 1978-104A A-2 Plesetsk	Total weight: 6300kg? Objective: "Continuation of outer space investigations." Description: Unavailable.	378	182	89.8	72.9	Probable Soviet military photoreconnaissance satellite; recovered Nov. 28.
Nov. 16	Cosmos 1048 USSR 1978-105A C-1 Plesetsk	Total weight: 750kg? Objective: "Continuation of outer space investigations." Description: Unavailable.	824	788	101	74	Probable Soviet military comsat using data-store/dump techniques.
Nov. 18	NATO 3C U.S. 1978-106A Delta ETR	Total weight: 701kg. Objective: Defense communications. Description: Drum 2.18m in dia, 2.23m long, cylindrical solar array; spin-stabilized.	35 814	35 460	1428.5	4.3	Third advanced NATO comsat, serving as backup for NATO 3A and 3B; on station at 50°W over Atlantic.

Launch Date	Spacecraft, Country, Int'l Designation, Vehicle, Launch Site	Payload Data	Apogee (km)	Perigee (km)	Period (min)	Inclination (degrees)	Remarks
Nov. 21	Cosmos 1049 USSR 1978-107A A-2 Plesetsk	Total weight: 6300kg? Objective: "Continuation of outer space investigations." Description: Unavailable.	375	183	89.7	72.9	Probable Soviet military photoreconnaissance satellite; recovered Dec. 4.
Nov. 28	Cosmos 1050 USSR 1978-108A A-2 Plesetsk	Total weight: 6300kg? Objective: "Continuation of outer space investigations." Description: Unavailable.	298	258	89.8	62.8	Probable Soviet military photoreconnaissance satellite; recovered Dec. 11.
Dec. 5	Cosmos 1051 USSR 1978-109A C-1 Plesetsk	Total weight: 50kg? Objective: "Continuation of outer space investigations." Description: Unavailable.	1484	1392	114.6	74.0	Multiple launch (8 comsats) on single launcher.
	Cosmos 1052 USSR 1978-109B C-1 Plesetsk	Total weight: 50kg? Objective: "Continuation of outer space investigations." Description: Unavailable.	1487	1409	114.8	74.0	
	Cosmos 1053 USSR 1978-109C C-1 Plesetsk	Total weight: 50kg? Objective: "Continuation of outer space investigations." Description: Unavailable.	1487	1427	115.0	74.0	
	Cosmos 1054 USSR 1978-109D C-1 Plesetsk	Total weight: 50kg? Objective: "Continuation of outer space investigations." Description: Unavailable.	1488	1444	115.2	74.0	

Date	Satellite	Description				Notes
Dec. 5	Cosmos 1055 USSR 1978-109E Plesetsk	Total weight: 50kg? Objective: "Continuation of outer space investigations." Description: Unavailable.	1489	1462	115.5	74.0	
	Cosmos 1056 USSR 1978-109F C-1 Plesetsk	Total weight: 50kg? Objective: "Continuation of outer space investigations." Description: Unavailable.	1502	1470	115.7	74.0	
	Cosmos 1057 USSR 1978-109G C-1 Plesetsk	Total weight: 50kg? Objective: "Continuation of outer space investigations." Description: Unavailable.	1513	1479	115.9	74.0	
	Cosmos 1058 USSR 1978-109H C-1 Plesetsk	Total weight: 50kg? Objective: "Continuation of outer space investigations." Description: Unavailable.	1536	1479	116.2	74.0	
Dec. 7	Cosmos 1059 USSR 1978-110A A-2 Plesetsk	Total weight: 6300kg? Objective: "Continuation of outer space investigations." Description: Unavailable.	360	188	89.7	62.8	Probable Soviet military photoreconnaissance satellite; recovered Dec. 20.
Dec. 8	Cosmos 1060 USSR 1978-111A A-2 Baykonur-Tyuratam	Total weight: 5700kg? Objective: "Continuation of outer space investigations." Description: Unavailable.	311	209	89.5	65	Probable Soviet military photoreconnaissance satellite; recovered Dec. 21.
Dec. 11	Navstar 4 U.S. 1978-112A Atlas F WTR	Total weight: 433kg. Objective: Support DOD's Global Positioning System. Description: Irregular cylinder, 4 extended solar panels, complex of antennas.	20 316	20 267	722.4	63.3	Fourth satellite in DOD Global Positioning System; launched by USAF.

Launch Date	Spacecraft, Country, Int'l Designation, Vehicle, Launch Site	Payload Data	Apogee (km)	Perigee (km)	Period (min)	Inclination (degrees)	Remarks
Dec. 14	Dscs 2-11 U.S. 1978-113A Titan IIIC ETR	Total weight: 590kg. Objective: "Develop spaceflight techniques and technology." Description: Unavailable.	35 796	35 726	1436	2.3	Pair of military comsats launched by USAF.
	Dscs 2-12 U.S. 1978-113B Titan IIIC ETR	Total weight: 590kg. Objective: "Develop spaceflight techniques and technology." Description: Unavailable.	36 413	36 261	1464.3	2.5	
Dec. 14	Cosmos 1061 USSR 1978-114A A-2 Plesetsk	Total weight: 200kg? Objective: "Continuation of outer space investigations." Description: Unavailable.	333	211	89.6	62.8	Probable Soviet military photoreconnaissance satellite, recovered Dec. 27.
Dec. 15	Cosmos 1062 USSR 1978-115A C-1 Plesetsk	Total weight: 875kg? Objective: "Continuation of outer space investigations." Description: Unavailable.	548	508	95.1	74.0	Probable Soviet military intelligence electronic ferret.
Dec. 16	Anik 4 (Telesat D) U.S. 1978-116A Delta ETR	Total weight: 922kg. Objective: Provide transmission of television, voice, and other data throughout Canada. Description: Box 325.9cm long, antenna 112cm, and 217.4cm dia.	35 794	35 782	1436.1	0.0	Second-generation comsat launched by NASA for Canada's Domestic Communications Satellite System.
Dec. 19	Cosmos 1063 USSR 1978-117A A-1 Plesetsk	Total weight: 3800kg? Objective: "Continuation of outer space investigations." Description: Unavailable.	661	632	97.5	81.2	Probable Soviet military intelligence electronic ferret.

Date	Name/Designation	Description	Col1	Col2	Col3	Col4	Notes
Dec. 19	Horizont 1 USSR 1978-118A D-1e Baykomur-Tyuratam	Total weight: 2880kg? Objective: Provide telephone and TV communications. Description: Unavailable.	48 365	22 581	1420	11.3	Planned orbit was inclined for better coverage of far north areas; apparently failed to reach planned orbit.
Dec. 20	Cosmos 1064 USSR 1978-119A C-1 Plesetsk	Total weight: 680kg? Objective: "Continuation of outer space investigations." Description: Unavailable.	991	435	98.7	83	Probable Soviet third-generation "B" nav-sat; went into wrong orbit.
Dec. 22	Cosmos 1065 USSR 1978-120A Kapustin Yar	Total weight: 90kg? Objective: "Continuation of outer space investigations." Description: Unavailable.	556	346	93.7	50.7	Probable Soviet radar-calibration satellite.
Dec. 23	Cosmos 1066 USSR 1978-121A A-1 Plesetsk	Total weight: 3300kg? Objective: "Continuation of outer space investigations." Description: Unavailable.	908	848	102.2	81.2	Possibly related to Meteor 2 series of weather satellites; press reported failure.
Dec. 26	Cosmos 1067 USSR 1978-122A C-1 Plesetsk	Total weight: 600kg? Objective: "Continuation of outer space investigations." Description: Unavailable.	1226	1184	109.2	83	Probable Soviet geodetic-research satellite.
Dec. 26	Cosmos 1068 USSR 1978-123A A-2 Plesetsk	Total weight: 6300kg? Objective: "Continuation of outer space investigations." Description: Unavailable.	401	187	90.2	62.8	Probable Soviet military photoreconnaissance satellite, recovered Jan. 8, 1979.
Dec. 28	Cosmos 1069 USSR 1978-124A A-2 Plesetsk	Total weight: 5900kg? Objective: "Continuation of outer space investigations." Description: Unavailable.	290	244	89.8	62.8	Probable Soviet military photoreconnaissance satellite, recovered Jan. 10, 1979.

Appendix B
MAJOR NASA LAUNCHES, 1978

During 1978, the U.S. had 32 launches (with 37 payloads), including two *Pioneer* launches to take pictures and conduct detailed scientific examinations of Venus. Of these, 12 (with 14 payloads) were launches by DOD. The remaining 20 with 23 payloads were NASA launches. Of these, 10 were scientific payloads (including the Cameo barium and lithium-cloud experiment payload carried in the rocket body of the *Nimbus 7* and the interplanetary *Pioneer* launchers).

Payloads launched for others in 1978 included two for DOD (a piggyback classified payload in March and FltSatCom 1 in Feb.); three comsats for ComSatCorp (Intelsat IV-A F-3 in Jan., Intelsat IV-A F-6 in Mar., and *Comstar 3* in June); two for NOAA (*Goes 3* in June and *Tiros N* in Oct.); and six international satellites (in Mar. *Oscar 8,* a cooperative effort by U.S., Canada, Japan, and West Germany; *BSE* for Japan in Apr.; *OTS 2* in May and *Geos 2* in July for ESA; *Nato IIIC* in Nov. for NATO; and *Anik 4* in Dec. for Canada).

These tables usually categorize vehicle and payload performance as S for successful, P for partially successful, or U for unsuccessful. A fourth category (Unk) would indicate payloads that did not operate because of vehicle failure. These categories, which are unofficial, do not take into account that U missions might produce valuable information, or that payloads with a long-life design might later fail to meet the design requirements and might then become officially unsuccessful. Futher information on these launches appears in Appendix A and in the indexed entries in the text.

MAJOR NASA LAUNCHES, 1978

Date	Name (NASA Code)	General Mission	Launch Vehicle (Site)	Performance Vehicle	Performance Payload	Remarks
Jan. 7	*Intelsat IVA F-3*	To launch satellite into transfer orbit. Satellite to provide 6250 two-way voice circuits plus two television channels simultaneously or a combination of telephone, TV, and other forms of communication traffic.	Atlas-Centaur ETR	S	S	Launched by NASA for ComSatCorp., manager of Intelsat; fifth in a series of improved Intelsat IVA spacecraft with almost two-thirds greater communications capacity than *Intelsat IV* type.
Jan. 26	*IUE 1*	To launch satellite into elliptical geosynchronous orbit; orbiting ultraviolet observatory to investigate stellar atmospheres and interplanetary medium, celestial objects of different galaxies, and quasars.	Delta ETR	S	S	Launched into successful transfer orbit; spacecraft placed above South America.
Feb. 9	*Fltsatcom 1*	To place DOD satellite in successful transfer orbit.	Atlas-Centaur ETR	S	S	First of four planned satellites launched by NASA for the Navy and the Dept. of Defense.
Mar. 5	*Landsat 3*	To acquire multispectral imagery over the U.S. and foreign countries to improve remote-sensing interpretive techniques.	Delta WTR	S	S	Third operational U.S. earth-resources survey satellite, an improved version of *Landsat 1* launched in 1972 and *Landsat 2* placed in orbit in 1975.
	Oscar 8	To place satellite into a sun-synchronous orbit; to provide continuous radio communications with small amateur ground terminals.	Delta WTR	S	S	Latest in a series of satellites for use by amateurs, replacing *Oscar 6*, and launched by NASA as secondary payload with *Landsat 3*.
	DOD spacecraft	Mission not announced.	Delta WTR	S	S	Another secondary payload with *Landsat 3*, launched for DOD.
Mar. 31	*Intelsat IVA F-6*	To launch satellite into successful transfer orbit from which synchronous orbit can be achieved.	Atlas-Centaur ETR	S	S	Sixth and last in a series of improved satellites launched by NASA for ComSatCorp, manager of Intelsat, as a domestic communications satellite in geostationary orbit.

Date	Name	Purpose	Launch Vehicle		Remarks
Apr. 7	Yuri (BSE)	To launch satellite into synchronous transfer orbit and is designed to conduct experiments in direct broadcast television transmission and reception technology.	Delta ETR	S	Japanese Broadcast Satellite-Experimental (Japan/BSE) launched by NASA for the National Space Development Agency (NASDA).
Apr. 26	HCMM (AEM 1)	To launch spacecraft into near-earth, circular sun-synchronous orbit to conduct research into the feasibility of using day/night thermal imagery to apparent thermal inertial values and temperature cycle data.	Scout WTR	S	Heat Capacity Mapping Mission (HCMM), also known as Applications Explorer Mission A (AEM-A), satellite is first in a series to operate in special orbits that satisfy mission-unique experimental data acquisition requirements.
May 11	OTS 2	To test concepts to be used in providing satellite links in the 1980s for routing of portions of the intra-European telephone, telegraph, and telex traffic as well as television relay for Western Europe.	Delta ETR	S	Launched by NASA for the European Space Agency (ESA) as a replacement for OTS-A, which was destroyed when its launcher exploded on Sept. 13, 1977.
May 20	Pioneer Venus 1	To determine the structure of upper atmosphere and ionosphere of Venus, to observe characteristics of atmosphere and surface, to measure gravitational field, and to detect gamma ray bursts.	Atlas-Centaur ETR	S	Spacecraft took pictures of Venus and conducted detailed scientific examination of the planet from orbit.
June 16	GOES 3	To launch spacecraft into synchronous orbit of sufficient accuracy to enable satellite to provide continuous observations of the earth's atmosphere on an operational basis.	Delta ETR	S	Third operational spacecraft of a series of Geostationary Operational Environmental Satellites launched by NASA for NOAA.
June 27	SeaSat 1	Experimental ocean satellite designed to determine whether microwave instruments are suitable for measuring the surface conditions of the seas.	Atlas Agena D or Atlas F WTR	U	All sensors turned on simultaneously July 7; spacecraft ceased transmitting data Oct. 9, and subsequent attempts to reestablish contact were unsuccessful.
June 29	Comstar 3	To provide 14 000 two-way, high-quality voice circuits.	Atlas-Centaur ETR	S	Launched by NASA for Comsat General Corp.; U.S. domestic telephone-communications satellite with a design life of seven years.
July 14	Geos 2	Designed to make scientific studies of the magnetosphere.	Delta ETR	S	Launched by NASA for ESA to perform same experiments as Geos 1, which failed to achieve satisfactory orbit.

APPENDIX B

Date	Name (NASA Code)	General Mission	Launch Vehicle (Site)	Performance Vehicle	Performance Payload	Remarks
Aug. 8	*Pioneer Venus 2*	To determine the nature and composition of clouds of Venus and the composition, structure, and general circulation pattern of the atmosphere.	Atlas-Centaur ETR	S	S	Spacecraft (cylindrical bus carrying four atmospheric entry probes, all of which reached Venus's surface) burned up in Venus atmosphere.
Aug. 12	*ISEE 3*	To obtain detailed measurements of the solar wind and its fluctuations.	Delta ETR	S	S	Third spacecraft in a series of three International Sun-Earth Explorers; inserted into a halo orbit.
Oct. 13	*Tiros N*	To measure temperature and humidity in earth's atmosphere, surface temperature, surface and cloud cover, water-ice-moisture boundaries, and proton and electron flux near earth.	Atlas F WTR	S	S	Launched by NASA for NOAA; first of eight third-generation operational meteorological polar-orbiting spacecraft; primary data source for Global Atmospheric Research Program's Global Weather Experiment, international cooperative program involving some 140 countries.
Oct. 24	*Nimbus 7*	Designed to study the atmosphere's physical characteristics and to search for evidence of both natural and manmade pollutants.	Delta WTR	S	S	Last in a series of Nimbus meteorological satellites first launched Aug. 1964.
	Cameo	To study solar energy and plasma flows and electric fields in Arctic regions through release of barium and lithium in and above the earth's ionosphere.	Delta WTR	S	S	Project Cameo (Chemically Active Materials Ejected in Orbit) launched as a piggyback payload with *Nimbus 7*.
Nov. 13	*HEAO 2*	A high-energy astronomical observatory intended to detect x-rays from objects two-thirds of the way toward the edge of the universe.	Atlas-Centaur ETR	S	S	Second in a series of three NASA high-energy astronomical observatories, made more detailed observations of x-ray sources identified by *HEAO 1*.
Nov. 19	*NATO III C*	To place satellite into transfer orbit that will allow ultimate placement into a stationary synchronous orbit for communications.	Delta ETR	S	S	Third and last of a series of NATO communications satellites launched by NASA, intended to provide backup services for earlier ones.
Dec. 16	*Anik 4 (Telesat D)*	Intended to provide transmission of television, voice, and other data throughout Canada.	Delta ETR	S	S	Second-generation satellite, fourth in a series orbited for Canada's domestic communications-satellite system.

Appendix C
MANNED SPACEFLIGHTS, 1978

Manned flight activity worldwide in 1978 consisted of five launches, all in the USSR Soyuz series. All crews docked successfully with *Salyut 6*. These missions carried the first Czech, Polish, and GDR cosmonauts. The crew of *Soyuz 29*, who returned in *Soyuz 31*, set a duration record of 3350hr. 48min.

At the end of 1978, the U.S. had made 31 manned space flights (2 suborbital, 20 in earth orbit, 3 in lunar orbit, and 6 lunar landings) with a total of 43 different crewmen. The USSR had made 41 manned flights, all in earth orbit, with 54 cosmonauts.

MANNED SPACEFLIGHTS, 1978

Date (Ship) Launched	Recovered	Designation; Crew	Weight (km)	Duration	Remarks
Jan. 10	Mar. 16	Soyuz 27 (Pamir) Vladimir Dzhanibekov Oleg Makarov	6570 (est.)	(ship) 1555h 53min (crew) c. 140h	Carried a second crew to enter Jan. 11 the *Salyut 6* orbiting laboratory, already occupied by the crew of *Soyuz 26* docked at the aft port. After 5 days on *Salyut 6*, the *Soyuz 27* crew returned to earth aboard *Soyuz 26*, releasing the aft docking unit for use by cargo ship *Progress 1* that arrived Jan. 22, left Feb. 6. (*Soyuz 26* undocked from *Salyut 6* Jan. 16, landed with the *Soyuz 27* crew the same day.)
Mar. 2	Mar. 10	Soyuz 28 (Zenit) Aleksey Gubarov Vladimir Remek	6570 (est.)	(ship and crew) 190h 17min	Carried the third crew to board the *Salyut 6* orbiting laboratory. The crew included cosmonaut researcher Captain Vladimir Remek of the Czech Army, first non-Soviet or non-American to fly into space. *Soyuz 28* docked with *Salyut 6* aft port Mar. 3; crew transferred on that same day; returned in *Soyuz 28* Mar. 10. (Record-breaking *Soyuz 26* crew returned in *Soyuz 27* Mar. 16.)
June 15	Sept. 3	Soyuz 29 (Foton) Vladimir Kovalenok Aleksandr Ivanchenkov	6570 (est.)	(ship) 1911h 23min (crew) 3350h 45min	Manned spacecraft continuing the occupation of *Salyut 6* begun earlier in the year. *Soyuz 29* crew docked with *Salyut 6* forward port June 16; hosted 2 other crews during 140-day stay; vehicle returned with *Soyuz 31* crew Sept. 3.
June 27	July 5	Soyuz 30 (Kavkaz) Pyotr Klimuk Miroslaw Hermaszewski	6570 (est.)	(ship and crew) 190h 4min	Carried the second international crew, with flight engineer Major Miroslaw Hermaszewski of the Polish Air Force, to *Salyut 6*, already occupied by *Soyuz 29* crew. *Soyuz 30* docked with aft port of *Salyut 6* June 28, undocked and landed July 5; port used by *Progress 2* cargo vessel July 9-Aug. 2, *Progress 3* Aug. 10-21.
Aug. 26	Nov. 2	Soyuz 31 (Yastreb) Valery Bykovsky Sigmund Jähn	6600 (est.)	(ship) 1628h 14min (crew) 188h 49min	Carried the third international crew, including cosmonaut researcher Sigmund Jähn of the German Democratic Republic, to *Salyut 6*. *Soyuz 31* docked Aug. 27 at aft port of *Salyut 6*; crew returned in *Soyuz 29* Sept. 3 after a flight of 7.9 days. (On Sept. 7, *Soyuz 29* crew entered *Soyuz 31*, undocked from *Salyut 6*, and (after the station had been rotated) redocked with the forward docking unit, leaving aft port free for *Progress* cargo-vessel use. *Progress 4* arrived Oct. 6 to refuel station and boost its trajectory; undocked Oct. 24, reentered Oct. 26. *Soyuz 29* crew returned in *Soyuz 31* Nov. 2.)

Appendix D
NASA SOUNDING ROCKET LAUNCHES, 1978

The accompanying table lists the 60 sounding rockets of the Arcas class and above launched by NASA in 1978. Launch sites were in Antarctica, Norway, Canada, and Sweden, as well as the United States. The rockets carried payloads for the Naval Research Laboratory, SAO, American Science & Engineering, Aerospace Corp., NOAA, GSFC, MSFC, WFC, Norway, and 17 universities. Types of onboard experiments included 28 in plasma physics, 4 in solar physics, 9 in aeronomy, 7 in astrophysics, 4 in astronomy, 4 in meteorology, 2 in stratosphere research, 1 in space processing, and 1 vehicle systems test.

Information for the table came from Goddard Space Flight Center's Quick-Look Sounding Rocket data sheets, issued after launches, with additional information from some of the experimenters concerned. Launch dates are in local time.

NASA SOUNDING ROCKET LAUNCHES, 1978

Launch Date	Rocket, NASA Designation, Launch Site	Apogee (km)	Remarks
Jan. 1	Super Arcas 15.161UE Siple Station, Antarctica	?	Univ. of Houston plasma-physics experiment.
	Super Arcas 15.162UE Siple Station, Antarctica	?	Univ. of Houston plasma-physics experiment.
	Nike Apache 14.533UE Wallops Flight Center, Va.	196.5	Univ. of Ill. plasma-physics experiment.
Jan. 4	Super Arcas 15.160UE Siple Station, Antarctica	250 ft.	Univ. of Houston plasma-physics experiment; failed.
Jan. 9	Astrobee-F 25.029GA White Sands Missile Range, N.M.	163	GSFC aeronomy experiment.
Jan. 10	Astrobee F 25.012UH White Sands Missile Range, N.M.	196.3	SAO high-energy astrophysics experiment; partial failure.
Jan. 12	Super Arcas 15.163UE Siple Station, Antarctica	?	Univ. of Houston plasma-physics experiment.
Jan. 30	Nike Tomahawk 18.211UE Andoya, Norway	220	Univ. of Md. plasma-physics experiment.
Jan. 31	Nike BBVC 27.029CS White Sands Missile Range, N.M.	273	American Science and Engineering solar-physics experiment.

Date	Vehicle	Altitude (km)	Experiment
Feb. 2	Terrier Malemute 29.008UE Poker Flat Range, Al.	616	Univ. of Wisc. plasma-physics (auroral) experiment.
Feb. 4	Aerobee 200A 26.063UH White Sands Missile Range, N.M.	28.5 (sm)	Univ. of Wisc. high-energy astrophysics (cosmic x-ray) experiment; failed.
Feb. 10	Black Brant VB 21.056UG White Sands Missile Range, N.M.	147.9 (sm)	Johns Hopkins Univ. galactic-astronomy experiment.
Feb. 13	Black Brant VC 21.042DS White Sands Missile Range, N.M.	210	NRL solar-physics experiment.
Feb. 27	Nike Tomahawk 18.1017UE Poker Flat Range, Al.	?	Univ. of Al. plasma-physics (barium ion injection) experiment.
Feb. 28	Aerobee 170 13.136DA White Sands Missile Range, N.M.	222.7	NRL aeronomy experiment.
Feb. 28	Nike Tomahawk 18.1015UE Poker Flat Range, Al.	?	Cornell Univ. plasma-physics experiment.
Mar. 2	Nike Tomahawk 18.1016UE Poker Flat Range, Al.	?	Cornell Univ. plasma-physics experiment.
Mar. 8	Astrobee F 25.023UH White Sands Missile Range, N.M.	189.5	MIT high-energy astrophysics (x-ray mapping) experiment.
Mar. 9	Terrier Malemute 29.007UE Poker Flat Range, Al.	336	Rice Univ. plasma-physics (magnetospheric) experiment.
Mar. 10	Nike Tomahawk 18.1018UE Poker Flat Range, Al.	?	Univ. of Al. plasma-physics experiment.

Launch Date	Rocket, NASA Designation, Launch Site	Apogee (km)	Remarks
Mar. 12	Taurus Orion 33.001UA Fort Churchill, Canada	204.4	Univ. of Pitt. plasma-physics (aeronomy) experiment.
Mar. 13	Astrobee F 25.031UE CRR	198	Univ. of Mich. plasma-physics (energetic-particle flux and UV radiation) experiment.
Mar. 21	Super Arcas 15.165GA Poker Flat Range, Al.	62	GSFC aeronomy experiment.
Mar. 27	Super Arcas 15.167GA Poker Flat Range, Al.	67	GSFC aeronomy experiment; partial failure.
	Super Arcas 15.169GA Poker Flat Range, Al.	67	GSFC aeronomy experiment.
	Nike Tomahawk 18.215GM Poker Flat Range, Al.	219	GSFC meteorology (auroral energetics) experiment.
	Super Arcas 15.170UE Poker Flat Range, Al.	75	Penn. State Univ. plasma-physics experiment.
Mar. 29	Super Arcas 15.164GA Poker Flat Range, Al.	63	GSFC aeronomy experiment.
	Super Arcas 15.166GA Poker Flat Range, Al.	62	GSFC aeronomy experiment.

Date	Vehicle / Location		Experiment
Mar. 29	Super Arcas 15.168GA Poker Flat Range, Al.	60	GSFC aeronomy experiment.
	Nike Tomahawk 18.214GM Poker Flat Range, Al.	?	GSFC meteorology (auroral energetics) experiment.
Mar. 30	Super Arcas 15.171UE Poker Flat Range, Al.	57.8	Penn. State Univ. plasma-physics (electric field measurements) experiment.
	Taurus Orion 33.002GA CRR	179	GSFC plasma-physics (aeronomy) experiment.
Apr. 9	Nike BBVB 27.010AE CRR	247	NOAA plasma-physics (electric fields) experiment.
Apr. 11	Nike BBVC 27.026UH White Sands Missile Range, N.M.	303	Univ. of Calif., Berkeley, high-energy astrophysics (spectroscopic) experiment.
Apr. 13	Nike Orion 31.006UE Esrange, Sweden	154	Johns Hopkins Univ. plasma-physics experiment.
May 6	Aerobee 170A 13.137UH White Sands Missile Range, N.M.	200.2	Univ. of Wisc. high-energy astrophysics (cosmic x-ray background) experiment.
May 15	Astrobee F 25.026DG White Sands Missile Range, N.M.	211	NRL galactic-astronomy (x-ray UV) experiment.
June 11	Nike Apache 14.539UE Wallops Flight Center, Va.	219 709 (m)	Cornell Univ. plasma-physics experiment.
June 15	Nike Tomahawk 18.1023UE White Sands Missile Range, N.M.	265.3	Univ. of Tex., Dallas, plasma-physics experiment.

APPENDIX D

Launch Date	Rocket, NASA Designation, Launch Site	Apogee (km)	Remarks
June 20	Nike Apache 14.543UE Wallops Flight Center, Va.	184	Univ. of Ill. plasma-physics experiment.
June 26	Nike Tomahawk 18.1019CE Wallops Flight Center, Va.	314.7	Aerospace Corp. plasma-physics experiment.
July 20	Astrobee F 25.037UH White Sands Missile Range, N.M.	190	SAO high-energy astrophysics experiment.
Aug. 14	Nike BBVC 27.034DS White Sands Missile Range, N.M.	220	NRL solar-physics (high-resolution telescope spectrograph) experiment; failed.
Aug. 15	Super Arcas 15.158UU Wallops Flight Center, Va.	61.5	Univ. of Mich. stratosphere research.
Aug. 29	Super Arcas 15.157UU Wallops Flight Center, Va.	63	Univ. of Mich. stratosphere (nitric oxide) research.
Sept. 11	Black Brant V 21.046NP White Sands Missile Range, N.M.	165	MSFC space-processing experiment.
Sept. 20	Astrobee F 25.028UL White Sands Missile Range, N.M.	224	Univ. of Colo. lunar and Venusian astronomy (UV) research.
Sept. 26	Astrobee F 25.042GG White Sands Missile Range, N.M.	231	GSFC galactic-astronomy (UV) research.

Sept. 27	Nike Apache 14.542UE Wallops Flight Center, Va.	182.6	Univ. of Ill. plasma-physics experiment.
Oct. 31	Taurus Tomahawk 12.1005WT Wallops Flight Center, Va.	554.7	WFC vehicle-system test.
Nov. 3	Aerobee 170A 13.135UE White Sands Missile Range, N.M.	151	Univ. of Mich. plasma-physics (atomic oxygen in night glow) experiment.
Nov. 16	Aerobee 170A 13.138GS White Sands Missile Range, N.M.	?	GSFC solar-physics experiment.
Nov. 27	Nike Tomahawk 18.207IE Andoya, Norway	?	GSFC/NDRE, Norway, plasma-physics (auroral) experiment.
	Nike Tomahawk 18.216IE Andoya, Norway	?	GSFC/NDRE, Norway, plasma-physics (auroral) experiment.
Dec. 1	Astrobee F 25.038UL White Sands Missile Range, N.M.	238	Johns Hopkins Univ. lunar and planetary (Jupiter) astronomy research.
Dec. 11	Nike Tomahawk 18.102UA White Sands Missile Range, N.M.	228	Univ. of Minn. plasma-physics (aeronomy) experiment.
	Astrobee F 25.001UH White Sands Missile Range, N.M.	163.5	Calif. Inst. of Tech. high-energy astrophysics (stellar x-ray sources) research.
Dec. 15	Super Arcas 15.149GM White Sands Missile Range, N.M.	70	GSFC meteorology (ozone) research.
	Super Arcas 15.197GM Wallops Flight Center, Va.	—	GSFC meteorology research.

Appendix E
ABBREVIATIONS OF REFERENCES

Listed here are the abbreviations used for citing sources in the text. Not all the sources are listed, only those that are abbreviated.

AAAS Bull	American Association for the Advancement of Science's *AAAS Bulletin*
A&A	American Institute of Aeronautics and Astronautics' magazine, *Astronautics & Aeronautics*
A&A 1977	NASA's *Astronautics and Aeronautics, 1977: A Chronology*
ABC	American Broadcasting Company
AEC Release	Atomic Energy Commission news release
Aero Daily	*Aerospace Daily* newsletter
Aero Med	*Aerospace Medicine* magazine
AF Mag	Air Force Association's *Air Force Magazine*
AFHF Newsletter	*Air Force Historical Foundation Newsletter*
AFJ	*Armed Forces Journal* magazine
AFSC *Newsreview*	Air Force Systems Command's *Newsreview*
AFSC Release	Air Force Systems Command news release
AIA Release	Aerospace Industries Association of America news release
AIAA *Facts*	American Institute of Aeronautics and Astronautics' *Facts*
AIAA Release	American Institute of Aeronautics and Astronautics news release
AIP *Newsletter*	American Institute of Physics *Newsletter*
AP	Associated Press news service
ARC *Astrogram*	NASA Ames Research Center's *Astrogram*
Astro Journ	American Astronomical Society's *Astrophysical Journal*
Atlanta JC	*Atlanta Journal Constitution* newspaper
AvWk	*Aviation Week & Space Technology* magazine
B News	*Birmingham News* newspaper
B Sun	*Baltimore Sun* newspaper
Bull Atom Sci	Education Foundation for Nuclear Science's *Bulletin of the Atomic Scientists*
Bus Wk	*Business Week* magazine
C Daily News	*Chicago Daily News* newspaper
C Trib	*Chicago Tribune* newspaper
Can Press	Canadian Press news service
CBS	Columbia Broadcasting System

C&E News	*Chemical & Engineering News* magazine
Cl *PD*	Cleveland *Plain Dealer* newspaper
Cl Press	*Cleveland Press* newspaper
Columbia J Rev	*Columbia Journalism Review* magazine
ComSatCorp Release	Communications Satellite Corporation news release
CQ	*Congressional Quarterly*
CR	*Congressional Record*
CSM	*Christian Science Monitor* newspaper
CTNS	Chicago Tribune News Service
D News	*Detroit News* newspaper
D Post	*Denver Post* newspaper
DASA Release	Defense Atomic Support Agency news release
DFRC	See FRC.
DJ	Dow Jones news service
DOC PIO	Department of Commerce Public Information Office
DOD Release	Department of Defense news release
DOT Release	Department of Transportation news release
EOP Release	Executive Office of the President news release
ESA Release	European Space Agency news release, use dated (not numbered)
FAA Release	Federal Aviation Administration news release
FBIS—Sov	Foreign Broadcast Information Service, Soviet number
FonF	*Facts on File*
FRC Release	Flight Research Center news release, after 8 Jan. 1976, became Dryden Flight Research Center (DFRC) news release
FRC *X-Press*	NASA Flight Research Center's *X-Press*
GE Forum	*General Electric Forum* magazine
Goddard News	NASA Goddard Space Flight Center's *Goddard News*
GSFC Release	NASA Goddard Space Flight Center news release
GSFC *SSR*	NASA Goddard Space Flight Center's *Satellite Situation Report*
GT&E Release	General Telephone & Electronics news release
H Chron	*Houston Chronicle* newspaper
H Post	*Houston Post* newspaper
INTELSAT Release	Intl. Telecommunications Satellite Org. news release
JA	*Journal of Aircraft* magazine
JPL *Lab-Oratory*	Jet Propulsion Laboratory's *Lab-Oratory*

JPL Release	Jet Propulsion Laboratory news release
JPRS	Department of Commerce Joint Publications Research Service
JSC Release	NASA Lyndon B. Johnson Space Center (Manned Spacecraft Center until 17 Feb. 1973) news release
JSC *Roundup*	NASA Lyndon B. Johnson Space Center's *Space News Roundup*
JSR	American Institute of Aeronautics and Astronautics' *Journal of Spacecraft and Rockets* magazine
KC Star	*Kansas City Star* newspaper
KC Times	*Kansas City Times* newspaper
KSC Release	NASA John F. Kennedy Space Center news release
LA *Her-Exam*	Los Angeles *Herald-Examiner* newspaper
LA Times	*Los Angeles Times* newspaper
Langley Researcher	NASA Langley Research Center's *Langley Researcher*
LaRC Release	NASA Langley Research Center news release
LATNS	Los Angeles Times News Service
LeRC Release	NASA Lewis Research Center news release
Lewis News	NASA Lewis Research Center's *Lewis News*
M HER	*Miami Herald* newspaper
M News	*Miami News* newspaper
M Trib	*Minneapolis Tribune* newspaper
Marshall Star	NASA George C. Marshall Space Flight Center's *Marshall Star*
MJ	*Milwaukee Journal* newspaper
MSFC Release	NASA George C. Marshall Space Flight Center news release
N Hav Reg	*New Haven Register* newspaper
N News	*Newark News* newspaper
N Va Sun	*Northern Virginia Sun* newspaper
NAA *News*	National Aeronautic Association *News*
NAA Record Book	National Aeronautic Association's *World and U.S.A. National World Aviation—Space Records*
NAC Release	National Aviation Club news release
NAE Release	National Academy of Engineering news release
NANA	North American Newspaper Alliance
NAS Release	National Academy of Sciences news release
NAS—NRC Release	National Academy of Sciences—National Research Council news release

NAS—NRC—NAE *News Rpt*	National Academy of Sciences—National Research Council—National Academy of Engineering *News Report*
NASA Actv	*NASA Activities*
NASA anno	NASA announcement
NASA GMR	NASA Headquarters "General Management Review Report"
NASA HHR—39	NASA Historical Report No. 39
NASA Hist Off	NASA History Office
NASA Hq *WB*	NASA Headquarters *Weekly Bulletin*
NASA Int Aff	NASA Office of International Affairs
NASA *LAR*, XIII/8	NASA *Legislative Activities Report*, Vol. XIII, No. 8
NASA Leg Off	NASA Office of Legislative Affairs
NASA MOR	NASA Headquarters Mission Operations Report, preliminary prelaunch and postlaunch report series (information may be revised and refined before publication)
NASA prog off	NASA program office (for the program reported)
NASA proj off	NASA project office (for the project reported)
NASA Release	NASA Headquarters news release
NASA Rpt SRL	NASA report of sounding rocket launching
NASA SP-4019	NASA Special Publication No. 4019
Natl Obs	*National Observer* magazine
Nature	*Nature Physical Science* magazine
NBC	National Broadcasting Company
NGS Release	National Geographic Society news release
NMI	NASA Management Instruction
NN	NASA Notice
NOAA Release	National Oceanic and Atmospheric Administration news release
NRL Release	Naval Research Laboratory news release
NSC Release	National Space Club news release
NSC *News*	National Space Club *News*
NSC *Letter*	National Space Club *Letter*
NSF *Highlights*	National Science Foundation's *Science Resources Studies Highlights*
NSF Release	National Science Foundation news release
NSTL Release	NASA National Space Technology Laboratories news release
NY News	*New York Daily News* newspaper
NYT	*New York Times* newspaper

NYTNS	New York Times News Service
O Sen Star	*Orlando Sentinel Star* newspaper
Oakland Trib	*Oakland Tribune* newspaper
Omaha W-H	*Omaha World-Herald* newspaper
ONR *Rev*	Navy's Office of Naval Research *Reviews*
P *Bull*	Philadelphia *Evening* and *Sunday Bulletin* newspaper
P Inq	*Philadelphia Inquirer* newspaper
PAO	Public Affairs Office
PD	National Archives and Records Service's *Weekly Compilation of Presidential Documents*
PIO	Public Information Office
PMR *Missile*	USN Pacific Missile Range's *Missile*
PMR Release	USN Pacific Missile Range news release
Pres Rpt 74	*Aeronautics and Space Report of the President: 1974 Activities*
SAO Release	Smithsonian Astrophysical Observatory news release
D/SBD	*Defense/Space Business Daily* newspaper
Sci Amer	*Scientific American* magazine
Sci & Govt Rpt	*Science & Government Report*, independent bulletin of science policy
SciServ	Science Service News service
SD	*Space Digest* magazine
SD Union	*San Diego Union* newspaper
SET Manpower Comments	Scientific Manpower Commission's *Scientific, Engineering, Technical Manpower Comments*
SF	British Interplanetary Society's *Spaceflight* magazine
SF Chron	*San Francisco Chronicle* newspaper
SF Exam	*San Franciso Examiner* newspaper
Sov Aero	*Soviet Aerospace* newsletter
Sov Rpt	Center for Foreign Technology's *Soviet Report* (translations)
SP	*Space Propulsion* newsletter
Spaceport News	NASA John F. Kennedy Space Center's *Spaceport News*
Spacewarn	IUWDS World Data Center A for Rockets and Satellites' *Spacewarn Bulletin*
SR list	NASA compendium of sounding rocket launches
SSN	*Soviet Sciences in the News*, publication of Electro-Optical Systems, Inc.

St Louis G-D	*St. Louis Globe-Democrat* newspaper
St Louis P-D	*St. Louis Post-Dispatch* newspaper
T-Picayune	New Orleans *Times-Picayune* newspaper
Tech Rev	Massachusetts Institute of Technology's *Technology Review*
Today	*Today* newspaper
testimony	Congressional testimony, prepared statement
text	Prepared report or speech text
transcript	Official transcript of news conference or congressional hearing
UN Reg	United Nations Public Registry of Space Flight
UPI	United Press International news service
USGS Release	U.S. Geological Survey news release
USPS Release	U.S. Postal Service news release
W Post	*Washington Post* newspaper
W Star-News	*Washington Star-News* newspaper
WFC Release	NASA Wallops Flight Center news release
WH Release	White House news release
WJT	*World Journal Tribune* newspaper
WSJ	*Wall Street Journal* newspaper

Index

Aborted flights, 106
Abourezk, James, 196
Abruzzo, Ben, 199
Acton, Loren, 94, 184
Acurex, Inc., 249
Adams, Brock, 32
AEG Telefunken, 171
Aeritalia, 33
Aero Club of Great Britain, 101
Aerospace Corp., 50
Aerospace Industries Association, 32, 100, 314
Africa, 3, 5
AFSatCom, 11, 26
Agency for International Development, 38, 58, 175
Agriculture, 69, 74, 97, 145, 187, 188, 272
Agriculture, Department of, 1
Air Force, 8, 11, 17, 22, 25, 26, 35, 82, 100, 113, 118, 153, 157, 203
   aborted mission, 64
   Aeronautical Systems Division, 312
   Avionics Laboratory, 286
   contractors, 28, 45, 46, 90
   contracts, 179
   Flight Dynamics Laboratory, 63
   Geophysics Laboratory, 311
   Project Blue Book, 74
   Space and Missile Systems Organization, 90, 135, 261, 286, 313
   Space and Missile Test Command, 144
Air Force Geophysical Laboratory, 235
Air-launched cruise missile, 5
Aircraft
   civilian, 8, 32, 65, 70, 72, 154
   design, 8, 43, 96, 209, 230
   experimental, 101, 139, 157, 174, 225, 255
   Gossamer Condor, 45
   military, 9
   research, 64, 87, 89, 92, 301
   supersonic, 43, 296, 311
   testing, 56
   transport, 5, 35, 41, 47, 200, 255, 301
Aircraft Energy Efficiency Program, 41
Airlines, U.S., 206
Airports, 39, 42, 59, 72, 121, 126, 173
   Gosselies, 5
   Kjevik, 5
   Nairobi, 5
   Yoff, 5

Albers, James, 296
Aleutian, Islands, 71
Alexander, Joseph, 192
Alfa 2 (launch vehicle), 33
Allen, Joseph, 178
Allenby, Richard, 21
Aller, Robert, 307
Allnutt, Robert, 135, 231
Alloys, 36
American Association for the Advancement of Science, 35, 91
American Astronautical Society, 91
American Astronomical Society, 268
American Indians, 81
American Institute of Aeronautics and Astronautics, 10, 35, 68, 91, 99, 108, 142
American Society of Engineering Education, 200
American Telephone & Telegraph Co., 113, 137
American Water Resources Association, 236
Ames Industrial Corp., 43
Ames Research Center, 17, 80
  contract, 111
  missions, 270, 282, 293
  people, 167, 263
  programs, 39, 43, 64, 86, 190, 233
Amsat, 206
Amsat Oscar, 53
Anderson, Arthur, 74
Anderson, Maxie, 199
Andrews, Bill, 68
Anik Al, 289
ANS (satellite, Netherlands), 163
Antarctic, 43
Antennas, 10, 108
  ground-based, 26, 71, 106, 155, 202, 246, 305
  space-based, 3, 109, 149, 158, 160
Anti-gravity forces protection, 312
Apollo 8, 279
Apollo 11, 79
Apollo 13, 98
Apollo 15, 179
Apollo 16, 168
Apollo-Soyuz, 10, 42, 79, 160
APPLE (India), 207
Apple (satellite), 304
Applications Explorer Missions, 95
Applications Systems Verification and Transfer, 54
Applications Technology Satellite 6, 192
Aquarius (command module), 98
Arco Solar, Inc., 240
Argentina, 12, 273
Ariane, 7, 20, 59, 93, 140, 153, 206, 216, 223, 280, 294, 297, 304
Ariel-5, 60
Armstrong, Neil, 77, 224
Army, 8, 17, 64, 86, 113, 235
  Ballistic Missile Agency, 16

ARO, Inc., 205
Association for the Space Science Exposition, 79
Asteroids, 211
Astrobee F (rocket), 311
Astrometry, 20
Astronauts, 7, 27, 35, 39, 42, 56, 61, 70, 74, 94, 98, 106, 129, 157, 172, 191, 209, 224, 231, 238
  pilots, 8
  scientists, 8
Astronomy, 15, 35, 70, 78, 95, 132, 145, 165, 167, 168, 184, 186, 190, 193, 214, 239, 243, 275
Atlas (rocket), 2, 5, 89, 94, 121, 123, 144, 210
Atlas-Agena, 82
Atlas-Centaur, 2, 3, 5, 25, 86, 94, 96, 136, 143, 168, 173, 194, 262, 268, 272, 297
ATMOSAT project, 50
Atmosphere Explorer C, 192
Atmospheric cloud physics laboratory, 95
Atmospheric variability experiments, 109
ATS-6 (satellite), 79
Australia, 5, 6, 92, 145, 272
Automatic pilot-advisory system, 97
Aviation Regulatory Reform Act, 311
Aviation Week, 2
Aviation/Space Writers Association, 32
AWACS (aircraft), 204
Awards, 26, 35, 45, 49, 67, 82, 99, 101, 103, 107, 136, 142, 152, 178, 185, 200, 216, 224, 228, 230, 235, 238, 243, 247, 252, 263, 299
B-1 bomber, 9, 75
BAe Dynamics Group (UK), 132
Bahcall, J. N., 109
Bainbridge, William, 154
balloons, 37, 109, 199
Barber, Paul, 199
Bartoe, John-David, 94, 184
Bell Helicopter Textron, 64
Bell-Northern Research Ltd., 39
Beloit College, 109
Bendix Corp., 202, 296
Benham, Harold, 154
Berezhnoy, Igor, 42
Berger, Harold, 292
Big Bird (satellite), 266
Bioengineering, 52
Biology, 39, 43, 49, 70, 138
Blackstock, Thomas, 103
Blamont, Jacques, 50
Blocker, Truman, 114
Blue Ridge Electrical Membership Corp., 171
Bluford, Guion, 8
Boeing Aerospace Co., 5, 35, 89, 90, 96, 156, 249, 255, 303, 313
Boeing Engineering and Construction Co., 249
Bond, Langhorne, 5, 20
Borman, Frank, 224
Botany, 43

Bowen, Fred, 107
Boyd, John 252
Boynton, Judd, 268
Brand, Vance, 61
Brandenstein, Daniel, 8
Brazil, 12, 119, 273
Bristow, Frank, 87
British Aerospace Dynamics Corp., 111, 143, 285
British Civil Aviation Authority, 108
British Interplanetary Society, 91, 131
British Science Research Council, 186
Brown, Harold, 1, 130
Brown, Richard, 110
Brzezinski, Zbigniew, 19
BSE satellite, 97
Buchli, James, 8
Buckner, Horst, 32
Bulgaria, 49, 263
Bunker Ramo Corp., 179
Bykovsky, Valery, 200
C-5 aircraft, 5
C-130 aircraft, 5
C-141 aircraft, 5
Cable & Wireless, Ltd. (UK), 68
Calder, Peter, 171
Caldwell, J. J., 109
California Institute of Technology, 152
Calio, Anthony, 214
Calspan Corp., 151
CalTech, 27, 122, 244
Canada, 4, 5, 12, 18, 38, 39, 94, 121, 202, 223, 272, 275, 294
Canaveral Council of Technical Societies, 195
Cape Canaveral, 5, 22, 62, 64, 82, 88, 93, 100, 268
Carlos, H.M. Juan, 109
Carr, Gerald, 224
Carter, Jimmy, 1, 8, 9, 28, 31, 44, 45, 46, 74, 75, 77, 92, 104, 113, 115, 122, 130, 142, 178, 224, 235, 239, 241, 245, 249, 260, 265, 274, 275
Caruso, Andrea, 15
Cassidy, William, 37, 38, 144
Ceallaigh, C. O., 32
Central Intelligence Agency, 1
Centre National d'Etudes Spatiales (France), 20, 50, 140, 210
Chaffee, Roger, 231
Chapman, Kenneth, 4, 116
Charles Stark Draper Laboratory, Inc., 189
Chase Econometrics, 63
Chemical clouds, 255
Cheston, T. Stephen, 79
Chimex Systems, Inc., 181
China, 33, 170
China Navigation Society, 27
Chirivella, Jose, 152
Chitre, NandKishore, 15
Christopher, Warren, 280

Civil Aeronautics Board, 40, 113
Civil Aviation Authority (UK), 20
Civil Service Commission, 44
Clanton, Uel, 187
Clapper, Peter, 46
Clouds, 95, 162, 255
Coast Guard, 121, 176, 293
Coats, Michael, 8
Cohen, Aaron, 191
Comets, 88
Communications
   global systems, 47, 57, 78, 81, 83, 108, 153, 246, 304
   global systems antenna, 10
   lasers, 286
   problems, 203
   submarines, 203
   systems, 14, 25, 39
Communications Technology Satellite, 81
Computer Sciences Corp., 155, 181, 211
Computer Systems, 2, 6, 43, 49, 51, 81, 85, 141, 167, 172, 254
   massive parallelism, 72
   spacecraft, 87, 167
Comsat General, 147
ComSatCorp, 2, 3, 5, 7, 15, 47, 58, 94, 108, 113, 136, 143, 171, 202, 280, 314
Comsats, 64, 80, 81, 83, 93, 108, 121, 136, 159, 164, 166, 261, 275, 277, 284, 291, 304
COMSTAR, 47, 136, 314
Concorde, 8, 43, 59, 171, 311
Condor, Gossamer, 70
Conferences. See meetings, 55
Congress, U.S.
   conference committees, 9
   House of Representatives, 38, 52, 92, 93, 103, 112, 132, 155, 289
   Senate, 75, 81, 92, 103, 112, 155, 289
Conrad, Jr., Charles, 224
Construction in space, 18, 84
Control systems, 6, 15, 43, 202, 263, 272, 277, 312
Copernicus (orbiter), 12, 16, 165
Cornell University, 268
Corning Glass Works, 6
Cosmic rays, 32, 95
Cosmology, 243
Cosmonauts, 7, 13, 31, 36, 49, 56, 85, 150, 176, 193, 200, 224, 254, 261
Cosmos 936, 14
Cosmos 954, 18, 38, 275
Cosmos 1031, 227
Cosmos 1047, 275
Cosmos program, 80, 123, 151, 187
Cost reductions, 1, 9, 35, 30, 42, 84, 105, 178, 201
Counts, Parker, 254
Covert, Eugene E., 75
Covey, Richard, 8
Cox, Roy, 108
Cranston, Alan, 9
Creighton, John, 8

Crippen, Robert, 61
Crow, Duward, 136
Cruise missiles, 100
Cts (satellite), 39, 121, 202
Cuba, 49, 69
Curien, Hubert, 117
Czechoslovakia, 49, 69, 125, 263, 277
Dana, Bill, 249
Data
  analysis, 2, 4, 6, 28, 35, 52, 88, 142, 166, 174, 271, 300
  collection, 4, 7, 34, 40, 42, 60, 77, 95, 96, 109, 143, 192, 198, 210
  management, 298
  processing, 179
  relay, 181, 196
  storage, 10, 179
  systems, 244
  transmission, 21, 57, 84, 145, 171, 223, 247
  users, 282
Data-relay satellite, 155
Day, Roy, 75
de Havilland Aircraft, 174
de Laboulaye, Francois, 280
DeBakey, Michael, 148
Deep Space Network, 67, 219, 274
Defense Advanced Research Projects Agency, 185
Defense Meteorological Satellite Program, 204
Delta (rocket), 5, 15, 16, 40, 88, 93, 94, 97, 110, 116, 146, 221, 254, 261, 275, 277
Department of Agriculture, 31, 57, 82, 188, 273
Department of Commerce, 4, 31
  National Telecommunications and Information Administration, 274
Department of Defense, 1, 11, 18, 21, 25, 26, 29, 31, 75, 77, 89, 115, 116, 122, 126, 155, 186, 203
  Advanced Research Project Agency, 169
Department of Energy, 13, 31, 45, 52, 60, 68, 87, 112, 167, 171, 196, 233, 249
Department of Interior, 1, 31, 38, 57, 82, 236
Department of State, 29
Department of Transportation, 40, 46, 99, 173, 226, 311
Design
  aircraft, 8
  engineering, 2, 19, 247, 291
  instrumentation, 50
  spacecraft, 2, 292, 302
Detroit Diesel Allison Division (GM Corp.), 60
Development Sciences, Inc., 300
Dhawan, S., 164
Dinneen, Gerald, 203
Divita, David, 88
Dixon, Robert J., 82
DNA and RNA, 138
Domingo, D.V., 32
Double Eagle II, 199
Dryden Flight Research Center, 29, 43, 56, 83, 225
Dscs 2 (satellite), 78
DuPont Corp., 50

Dupree, Andrea, 186
Dzhanibekov, Vladimir, 7
E-Systems, Inc., 159
Earhart, Amelia, 83, 107
Eastern Test Range, 86, 169, 286
Economics
  aerospace industry, 4, 314
  aviation industry, 72, 206
  B-1 bomber, 9
  data storage, 10
  European space program, 278
  satellite launch, 245
  see also expenditures, 3
Ecuador, 21
Edelson, Edward, 137
Education, 310
Edwards Air Force Base, 28
Egypt, 164
El-Baz, Farouk, 164
Electric Space Division, 249
Electricity, generated from space, 80
Electronic message systems, 279
Elliott, David, 87
Elliot, Jim, 268
Emme, Eugene, 91
EMR Telemetry, 227
Encke comet, 229
Energy
  conservation, 31, 35, 43, 68, 71, 103, 131, 137, 201, 215, 259, 291, 298
  costs, 220
  efficiency, 41
Energy Materials Corp., 240
England, 4
Engle, Joe, 61
ENTEL, 202
Enterprise (orbiter), 6, 11, 29, 121
Environmental Protection Agency, 31, 58, 145
Environmental Research & Technology, Inc., 314
Environmental Research Laboratory, 17
Eppley Laboratory, Inc., 13
ERNO (W. Germany), 27, 251
Erythrocytes, 298
Escuela Politecnia National (Ecuador), 21
European Broadcast Union, 110
European Space Agency, 5, 7, 11, 15, 16, 20, 27, 33, 34, 57, 59, 83, 84, 88, 94, 109, 111, 129, 140, 143, 150, 153, 159, 165, 177, 183, 186, 198, 206, 212, 218, 254, 265, 278
  expenditures, 93
  Ots 2, 110
European Space Operations Centre, 85, 109, 143, 177, 282
Eutelsat, 160, 198
Evans, John, 67
Exos B, 212
Exosat, 93, 216
Expenditures

comsats, 291
climate research, 31
European Space Agency, 111
Intelsat program, 3, 119
meteosat, 213
MX-missile, 1
National Aeronautics and Space Administration, 152
research and development, total U.S., 103
Space Shuttle, 44, 117, 218, 251
See also Economics and Funding
Experiments
  atmospheric, 111
  Avefria, 111
  biological, 39
  chemical clouds, 255
  GARP, 282
  life sciences, 91
  materials processing, 30, 124, 160, 219, 233
  meteorological, 271, 283
  scientific, 29, 30, 32, 53, 70, 110, 134, 177, 184, 194, 269
Explorer, 166
Explorer 1, 16, 17, 62
Explorer 52, 147
Extraterrestrial life, 116, 144
Extravehicular activity, 50
Fabian, John, 8
Fairchild Space and Electronics Co., 30
Faise, Jr., Fred, 98
Farnham, Lee, 147
Farquhar, Robert, 276
Fastie, W. G., 110
Federal Aviation Administration, 5, 8, 20, 46, 113, 185, 202, 206, 215, 226, 298
Federal Communications Commission, 58, 113
Federal Electric Corp., 286
Federal Republic of Germany, 27, 29, 58, 85, 92, 93, 106, 148, 171, 206. 218, 251
Feoktistov, Konstantin, 85, 193, 227
Ferranti Electronics, Ltd., 246
Fichtel, Carl, 215
Finke, Wolfgang, 117
Finn, Terence, 199
Firewheel (Germany), 206
Fisher, Anna, 8
Fishman, Gerald, 215
Fletcher, James C., 26, 35
Flight safety, 18
Flippo, Ronnie, 45, 251, 289
FltSatCom, 11, 25, 52, 86, 183
Fluorocarbons, 37
Food, 278
Ford, Gerald, 260
Ford Aerospace and Communications Corp., 223, 261, 275
Forest Service, 167
Fowler, James, 204
Fox, Kenneth, 182

France, 4, 20, 39, 50, 92, 119, 140, 210
Friedmann, E. Imre, 43
Frietag, Robert F., 79
Frosch, Robert 11, 12, 38, 44, 74, 75, 81, 97, 142, 164, 191, 223, 231, 238, 247, 255, 259, 292, 311
Frutkin, Arnold, 49, 270
Fullerton, Charles, 61
Fulton, Fitzhugh, 67
Funding, 226
   government, 1, 3, 9, 13, 16, 27, 31, 44, 48, 52, 103, 196, 211, 220, 285
   research and development, 48, 103, 259, 260, 313
   See also Expenditures.
Fuqua, Don, 133, 214, 251, 289
Fury, Joe, 203
Gagarin, Yuri, 85
Galileo project, 12, 91, 148, 161
Gamma rays, 34, 95, 120, 233
   observatory, 214
Garber, Paul, 297
Gardner, Dale, 8
GARP, 282
Garrett Corp., 296
Garriott, Owen, 183
Gazenko, Oleg, 298
Geddes, William, 15
Gemini, 79
Gemini-Titan, 79
General Dynamics Corp., 5, 89, 113, 135, 137, 169, 205, 246
General Electric Co., 40, 52, 96, 114, 147, 148, 159, 171, 213, 215, 219, 225, 277, 298, 305
Geological Survey (U.S.), 187, 314
Geophysics Labs and Research, 268
George Washington University, 107
Geos (satellite), 143, 177, 194
Geos 1 and *2*, 254
Geos 2, 239
Geos 3, 284
German Democratic Republic, 49, 200
Giacconi, Riccardo, 275
Giberson, Gene, 248
Gibson, Edward, 224
Gibson, Robert, 8
Gibson, Roy, 247, 294
Gilliam, Issac, 136
Ginter, R. D. "Duff", 99
Glaser, Peter, 70
Glazer, J. Henry, 79
Glenn, Jr., John, 224
Glissada (USSR), 41
Global Associates, 303
Global Atmospheric Research Program, 116, 146, 182, 210, 299
Global communication systems, See Communications.
Global Weather Experiment, 146, 299
Goddard Space Flight Center, 16
   contractors, 30

missions, 111
people, 21, 192
programs, 40, 70, 96, 110, 181, 205, 210, 213
Godfrey, Frank, 161
Goes (satellite), 109, 161, 146, 181
Goes 1, 142, 282
Goes 3, 283
Goes C, 116
Gold (presence in space), 166
Goldreich, Richard, 268
Gomez, Louis, 195
Gossamer Condor, 45, 101
Grechko, Georgy, 31, 56, 151, 224
Gregory, Frederick, 8, 39
Grey, Jerry, 132
Griffith, Cecilia, 162
Griggs, Stanley, 8
Grissom, Virgil (Gus), 225, 231
Grumman Aerospace Corp., 84, 98, 135
Gubarev, Aleksey, 49
Gurovsky, Nikolai, 14
Gursky, Herbert, 186
H-Sat, 59
Haigh, Henry, 126
Haile, Frank, 154
Haines, Bill, 303
Haise, Fred, 61
Hall, Minard, 21
Halley's comet, 88, 229
HALO experiments, 11
Hannigan, James, 254
Harr, Jr., Karl, 100
Harris Corp., 10, 204
Hart, Terry, 8
Hartsfield, Henry, 139
Harvey, John, 94
Haskins, Glenn, 87
Hauck, Frederick, 8
Hawley, Steven, 8
Hayflick, Leonard, 213
Heacox, William, 166
Head, Sara, 70
Heao, 132
Heao 1, 34, 77, 120, 239
Heao 2, 268, 272, 275, 283
Heao B, 94, 262
Hearth, Donald, 35, 107, 163, 200
Heat-capacity mapping mission, 95, 187
Helios 1, 192
Helios B, 233
Helms, Curtis, 178
Hermaszewski, Miroslaw, 150, 261
Hessberg, Rufus, 14
Heyman, Joseph, 228

HiMat (aircraft), 157
Hinners, Noel, 14, 120, 214, 229, 243, 273, 285, 290
Hoffman, Jeffrey, 8
Honeywell, Inc., 2
Horizont (satellite), 304
Horton, Vic, 68
Horz, Fred, 32
House of Representatives, U.S. See Congress, 75
Hovell, C. Ronald, 23
How-Ren, Pan, 170
Hsnmin, Jen, 284
Hughes Aircraft Co., 38, 45, 47, 96, 124, 148, 186, 312
Hughes Communications Services, Inc., 216, 280
Hughes, Thomas, 91
Hungary, 49, 263
Hussain, Farooq, 17
Hutchinson, Neil, 62
Hydrazine engine, 300
Icing, 185
Ifft, Edward, 179
ILLIAC IV (computer), 167
Imagery, 40, 47, 178
India, 7, 164, 207, 263, 273
Indian Space Research Organization, 304
Industralization in space, 84, 231
Infrared
    astronomical satellite, 144
    imaging, 47
    spectrum, 95, 146
    surveillance, 45
    water-vapor radiometer, 17
INSAT-1, 164
Institute of Electrical and Electronics Engineers, 88
Institute of Space and Aeronautical Science (Japan), 41, 212
Intelsat, 3, 10, 15, 33, 47, 58, 68, 113, 118, 143, 145, 162, 166, 171, 198, 217, 223, 264, 277, 280, 297, 301, 314
    expenditures, 3, 119
Intelsat IV-A, 2
Intelsat IVA F-3, 2, 3
Intelsat IVA F-6, 3
Interceptor satellites, 28, 64, 77, 89, 123, 130, 265, 275
Intercosmos 18, 277
International Academy of Astronautics, 91
International Aeronautic Federation, 126
International Civil Aviation Organization, 6, 20, 92, 107, 173, 203
International cooperation, 92, 97, 111, 117, 146, 165, 177, 183, 188, 198, 202, 210, 238, 280, 294, 299, 307
    U.S. and Europe, 15, 83, 107, 143
International cooperation. See also USSR, 92
International Council of Scientific Unions, 146, 299
International Geophysical Year, 17, 62
International Hydrodynamics Co., 83
International Magnetospheric Study, 143
International Maritime Satellite Observation (INMARSAT), 223

International Telephone and Telegraph, 96
International Ultraviolet Explorer, 15, 16, 34, 70, 186, 243
International Union of Geodesy and Geophysics, 112
Interstellar wind, 32
Ionosphere, 221
Iran, 12
Ireland, 92
ISEE (satellite), 177, 192, 198
ISEE 3, 276
ISEE C, 165
Italy, 12, 33
Itek Optical Systems Division, 222
ITT World Communications, 114
IUE (satellite), 166
Ivanchenkov, Alexander, 150, 176, 254, 261, 298
Jackson State University, 39
Jaehn, Sigmund, 261
Jaffe, Leonard, 106
Jähn, Sigmund, 200
Japan, 3, 5, 12, 41, 51, 94, 97, 119, 149, 198, 212
Japan Science Society, 79
Japanese Broadcasting Corp., 52
Japanese Defense Agency, 28
Japanese Maritime Science Promotion Foundation, 79
Jay, Peter, 280
Jefferys, William, 110
Jennings, Donald, 183
Jet Propulsion Laboratory, 16, 300
  contracts, 249
  missions, 50, 57, 86, 96
  people, 87, 88, 169, 197
  programs, 158, 240
Jian-Cheng, Wu, 170
Johns Hopkins University, 110
Johnson, C. Curtis, 155
Johnson, David, 243
Johnson, Lyndon, 231
Johnson, Marshall, 123
Johnson Space Center, 59
  contractors, 50
  contracts, 222
  people, 8, 12, 172
  programs, 42, 56, 87
Johnston, David, 114
Joint American-Soviet Particle Intercalibration Project, 228
Joint Chiefs of Staff, 1
Jupiter (planet), 86, 91, 105, 148, 191
Jupiter (launch vehicle), 2, 12, 16, 84, 210, 241
Kayex, Corp., 240
KC-135 (aircraft), 35, 225
Keldysh, Mstislav, 148
Kennedy Space Center, 2, 11
  contractors, 6
  contracts, 83, 129, 136, 156

missions, 4, 15, 40, 93, 96, 262
  people, 12
  programs, 7, 27, 33, 49, 111, 295
  Spaceport, 130, 136
Kenya, 90
Kevlar, 50
Keys, John, 223
Killer satellites, 17, 28
Kinberg, Robert, 255
Kirhofer, William, 96
Kirk, Robert, 32
Kissel, Ralph, *120*
Klein, M. J., 197
Klimuk, Pyotr, 150
Kline, Raymond, 214
Kluicke & Soffa Industries, Inc., 240
Kovalenok, Vladimir, 150, 176, 254, 261, 298
Kraft, Jr., Christopher, 114, 165, 191, 231
Kramer, James, 214
Kresge Eye Institute, 14
Kubasov, Valery, 160
Kuhn, Peter, 17
Kuo-feng, Hua, 19
Kurfess, James, 215
L-5 Society, 276
Lambert, K. L., 110
Lampton, Michael, 129, 188
Landing system, 173
Landsat, 57, 67, 68, 166, 187, 205, 244, 272, 298
  contractors, 52
Landsat 1, 12
Landsat 2, 7, 12
Landsat 2 and *3*, 247
Landsat 3, 53, 97
Landsat C, 7, 11, 40
Landsat D, 7, 11, 30, 159
Langley Research Center, 3, 18, 66
  contracts, 85, 189, 237, 277
  people, 103, 107, 161, 171, 213, 217, 299
  programs, 201
Large Area Crop Inventory Experiment, 188, 272
Large Space Systems Technology, 18
Laser-heterodyne radiometer, 37
LaserCom, 287
Lasers, 17, 29, 41, 71, 77, 83, 141, 185, 203, 206, 286
Latapie, Francis, 15
Lawless, James, 137
Learjet, 18
Leasat, 147, 216, 280
Leckrone, David, 166
Leonov, Aleksey, 69
LES 8 and *9* (satellites), 22
Lewandowski, Bohdan, 247
Lewis, Charles, 62, 238

Lewis Research Center, 14
  contractors, 60
  contracts, 2
  missions, 53
  people, 19
  programs, 39, 82, 105, 156
Lichtenberg, Byron, 129
Lilly, William, 75
Lind, Don, 32
Lindberg, Anne Morrow, 107
Lindberg, Charles A., 311
Linsky, Jeffrey, 70
Liquid-fuel rocket, 2, 6, 104, 169
Lissaman, P.B.S., 70
Lixiscope, 190
Lockheed International, 89, 114, 145
Lockheed Missiles and Space Co., 6, 111, 158, 184, 241
Lockheed-Georgia Co., 5
Lomas, Lucius, 264
Lomax, Harvard, 35
Long Duration Exposure Facility, 3, 31, 133
Loomis, Vincent, 83
Lorre, Jean, 169
Los Alamos Scientific Laboratory, 64
Loudenslager, Leo, 126
Louis Pasteur University (France), 278
Lousma, Jack, 161
Lovelace, Alan, 44, 133, 248, 252, 255
Lovell, Jim, 99
Lucas, William, 35
Lucid, Shannon, 8
Lunar and Planetary Institute (Texas), 55
Lunar Curatorial facility, 38
Lundin, Bruce, 248, 254
Lunney, Glynn, 191
Lushina, Louis, 33
Lynn, Donn, 169
MacCready, Paul, 46
Magion, 277
Magnavox, 130
Magnetosphere, 143, 147, 177, 212
Makarov, Oleg, 7
Malaga, Joseph, 195
Malfunctions, 80, 83, 88, 111, 187, 241, 248, 271
Malkin, Mike, 75
Management Services, Inc., 136
Mannen, James, 254
Manring, Lewis, 137
Mapping, 187, 233, 285, 298
Marecs (satellite), 198, 223, 280
Marine Corps, 8
Mariner 9, 266
Marisat, 280, 314
Marks, Alvin, 154

Marots-A, 207
Mars, 44, 50, 56, 88, 163, 174, 266
    White Rock, 267
Mars airplane, 300
Marshall Islands, 83
Marshall Space Flight Center, 10, 11, 16, 22, 25
    contractors, 30, 193
    contracts, 135, 184, 296
    ground control laboratory, 31
    missions, 71
    people, 21, 49, 178
    programs, 6, 7, 13, 15, 27, 34, 41, 109, 141, 160
Martin Marietta Aerospace, 313
Martin Marietta Corp., 13, 45, 65, 78, 90, 113, 119, 156, 159, 294, 296
Mason, Brian, 144
Massachusetts Institute of Technology, 193
Materials processing (in space), 30, 150, 219, 233
Maverick (missile), 312
MB Associates, 240
McBride, Jon, 8
McCall, Robert, 163
McCarthy, John, 19
McDonnell Douglas Astronautics Co., 65
McDonnell Douglas Corp., 17, 46, 54, 89, 247, 296, 313
McMurtry, Tom, 68
McNair, Ronald, 8
McRee, Griffith, 200
McTigue, John, 68
Mechanical Technology, Inc., 68
Media, 249
Medicine, 87, 105, 107
Meetings, 10, 14, 18, 21, 32, 41, 55, 66, 70, 81, 88, 107, 114, 122, 142, 161, 171, 185, 188, 195, 213, 216, 236, 243, 246, 247, 265, 268, 273, 292, 296
Menrad, Wesley, 152
Merbold, Ulf, 188
Mercury (planet), 56
Mercury (spacecraft), 42, 79
Mercury Co., 237
Mercury-Atlas, 79
Mercury-Redstone, 79
Messerschmitt-Bolkow-Blohm, 171
Meteorites, 37, 144
Meteorology, 31, 49, 60, 109, 142, 161, 177, 189, 204, 209, 252, 271, 283, 299
Meteosat system, 129, 213
Meteosat 1, 84, 282
Meteosat 2, 206, 304
Methane (in space), 182
Mexico, 264
Micro Craft, Inc., 66
Micrometeorites, 4, 13, 32, 176
Microwave scanning-beam landing system, 83, 92, 158, 173, 202. See also Time reference scanning-beam/microwave landing system.
Midwest Research Institute, 63
Miller, R. B., 96

Miller, R. H., 35
Miller, Richard, 167
Mineral exploration in space, 80, 166
Mini-Sniffer (aircraft), 300
Minority employment, 7, 8, 39
Mission specialists, 8, 41
Models, 6, 53, 56, 66, 205, 233, 286, 301
Molniya 3 comsat, 13
Molniya program, 10
Mongolia, 49, 69
Monitoring
    earth resources, 40, 53, 95, 121, 145, 158, 159, 176, 187, 205, 212, 241, 271, 298
    space-based, 31, 82, 159, 188, 230, 239, 266
Morgan, Walter, 108
Morrison, James, 33
Moseson, Merland, 255
Motorola, Inc., 240
Muckelstone, Robert, 154
Mueller, George, 68
Mullane, Richard, 8
Multispectral scanner system, 40, 53
Murray, Bruce, 300
MX (missile), 1, 2, 126, 153
Myers, Dale, 45
Mylar, 50
Nagel, Steven, 8
Nagoya University, 42
Narimanov, Georgy, 80
NASA-928 (aircraft), 88
National Academy of Engineering, 77
National Academy of Sciences, 75, 238
National Aeronautic Association, 46, 56, 82, 125
National Aeronautics and Space Administration
    budget, 10, 11, 30, 52, 57, 67, 92, 93, 103, 106, 188, 250
        impact on GNP, 63
    contractors, 30, 90, 114, 153, 159, 303, 305
    contracts, 151, 156
    End-to-End Data Systems (NEEDS) project, 10
    expenditures, 152, 155
    ground facilities, 50
    launch schedule, 308
    management, 4, 5, 10, 33, 44, 106, 172, 231, 250, 252, 270
    minority businesses, 181
    minority employment, 39
    missions, 49, 57, 256
    Office of Aeronautics and Space Technology, 28, 99, 152, 185
    Office of Space and Terrestrial Applications, 23, 25, 106, 142
    Office of Space Science, 122, 285, 290
    Office of Space Transportation Systems, 52
    people, 11, 20, 22, 33, 49, 91, 96, 99, 135, 178, 199, 213
    policy, 104, 142
    programs, 67
    research, 14, 15, 17, 31, 38, 63, 97, 142
    small business activity, 310

National Aeronautics and Space Administration Communications Network (NASCOM), 219
National Business Aircraft Association, 185
National Center for Atmospheric Research, 87
National Environmental Satellite Service, 176, 209, 243
National Eye Institute, 105
National Investigations Committee on Aerial Phenomena, 74
National Medical Association, 190
National Oceanic and Atmospheric Administration, 1, 5, 13, 17, 18, 31, 68, 94, 109, 116, 122, 142, 146, 161, 176, 178, 210, 273
   Wallops Flight Center, 86, 90, 181, 201, 211, 235, 260
National Parachute Test Range, 55, 90, 106, 222
National Radio Astronomy Observatory, 183
National Remote Sensing Agency (India), 7
National Research Council, 74, 81, 235, 302
National Rocket Club, 91
National Science Foundation, 30, 31, 37, 43, 44, 144, 211, 270
National Security Council, 161
National Space and Technology Laboratory, 96, 261
National Space Club, 200
National Space Development Agency (Japan)
   Broadcasting Satellite for Experiment Purposes, 51
National Space Technology Laboratories, 12, 119, 182, 252
National Taxpayers Union, 196
National Transonic Facility, 253, 277, 302
Nato 3C, 260, 275
Naval Electronic Systems Command, 26
Naval Research Laboratory, 159, 168, 296
Navigation systems, 177, 286
NavStar, 100, 130, 134, 177, 286
NavStar 2, 118, 203
Navy, 5, 8, 17, 21, 25, 26, 52, 94, 99, 121, 176, 270
Nellis Air Force Base, 83
Nelson, George, 8
Nelson, Jr., David, 268
Nesterov, Peter, 126
Netherlands, 144, 163
Newcomb Cleveland Award, 35
Newman, Larry, 199
Nicollier, Claude, 188
Nike-Apache (rocket), 228
Nimbus 6, 42
Nimbus B, 212
Nimbus G, 11, 201, 213, 220, 230, 245, 247, 253
Nippon Electric Co., 10
Noise standards (aviation), 32
NOISEMAP, 204
Nola, Frank, 105, 137, 259
NORAD, 218
Norris, T. Bland, 255
North American Rockwell, 98
North Carolina A&T State University, 39
Northern Research and Engineering Corp., 293
Nosaka, K., 145

O'Conner, Joseph, 147
O'Leary, Brian, 80
O'Neill, Gerald, 116
O'Sullivan, D., 32
OAO 2, 16
Obayashi, Tatsugo, 41
Observances, 175, 221, 227, 231, 249, 255, 311
Observations
  space-based, 34
Observatories
  ground-based, 36, 277
  space-based, 15, 20, 70, 85, 94, 109, 120, 132, 214, 239, 247, 262, 269, 276, 283
Ocampo-Friedman, Roseli, 43
Ockels, Wubbo, 188
Odetics Corp., 145
Odyssey (Apollo mothership), 98
Ogarkov, Nikoli V., 77
Onizuka, Ellison, 8
Orbit 81, 310
Orbita network, 14
Orbiting Astronomical Observatory 3, 12
Orbiting relay stations, 22
Oscar D, 53
Ots (ESA satellite), 59
Ots 2, 109, 149, 246
OTS A, 88, 149
Out of Ecliptic program, 11
Oxford University, 61
Ozone monitoring, 88, 212
P80-2 mission, 45
Paarlberg, Don, 273
Pacific Northwest Regional Commission, 67
Page, Thornton, 170
Pan American World Airways, 151
Papago Construction Co., 112
Parachutes, 55, 90, 100, 106, 125, 209, 222, 266, 300
Parker, Robert, 183
Particle-intercollabertion project, 112
Patchett, Bruce, 94
Patterson, N. Paul, 94
Patton, Jr., James, 216
Payload specialists, 184, 188, 238
Payne, Randolph, 145
Pegasus program, 218, 286
Penaranda, Frank, 22
People's Republic of China, 17, 19, 26, 81, 227, 245, 272, 284, 307
Perkin-Elmer Corp., 6
Perras, Marcel, 145, 301
Peterson, Laurence, 215
Petrov, Boris, 14
Phobos, 2
Pierce, John, 27
Pinson, Larry, 230
Pioneer, 96, 117, 123, 138, 173, 194, 233, 240, 270, 280, 293, 299, 305

Pioneer 11, 1, 84, 170
Pioneer Venus, 61
Planetary exploration, 50
Planning Research Corporation, 129
Plasma-interaction experiment, 53
Plessey Corporation, 20
Podraczky, Emeric, 15
Pogue, William, 224
Poland, 49
Policy
  research, 21
  science, 260
  social, 80
  space, 1, 33, 241, 249, 274, 289
Pollution, 11, 88, 97, 145, 174, 212, 215, 230, 245
Porter, George, 137
Poseidon (missile), 82
Postal Service, U.S., 163, 279
Power sources
  batteries, 96
  electric, 103, 105
  microwave, 80, 107, 116
  nuclear, 115
  renewable, 103, 156, 265
  solar, 16, 26, 80, 89, 115, 156, 175, 240, 249
Power systems
  nuclear, 275
Powersats, 93, 196
Prague Institute of Geophysics, 277
Presidential Command Network, 25
Press, Frank, 26, 30, 122, 161, 260, 307
Price, Melvin, 46
Princeton University, 109
Prinz, Dianne, 94, 184
Progress 1, 12, 29, 31, 36, 104
Progress 2, 176, 187
Progress 2, 3, and *4*, 261
Progress 3, 193
Project Cameo, 255
Propulsion systems, 196
Proxmire, William, 183, 188, 211
Public opinion (space program), 154
Puddy, Donald, 62
Puerto Rico, 164
Pyrheliometers, 13
Quiet short-haul research aircraft, 174, 190
Radar, 155, 199, 271, 273
Radio Research Laboratories of Japan, 52
Radio systems, 42, 86
Radio waves, 191, 197
Radiometers, 17, 37, 74, 95, 146, 159, 181, 210, 252, 271
Rand, Robert, 87
Rapp, Rita, 245
Rasool, Ichtiaque, 23, 148

Ray, Dixy Lee, 67
RCA Corp., 49, 89, 196, 219, 221, 240, 279, 283, 291, 310
RCA Global Communications, 114
Recovery systems, 90
Redstone (launch vehicle), 16
Reed, Dale, 300
Reeder, John, 217
Reentry, 218
Reentry flight dynamics simulator, 139
Remek, Vladimir, 49
Remote sensing, 7, 12, 34, 81, 95, 123, 166, 295
Research
  atmospheric, 263, 300
  aviation, 96
  biomedical, 114
  climate, 31
  computer systems, 72
  design, 51
  expenditures, 48
  funding, 259
  grants, 35, 44
  high energy, 291
  high pressure, 269
  human, 295
  lunar, 211
  magnetic field, 80
  medical, 49
  ocean, 89
  scientific, 306
  solar, 184, 247
  space colonization, 276
Research and development, 4, 142, 158, 185, 188, 195, 244, 289
  budget, 260
  funding, 103, 313
Resnick, Judith, 8
Resources monitoring, 21
Ride, Sally, 8
Riggins Company, Inc., 253
Robinson, Henry, 166
Rockwell International Corp., 39, 56, 63, 114, 119, 125, 134, 139, 158, 182, 286, 301
  Rocketdyne Division, 75
Rolls Royce Ltd., 171
Romanenko, Yuri, 31, 56, 151, 224
Romania, 49, 263
Rood, Jack, 154
Rosenblum, Louis, 175
Rostafinski, Wojciech, 62
Roth, William, 188
Rotor systems research aircraft, 86, 229
Rowe, Herbert, 22
Royal Aeronautical Society, 101
Ruloff, Arthur, 268
Rupp, Charles, 120
Safety

aviation, 32
Saito di Quirra, 33
SALT, 2
Salyut program, 30
Salyut 2, 291
Salyut 3, 290
Salyut 4, 70
Salyut 5, 14, 290
Salyut 6, 7, 12, 13, 30, 31, 36, 39, 56, 70, 104, 176, 193, 200, 224, 227, 254, 261
Salyut 6/Soyuz 29, 150
Salyut 27, 30
SAMSO, 127
San Andreas Fault Experiment, 141
Sanchini, Dominic, 75
Sas 3, 193
Satcom-3, 291
Satellite charing at high altitudes (SCATHA), 294
Satellite Communications Terminal Program Office, 22
Saturn (planet), 1, 2, 84, 87, 170, 210, 241
Saturn (rocket), 4, 72, 78, 79, 218
Saudi Arabia, 33
Scanning beam microwave landing system. See Microwave scanning beam landing system.
Scatterometers, 271
Schjeldahl Corp., 50
Schmitt, Harrison, 35, 75, 104, 232, 251, 289
Schneider, William, 99, 214
Schoenfelder, V., 215
Schroeder, D. J., 109
Science Research Council (UK), 15, 60
Scobee, Francis, 8
Scott, David, 35, 136, 300
Scott, Sheila, 154
Scout (launch vehicle), 90, 95, 99, 163, 262
SDC Integrated Services, Inc., 85
Search for extraterrestrial intelligence, 66, 106
Seasat, 57, 89, 100, 121, 134, 144, 176, 223, 241, 247, 248, 254, 271
SeaSat 1, 158
Seasat A, 11
SeaSat Users Research Group Europe, 247
Seddon, Margaret, 8
Senate, U.S. See Congress, U.S., 75
Sensor systems, 40, 58
  spacecraft, 34, 74, 77, 95, 177, 212, 236
Sensor Technology, Inc., 241
SEPAC (space experiments with particle accelerators), 41
SETI program, 220
Shakar, Mohamed, 164
Shaw, Jr., Brewster, 8
Shepard, Alan, 225
Shklovsky, Iosif, 115
Short, Nicholas, 166
Shramo, Daniel, 255
Shriver, Loren, 8
Shroud of Turin, 169

Sikorsky Aircraft, 86
Siltec Corp., 241
Simokaitis, Frank, 213
Simon, George, 94, 184
Simulation, 139, 156, 167, 174
Sky Van (aircraft), 201
Skylab, 12, 36, 61, 65, 71, 78, 82, 106, 115, 120, 140, 156, 164, 224, 249, 259, 302
Skylab 4, 56, 168
Slayton, Donald "Deke", 42, 221, 231
Smith, Aubrey, 260, 293
Smith, Bruce, 167
Smith, Richard, 178
Smithsonian Institution, 38, 42, 45, 144, 296
   Center for Earth and Planetary Studies, 164
   National Air and Space Museum, 79, 199, 263, 297
Snap 10A test vehicle, 18
Snow, Harold, 87
Societe Europeenee de Propulsion (France), 7, 294
Society for Automotive Engineers, 152
Solar activity, 32, 37, 57, 122, 147, 165, 192, 198, 214, 235, 276
Solar conversion devices, 16
  arrays, 26, 66, 78, 112, 132, 172, 184, 194, 219, 239, 248, 272
  cells, 132, 168, 171
  collectors, 194 See also Power sources, solar.
Solar system
  primordial matter, 37
Solar wind, 84
Solar-Polar
  see Out of Eliptic program, 11
Solid-fuel rocket, 4, 10, 16, 49, 54, 83, 88, 90, 100, 106, 111, 260, 271, 313
Sollos, Inc., 241
Sonnemann, Harry, 178
Soviet Academy of Sciences, 14, 80, 148
Soyuz 18, 56
Soyuz 21, 14
Soyuz 24, 14
Soyuz 26, 7, 13, 56, 224
Soyuz 27, 7, 12, 13, 31, 39
Soyuz 28, 49
Soyuz 29, 176, 200, 224, 261
Soyuz 30, 150, 176
Souyz 31, 200, 254, 261
Space
  construction, 18, 84
  food, 245
  medicine, 13, 91, 111, 151, 262
Space and Missile Systems Organizations (Air Force), 45
Space and Missile Test Center, 286
Space Computational Center, 179
Space debris hazards, 18
Space processing applications rocket, 219
Space Shuttle, 25
  costs, 44, 117, 218, 251
  crews, 7, 8, 39, 61, 278

designation system, 197
experiments, 3
external fuel tank, 6, 178
funding, 44, 220
launches, 115, 195, 277
management, 295
missions, 12
orbiter, 4, 6, 11, 44, 50, 52, 69, 91, 93, 104, 114, 158, 183
payloads, 3, 12, 16, 18, 20, 29, 30, 32, 36, 40, 38, 50, 57, 82, 89, 91, 106, 110, 111, 124, 156, 161, 164, 222, 273, 285, 297
policy, 90
problems, 226
propulsion, 6, 10, 19, 54, 72, 75, 80, 84, 90, 96, 182, 223, 253, 261, 272, 292
safety, 107
systems, 120, 138, 155
testing, 7, 21, 42, 54, 60, 69, 72, 81, 83, 90, 96, 103, 106, 119, 121, 132, 149, 156, 182, 222, 232, 242, 252, 253, 271, 292, 302
upper stages, 90
Space stations, 80, 85, 149, 161, 232
Space Telescope, 6, 11, 16, 20, 21, 109, 110
Space Transportation System, 41, 133
Space walk, record, 176
Spacecraft charging at high altitudes (SCATHA), 312
Spaceflight Tracking and Data Network, 51, 205
Spaceflight Tracking and Data Network station, 21
Spacelab, 21, 30, 95, 117, 124, 129, 134, 168, 183, 188, 296
  costs, 285
  crew, 308
  experiments, 94, 309
  funding, 290
  management, 25, 41
  payloads, 29, 160, 284
  preparations, 27, 33, 217, 238, 264
SPAR V, 219
Spectrometers, 283, 293
Speer, Fred, 34
Sperry Rand Corp., 296
Spin research vehicle, 225
Sports coverage by satellite, 146
SRI International, 189
Stafford, Thomas, 10, 45, 191
Stanford University, 39
State University of New York, 109
Steffen, C. J., 145
Stennis, John, 9
Stereosat program, 93, 220
Stevenson, III, Adlai E., 75, 229, 251, 289
Steward, Robert, 8
Stitt, James, 255
Strategic Air Command, 11, 26
Straub, Robert, 67
Strong, Keith, 94
Submarines, 131, 203
Sullivan, Kathryn, 8

Super Guppy (aircraft), 301
Supersonic transport, 216
Surveillance, 223, 265, 275, 290
    infrared, 45
    ocean, 270
    submarine, 131
Swain, Charles, 201
Sweden, 92
Swigert, Jack, 98
Symposium. See meetings, 59
Syncom-4, 38
Synthetic-aperture radar, 199
System for nuclear auxiliary power (SNAP), 115
Systems, computer. See Computer systems.
Systems Development Corp., 68
Syvertson, Clarence, 99, 263
Taurus-Tomahawk, 260
TD 1 satellite, 16
Teague, Olin, 93, 213, 230, 238
Technische Universitat (Austria), 264
Technology
    applications, 12, 27, 41, 52, 79, 236, 314
    future, 107
    transfer, 7, 62, 100, 122, 161, 172, 245, 284, 307
    utilization, 190, 201, 260, 286
Technology Utilization House, 201
Teledyne Brown Engineering, 40
Teleoperator retrieval system, 71, 156, 183, 220, 259, 302
Telesat, 314
Telesat D, 289
Telescopes, 35, 36, 145, 168, 186, 216, 243, 262, 276, 283
    radio, 197
    satellite's, 15
Television systems, 49, 50
Telos Computing, Inc., 302
Temple 2 comet, 88, 229
Terhune, Charles, 255
Terrell, Norman, 49
Testing
    aircraft, 139
    equipment, 30, 105
Texas A & M, 109
Thagard, Norman, 8
    3B (Spacecraft), 78
Thermal protection system tiles, 252
Thiokol Corp., 10, 19, 55, 153, 252, 253, 271
Thomas, Roger, 309
Thompson, A., 32
Thompson, J.R., 75
Thompson, Robert, 42, 191
Thor (missile), 17
Thor/Agena, 265
Time reference scanning-beam/microwave landing system, 5, 6, 20, 107. See also microwave scanning-beam landing system.

Timothy, Adrienne, 148
Tiros-N, 11, 89, 100, 209, 221, 230
Titan, 90, 313
Titan (rocket), 2, 64, 78, 82, 134, 169, 265
Tokyo Shibaura Electric Co., 52
Tokyo University, 212
Tomahawk (missile), 5, 100, 113
Tracking, 87, 91, 159, 171, 196, 205, 246, 286
Training, 172, 188
Treaties, 33, 79, 179
Trident (missile), 82
Triska, P., 277
Truly, Richard, 61
TRW Systems, Inc., 25, 95, 98, 160, 185, 132, 268
Tuckwell, N., 145
Turegano, J. A., 197
Twiggs, Tom, 174
Uemura, Naomi, 42
UFOs, 74
UK-5, 60
Ullman, Al, 32
Ultraviolet spectrum, 15, 34, 37, 168, 186
United Kingdom, 5, 15, 16, 20, 92, 94, 119, 145, 210, 246
United Nations, 33, 39, 202, 247
United Space Boosters, Inc., 49
United Stirling (Sweden), 68
United Technologies Corp., 55, 114, 209, 296, 303, 313
University of Chicago, 45, 296
University of Florida, 68
University of Iowa, 296
University of Pittsburgh, 37
University of Texas, 110, 114
University of Tokyo, 41
Uranus, 1, 197, 268
USS Barb, 100
USSR, 12, 198
  Academy of Sciences, 306
  Cosmos program, 80, 123
  Gagarin Space Centre, 69
  Glissada, 41
  Intercosmos program, 49, 150
  international cooperation, 9, 14, 39, 49, 55, 112, 263, 303
  lunokhod moon-rover (USSR), 80
  manned space flights, 7, 13, 65, 176, 227, 261, 290, 298
  space activities, 29, 39, 69, 80, 187, 214, 224, 227, 254, 263, 275, 304, 306
  space stations, 56
  SST, 43
  technology, 28, 41
  Tupolev TU-95, 65
  weapons systems, 17, 29, 77
  See also Salut, Soyuz, and Venera
Van Allen, James 17
Van Allen radiation belts, 16
Van Hoften, James, 8

Vandenberg Air Force Base, 5, 6, 7, 11, 89, 90, 95, 97, 100, 115, 159, 227
Vanguard project, 17, 62
Vaughn, Jr., Otha, 60
Vaughn, Pat, 67
Vela, 233
Venera missions, 117, 273
Venera 11 and *12*, 214, 306
Venus, 61, 80, 117, 173, 194, 214, 273, 293, 299, 306
Venus (planet), 96, 123, 138, 233, 240, 270, 280
Venus-orbiting imaging radar, 199
Verbeek, Earl, 187
Viking
  lander, 79
  orbiter, 2, 174
  project, 35, 44, 89, 267
Viking 1, 163
Viking 2, 163
Villafranca station (Spain), 16
Voight Corp., 32
Von Braun, Werner, 10
VORTEX (Venus orbiter radiometric temperature experiment), 61
Vought Corp., 17, 90, 116, 262
Voyager project, 158, 192, 210, 219, 241
Voyager 1, 2, 105
Voyager 1 and *2*, 306
Voyager 2, 1, 2, 86
Waldheim, Kurt, 247
Walker, David, 8
Walker, Joe, 301
Wallops Flight Center, 86, 90, 181, 201, 211, 235, 260
WB-57F (aircraft), 87
Weapons systems, 28
  aircraft, 9
  lasers, 29, 71, 77, 185
  missiles, 1, 5, 46
  nuclear, 17
  space-based, 17, 64, 77
Weiss, Ted, 104
Wenzel, K. P., 32
Western Test Range, 5, 286
Western Union Corp., 155, 196
Western Union International, 114
Western Union Space Communications, Inc., 307
Westlake, Reginald, 15
Whitcomb, Richard, 185
White, II, Edward, 231
White Sands Missile Range, 286
White Sands Test Facility, 195, 228
Williams, Dell, 255
Williams, Donald, 8
Wind sheer, 227
Wind tunnels, 35, 51, 64, 66, 99, 103, 205, 209, 233, 253, 277, 302
Wind-turbine electric generators, 164, 171, 265

Winglets, 35, 255
Winstead, Thomas, 178
Winter, David, 14, 114
Wisner, William, 154
Women, 7, 8, 83, 94, 107, 148, 184, 295
Wood, H. William, 15
Woodley, William, 162
Workshop. See meetings, 66
World Data Centers, 300
World Meteorological Organization, 229
World Weather Watch, 299
Wright, Wilbur and Orville, 206, 297
X-15 (aircraft), 249
X-rays, 34, 35, 60, 78, 95, 120, 163, 186, 190, 193, 239, 243, 262, 268, 275, 283
Xenon, 268
XFV-12A (aircraft), 139
XV-15 (aircraft), 64
Yardley, John, 75, 191, 197, 214, 225, 289, 303
YF-12 (aircraft), 92, 296
Yin, Lo I., 190
Young, John, 61, 191
Young, Richard, 43
Young, Thomas A., 263
Zaire, 33
Zisfein, Melvin, 199

The NASA History Series

HISTORIES

Anderson, Frank W., Jr., *Orders of Magnitude: A History of NACA and NASA, 1915-1980* (NASA SP-4403, 2d ed., 1981).

Benson, Charles D., and William Barnaby Faherty, *Moonport: A History of Apollo Launch Facilities and Operations* (NASA SP-4204, 1978).

Bilstein, Roger E., *Stages to Saturn: A Technological History of the Apollo/Saturn Launch Vehicles* (NASA SP-4206, 1980).

Boone, W. Fred, *NASA Office of Defense Affairs: The First Five Years* (NASA HHR-32, 1970, multilith).

Brooks, Courtney G., James M. Grimwood, and Loyd S. Swenson, Jr., *Chariots for Apollo: A History of Manned Lunar Spacecraft* (NASA SP-4205, 1979).

Byers, Bruce K., *Destination Moon: A History of the Lunar Orbiter Program* (NASA TM X-3487, 1977, multilith).

Compton, W. David, and Charles D. Benson, *Living and Working in Space: A History of Skylab* (NASA SP-4208, 1983).

Corliss, William R., *NASA Sounding Rockets, 1958-1968: A Historical Summary* (NASA SP-4401, 1971).

Ezell, Edward Clinton, and Linda Neuman Ezell, *On Mars: Exploration of the Red Planet, 1958-1978* (NASA SP-4212, 1984).

Ezell, Edward Clinton, and Linda Neuman Ezell, *The Partnership: A History of the Apollo-Soyuz Test Project* (NASA SP-4209, 1978).

Green, Constance McL., and Milton Lomask, *Vanguard: A History* (NASA SP-4202, 1970; also Washington: Smithsonian Institution Press, 1971).

Hacker, Barton C., and James W. Grimwood, *On the Shoulders of Titans: A History of Project Gemini* (NASA SP-4203, 1977).

Hall, R. Cargill, *Lunar Impact: A History of Project Ranger* (NASA SP-4210, 1977).

Hallion, Richard P., *On the Frontier: Flight Research at Dryden, 1946-1981* (NASA SP-4303, 1984).

Hartman, Edwin P., *Adventures in Research: A History of Ames Research Center, 1940-1965* (NASA SP-4302, 1970).

Levine, Arnold, *Managing NASA in the Apollo Era* (NASA SP-4102, 1982).

Muenger, Elizabeth A., *Searching the Horizon: A History of Ames Research Center, 1940–1976* (NASA SP-4304, 1985).

Newell, Homer E., *Beyond the Atmosphere: Early Years of Space Science* (NASA SP-4211, 1980).

Pitt, John A., *The Human Factor: Biomedicine in the Manned Space Program to 1980* (NASA SP-4213, 1985).

Roland, Alex, *Model Research: The National Advisory Committee for Aeronautics, 1915-1958* (NASA SP-4103, 1985).

Rosenthal, Alfred, *Venture into Space: Early Years of Goddard Space Flight Center* (NASA SP-4301, 1968).

Rosholt, Robert L., *An Administrative History of NASA, 1958-1963* (NASA SP-4101, 1966).

Sloop, John L., *Liquid Hydrogen as a Propulsion Fuel, 1945-1959* (NASA SP-4404, 1978).

Swenson, Loyd S., Jr., James M. Grimwood, and Charles C. Alexander, *This New Ocean: A History of Project Mercury* (NASA SP-4201, 1966).

REFERENCE WORKS

Aeronautics and Space Report of the President, annual volumes for 1975-1982.

The Apollo Spacecraft: A Chronology (NASA SP-4009, vol. 1, 1969; vol. 2, 1973; vol. 3, 1976; vol. 4, 1978).

Astronautics and Aeronautics: A Chronology of Science, Technology, and Policy, annual volumes 1961-1977, with an earlier summary volume, *Aeronautics and Astronautics, 1951-1960.*

Dickson, Katherine M., ed., *History of Aeronautics and Astronautics: A Preliminary Bibliography* (NASA HHR-29, 1968, multilith).

Hall, R. Cargill, ed., *Essays on the History of Rocketry and Astronautics: Proceedings of the Third through the Sixth History Symposia of the International Academy of Astronautics* (NASA CP-2014, 2 vols., 1977).

Hall, R. Cargill, *Project Ranger: A Chronology* (JPL/HR-2, 1971, multilith).

Looney, John J., ed., *Bibliography of Space Books and Articles from Non-Aerospace Journals, 1957-1977* (NASA HHR-51, 1979, multilith).

Roland, Alex F., *A Guide to Research in NASA History* (NASA HHR-50, 6th ed., 1982, available from NASA History Office).

Skylab: A Chronology (NASA SP-4011, 1977).

Van Nimmen, Jane, and Leonard C. Bruno, with Robert L. Rosholt, *NASA Historical Data Book, 1958-1968,* vol. 1, *NASA Resources* (NASA SP-4012, 1976).

Wells, Helen T., Susan H. Whiteley, and Carrie E. Karegeannes, *Origins of NASA Names* (NASA SP-4402, 1976).

Recent volumes are available from Superintendent of Documents, Government Printing Office, Washington, DC 20402; early volumes from National Technical Information Service, Springfield, VA 22161.

www.ingramcontent.com/pod-product-compliance
Lightning Source LLC
Chambersburg PA
CBHW081716170526
45167CB00009B/3597